Ergebnisse der Mathematik und ihrer Grenzgebiete

Band 76

Herausgegeben von P. R. Halmos · P. J. Hilton
R. Remmert · B. Szőkefalvi-Nagy

Unter Mitwirkung von L. V. Ahlfors · R. Baer
F. L. Bauer · A. Dold · J. L. Doob · S. Eilenberg
K. W. Gruenberg · M. Kneser · G. H. Müller
M. M. Postnikov · B. Segre · E. Sperner

Geschäftsführender Herausgeber: P. J. Hilton

B. A. F. Wehrfritz

Infinite Linear Groups

An Account of the Group-theoretic Properties
of Infinite Groups of Matrices

Springer-Verlag Berlin Heidelberg GmbH
1973

Bertram A. F. Wehrfritz

Queen Mary College, London University, London/England

AMS Subject Classifications (1970):
Primary   20 E 99, 20 H 20 · Secondary   15 A 30, 20 F 99, 20 G 15, 20 H 25

ISBN 978-3-642-87083-5   ISBN 978-3-642-87081-1 (eBook)
DOI 10.1007/978-3-642-87081-1

Softcover reprint of the hardcover 1st edition 1973

Preface

By a linear group we mean essentially a group of invertible matrices with entries in some commutative field. A phenomenon of the last twenty years or so has been the increasing use of properties of infinite linear groups in the theory of (abstract) groups, although the story of infinite linear groups as such goes back to the early years of this century with the work of Burnside and Schur particularly.

Infinite linear groups arise in group theory in a number of contexts. One of the most common is via the automorphism groups of certain types of abelian groups, such as free abelian groups of finite rank, torsion-free abelian groups of finite rank and divisible abelian p-groups of finite rank. Following pioneering work of Mal'cev many authors have studied soluble groups satisfying various rank restrictions and their automorphism groups in this way, and properties of infinite linear groups now play the central role in the theory of these groups. It has recently been realized that the automorphism groups of certain finitely generated soluble (in particular finitely generated metabelian) groups contain significant factors isomorphic to groups of automorphisms of finitely generated modules over certain commutative Noetherian rings. The results of our Chapter 13, which studies such groups of automorphisms, can be used to give much information here.

A third way that linear groups arise is via an earlier theorem of Mal'cev which states that a group G is isomorphic to some linear group of degree n if each of its finitely generated subgroups is isomorphic to a linear group of degree n (see 2.7 for a proof). It is principally via this result and its generalizations that infinite linear groups have influenced the theory of locally finite groups. With more information about which linear groups the finitely generated subgroups of G are isomorphic to, one can sometimes specify precisely a linear group isomorphic to G. This led to very important characterizations of certain groups such as $PSL(2, F)$ over locally finite fields F, which now play a crucial role in the theory of locally finite groups, see reference [31a]. I suspect that to date we have only scratched the surface of the applications of infinite linear groups to locally finite groups.

The class of linear groups is one of the few examples of a highly non-trivial, highly non-soluble, but still relatively accessible class of groups on which to test conjectures. It is quite common for counter examples to be constructed as or out of linear groups, and for a problem to be solved first for the linear case where the situation should be somewhat simpler. The obvious examples of this are the Burnside Problem and the minimal and maximal condition conjectures (see 1.23, 9.8 and 10.18). Finally it sometimes happens for purely ad-hoc reasons that one knows that a particular group is isomorphic (or at least related) to a certain linear group. When this happens one can often use this information to obtain painlessly properties of the original group. Examples of this phenomenon that come to mind are free groups, certain relatively free groups and certain generalized free products (see Chapter 2).

Much of the work on linear groups has been done by people with a sufficiently non-group-theoretic background to make their articles difficult for a group theorist to read. Also many other results on linear groups are tucked away in papers with purely group-theoretic titles. Thus there seemed to be a need to gather all this material together.

This book is an attempt to give an account of those properties of linear groups that seem relevant to infinite group theory, assuming by and large no more than a postgraduate student of group theory might reasonably be expected to know. This has not always been possible of course. In the main we require the reader to know no more linear algebra than usually appears in undergraduate courses, and little more commutative algebra. We assume no knowledge of algebraic geometry (with the exception of the final chapter which is an appendix on algebraic groups) developing what little we need as we go along. In contrast we assume a considerable quantity of (infinite) group theory. For the most part we need no more than might reasonably be taught in M. Sc. courses, but occasionally we need considerably more. I have tried to give full (and often alternative) references whenever this occurs and it is very rare for me not to be able to refer the reader to a textbook.

A word about the arrangement of this book; the basic chapters are 1, 5, 6 and to some extent 2, which provides the reader with some examples of linear groups to bear in mind. These should in a way have been labelled Chapters 1 to 4. However, assuming the reader to be interested in the group theoretic properties of linear groups there seemed to be a considerable risk of him abandoning the book through boredom before he reached the meat. I have therefore split the introduction into very roughly the ring theoretic part in Chapter 1 and the geometric part in Chapters 5 and 6. It is possible to go a good way with the theories of soluble linear groups and finitely generated linear groups without the latter part and this we do in Chapters 3 and 4. The determined reader is welcome to read Chapters 5 and 6 immediately after Chapter 1.

The remainder of the book is composed as follows. Chapter 7 studies Jordan decomposition in linear groups and contains some structure theorems for locally nilpotent linear groups, results that are the basis of much of Chapter 8 on upper central series and a little of Chapter 11 on locally supersoluble linear groups. The bulk of the material on periodic linear groups is in Chapter 9, some further material being in the second part of Chapter 12. Chapter 10 is devoted to properties of linear groups with a (sometimes very vague) varietal flavour. In Chapter 13 we show how properties of groups of automorphisms of finitely generated modules over commutative rings may be derived from properties of linear groups and Chapter 14 is little more than a description of the main properties of algebraic groups over algebraically closed fields with very few of the proofs.

This book has been written with the requirements of the reader wishing to make a serious study of the subject much to the forefront of my mind. However the needs of those wishing to use it as a reference work have not entirely been ignored. A reader who desires to look up a particular type of result or property should look first in the index. If this is unhelpful, as of course it often will be if for example he does not know precisely what he is looking for, he should turn to the earliest chapter with the most promising looking title. In the first paragraph or two he will find a brief summary of the properties of linear groups discussed in that chapter (apart from Chapters 1, 5 and 7 where this is not really practicable). If this is still unhelpful he should turn to the end of the chapter where, when relevant, he will find a brief summary of where the main types of groups studied in the chapter appear later in the book.

Approximately half of the material in this book has been available for some time as lecture notes issued in the Queen Mary College Mathematics Notes series. Chapters 10, 11, 12 and 14 have no counterpart in the Q.M.C. notes. Chapter 13 is a largely rewritten and considerably extended version of the Q.M.C. Chapter 10. Chapters 2, 4 and 9 are greatly extended versions of the corresponding Q.M.C. chapters. The remaining six chapters are only minor variants of their Q.M.C. counterparts, the main changes being correction of mistakes, the insertion of more examples, exercises and references to the literature, and the inclusion of the proofs of certain results that had only been stated before.

It had been my intention to give everybody his due credit. This has in practice proved quite impossible, particularly where the idea occurs as part of a proof, and the book is sprinkled with uncredited ideas and tricks that I have lifted from various people and places. Much of the material I have given in courses and seminars in London at one time or another and my audiences are responsible for many improvements in content, presentation and notation. I am especially indebted to Karl Gruenberg and Otto Kegel in this respect. My thanks are also due to Kurt Hirsch, who, among other contributions, originally stimulated my interest in linear groups with a course he gave on the subject during the academic year 1963–64. Finally Norman Massey helped me with the proof reading; I claim full credit however for all the errors left behind.

London, January 1973

Bertram A. F. Wehrfritz

Table of Contents

Notation

If G is a group, then

G' denotes the derived group of G,

$G = \gamma^1 G \supseteq \gamma^2 G \supseteq \cdots$ the lower central series of G,

$\{1\} = \zeta_0(G) \subseteq \zeta_1(G) \subseteq \cdots$ the upper central series of G,

$\zeta(G)$ the hypercentre of G,

$\eta(G)$ the Hirsch-Plotkin radical of G,

$\eta_1(G)$ the Fitting subgroup of G,

$\lambda(G)$ the product of all the G-hypercyclic normal subgroups of G (see p. 156),

$\phi(G)$ the Frattini subgroup of G,

$\delta(G)$ the intersection of the non-normal maximal subgroups of G, or G itself if none such exist,

$\psi(G)$ the intersection of the centralizers of the chief factors of G,

$\psi_1(G)$ the intersection of the centralizers of the finite chief factors of G, or G itself if none such exist,

$\pi(G)$ the set of primes p such that G contains an element of order p,

and $O^\pi(G)$ the intersection of the normal subgroups N of G such that $\pi(G/N) \subseteq \pi$.

If S is any subset of G then

$|S|$ denotes the cardinality of S,

$N_G(S)$ the normalizer of S in G,

$C_G(S)$ the centralizer of S in G,

$\langle S \rangle$ the subgroup of G generated by S,

and $\langle S^G \rangle$ the normal subgroup of G generated by S.

If a is any element of G, then a^G denotes the set of conjugates of a in G.

If H and K are subgroups of G, then

$(G:H)$ denotes the index of H in G,

$[H, K] = \langle [h, k] = h^{-1} k^{-1} h k : h \in H, k \in K \rangle = [H, {}_1 K]$,

$[H, {}_{s+1} K] = [[H, {}_s K], K] = [H, K, K, \ldots, K]$ where the K is repeated $s+1$ times,

and $H \mathbf{e} K$ means that for each h in H and k in K there exists an integer $r = r(h, k)$ such that $[h, {}_r k] = [h, k, k, \ldots, k] = 1$ where k is repeated r times.

If G acts as a group of operators on the group A, then either A]G or G[A denotes the split extension of A by G.

If H and K are groups and $\{H_i\colon i \in I\}$ is a family of groups then

$H \times K$ and $\underset{i \in I}{\times} H_i$ denote the direct products of the corresponding groups,

$H * K$ and $\underset{i \in I}{\times} H_i$ the free products,

$\underset{i \in I}{\prod} H_i$ the cartesian product,

$H \wr K$ the (restricted) wreath product of H by K

and $H \bar\wr K$ the complete wreath product of H by K.

$\mathfrak{F}, \mathfrak{G}, \mathfrak{A}, \mathfrak{N}, \mathfrak{S}$ denote respectively the classes of all finite groups, finitely generated groups, abelian groups, nilpotent groups and soluble groups. S, Q, L, R denote respectively the subgroup, quotient, local and residual operators.

If R is a commutative ring and V an R-module, then

$\operatorname{End}_R(V)$ denotes the ring of R-endomorphisms of V,

$\operatorname{Aut}_R(V)$ the group of R-automorphisms of V,

R_n the ring of $n \times n$ matrices over R, and if $x \in R_n$,

$\operatorname{tr}(x)$ denotes the trace of x,

and $\det(x)$ the determinant of x.

F always denotes a field.

$\bar{F}$ is an algebraic closure of F,

F^* the multiplicative group of F,

and $\operatorname{char} F$ the characteristic of F.

$\mathbf{S}_n$ is the symmetric group on n-symbols,

$\mathbf{C}_n$ a cyclic group of order n,

$\mathbf{C}_{p^\infty}$ a Prüfer p^∞-group,

$GL(n, F)$ the general linear group of degree n over the field F,

$SL(n, F)$ the special linear group, $= \{g \in GL(n, F)\colon \det g = 1\}$,

$Tr(n, F)$ the triangular group, $= \{(g_{ij}) \in GL(n, F); g_{ij} = 0 \text{ for } i < j\}$,

$Tr_1(n, F)$ the unitriangular group, $= \{(g_{ij}) \in Tr(n, F)\colon g_{ii} = 1 \text{ for each } i\}$,

$D(n, F)$ the diagonal group, $= \{(g_{ij}) \in GL(n, F)\colon g_{ij} = 0 \text{ for } i \neq j\}$.

If S is a subset of an F-algebra A then $F\{S\}$ denotes the subalgebra of A with identity generated by S.

If S is a subset of $GL(n, F)$ then $\mathscr{A}_F(S)$ denotes the minimal closed subgroup of $GL(n, F)$ containing S; see p. 74.

If g is an element of $GL(n, F)$ then $g = g_u g_d$ is the (multiplicative) Jordan decomposition of g over $\bar{F}$; see p. 91.

If G is a subgroup of $GL(n, F)$ then G^0 denotes the connected component of G containing 1.

If r is a real number $[r]$ denotes the largest integer not exceeding r. If r is a natural number $\pi(r)$ denotes the set of (positive) prime divisors of r.

Usually the zero of rings and additive groups is written 0. Occasionally the zero $n \times n$ matrix is written 0_n.

The identity of multiplicative groups is written 1. The identity $n \times n$ matrix is written 1_n. The identity of rings and fields is usually written 1. Occasionally the identity of the ring R is written 1_R.

Throughout mappings and representations have been written on the right. Module means right module in general. However, where the ring is commutative we allow ourselves to put the scalars on any convenient side.

Interdependencies Between the Chapters

In the following diagram a continuous line indicates a heavy dependence, while a broken line denotes only a partial dependence. Incidental references are not indicated in the diagram.

Chapter 13 partially depends on *every* previous chapter, and is not included below.

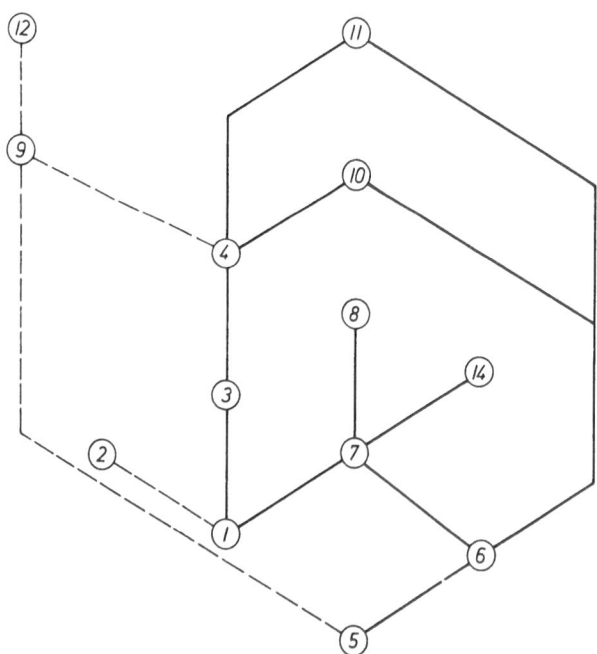

1. Basic Concepts

All rings will have identity elements and all modules will be unital. Let R be a commutative ring (in our applications R will always be an integral domain and will usually be either a field or $\mathbb{Z}$). R_n denotes the R-algebra of $n \times n$ matrices and $\mathrm{GL}(n, R)$ the group of units of R_n. By definition a *linear group* is a subgroup of $\mathrm{GL}(n, F)$ for some positive integer n and some (commutative) *field F*.

Let V be a free R-module of dimension n. Then

$$R_n \cong_R \mathrm{End}_R(V) \quad \text{and} \quad \mathrm{GL}(n, R) \cong \mathrm{Aut}_R(V).$$

If X is any subset of $\mathrm{End}_R(V)$, $R\{X\}$ denotes the R-subalgebra (with-1) of $\mathrm{End}_R(V)$ generated by X. We can regard V as an $R\{X\}$-module in a natural way. We say that X is irreducible, completely reducible, etc. (as a subset of $\mathrm{End}_R(V)$) according to whether V is irreducible, completely reducible, etc. as an $R\{X\}$-module. Identifying R_n with the R-endomorphism ring of the space of n-row vectors over R we transfer this terminology to subsets of R_n.

If G is a group, RG denotes the group ring of G over R. Let V be a free R-module of dimension n. A (group) homomorphism ρ of G into $\mathrm{Aut}_R(V)$ is called a *representation* of G over R of dimension n (some say of degree n). If $v \in V$ and $g \in G$, defining $v \cdot g = v(g\rho)$ makes V into an RG-module, an RG-module which we denote by (V, ρ) for the moment. Conversely if V is an RG-module such that V is a free R-module of finite dimension under the action of $R \subseteq RG$, then V determines a representation ρ of G such that $V = (V, \rho)$.

Let V_1 be a second free R-module of finite dimension and ρ_1 a homomorphism of G into $\mathrm{Aut}_R(V_1)$. ρ and ρ_1 are called equivalent representations of G over R if (V, ρ) and (V_1, ρ_1) are isomorphic as RG-modules.

A matrix representation of G over R of degree n is a homomorphism of G into $\mathrm{GL}(n, R)$. If ϕ and ψ are any two homomorphisms of G into $\mathrm{GL}(n, R)$ we call ϕ and ψ equivalent if there exists an element x of $\mathrm{GL}(n, R)$ such that

$$g\phi = x^{-1}(g\psi)x \quad \text{for all } g \text{ in } G.$$

1

These two notions of equivalence are essentially the same, for clearly if (V, ρ) and (V_1, ρ_1) are isomorphic as RG-modules then $\dim V = \dim V_1$. Note that two matrix representations of degree 1 are equivalent if and only if they are equal. A representation (of either type) is called irreducible, completely reducible, etc. whenever its image is irreducible, completely reducible, etc.

F will always denote a field. Let G be a subgroup of $\mathrm{GL}(n, F)$ and V a free F-module of dimension n. Let $u_1, \ldots, u_n$ be a basis of V. Make V into an FG-module relative to this basis in the natural way. If g is an element of G denote by ϕ_g the linear mapping of V into itself determined by g, i.e. $g = (\phi_g; u_i, u_j)$. Suppose that $\{0\} = V_0 \subseteq V_1 \subseteq \cdots \subseteq V_r = V$ is a series of FG-submodules of V. Since F is a field there exists a basis $v_1, \ldots, v_n$ of V such that

$$v_{n_1 + \cdots + n_{i-1} + 1}, \ldots, v_{n_1 + \cdots + n_i}$$

is a basis of V_i modulo V_{i-1} for $i = 1, 2, \ldots, r$. Because V_i/V_{i-1} is an FG-module this basis determines a representation ρ_i of G into $\mathrm{GL}(n_i, F)$. If $x^{-1} = (1; v_i, u_j)$ is the matrix that represents the change of basis from the v_i to the v_j, then

$$x^{-1} g x = (\phi_g; v_i, v_j) = \begin{pmatrix} g\rho_1 & & 0 \\ & \ddots & \\ * & & g\rho_r \end{pmatrix} \qquad (*)$$

(where there is a series of blocks down the diagonal, all zeros above these blocks and something unknown below). Whenever the above series is a composition series each ρ_i is irreducible. Thus if G is a subgroup of $\mathrm{GL}(n, F)$ there exist positive integers $r, n_1, n_2, \ldots, n_r$, for each $i = 1, 2, \ldots, r$ an irreducible representation ρ_i of G in $\mathrm{GL}(n_i, F)$, and an element x of $\mathrm{GL}(n, F)$ such that $n_1 + n_2 + \cdots + n_r = n$ and $(*)$ holds for all g in G.

Let K be an extension field of the field F. If A is an F-algebra and V an A-module of finite F-dimension, then $A^K = K \otimes_F A$ is a K-algebra and $V^K = K \otimes_F V$ is an A^K-module. In particular, each element x of

$$K \otimes_F \mathrm{End}_F(V)$$

determines a K-endomorphism $x\phi$ of V^K and

$$\phi: K \otimes_F \mathrm{End}_F(V) \to \mathrm{End}_K(V^K)$$

is a K-algebra isomorphism.

Now $F_n \subseteq K_n$ and $\mathrm{End}_F V$ can be embedded in $K \otimes_F \mathrm{End}_F(V)$ by $x \to 1 \otimes x$. Let $v_1, \ldots, v_n$ be a basis of V. Then $1 \otimes v_1, \ldots, 1 \otimes v_n$ is a K-basis of V^K. These bases determine an F-algebra isomorphism $\theta: F_n \to \mathrm{End}_F(V)$ and a K-algebra isomorphism $\psi: K_n \to \mathrm{End}_K(V^K)$. It is simple to check that the following diagram commutes.

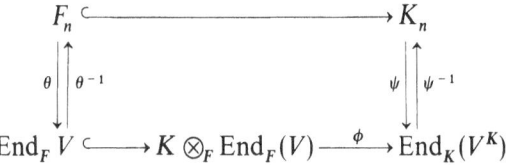

1.1. Schur's Lemma. *If R is a ring and U and V are irreducible R-modules, then any R-homomorphism of U to V is either an isomorphism or the zero map.* $\mathrm{End}_R(U)$ *is a division ring.* ☐

1.2. Corollary. *If F is algebraically closed and G an irreducible subgroup of* $\mathrm{GL}(n, F)$ *then* $\zeta_1(G) = G \cap F^* 1_n$.

Proof. For $C_{F_n}(G)$ is a finite dimensional division algebra over F and thus is $F1_n$. Hence $\zeta_1(G) = G \cap F1_n$. ☐

1.3. Corollary. *If F is algebraically closed and G is an abelian irreducible subgroup of* $\mathrm{GL}(n, F)$ *then* $n = 1$. ☐

We need the following well-known results on completely reducible modules. See [10] §15 and [26b] §§4.1, 4.2.

1.4. Theorem. *Let R be a ring and M an R-module. Then M is completely reducible if and only if M is generated by its irreducible submodules, which happens if and only if every submodule of M is a direct summand. If M is completely reducible and N is any submodule of M, then N and M/N are completely reducible. Also M is a direct sum of its homogeneous components,* $M = \bigoplus H_i$, *all the irreducible submodules of each* H_i *are isomorphic,* H_i *is fully invariant in M and every fully invariant submodule of M is a direct sum of some of the* H_i.

1.5. Theorem. *Let G be a group and H a subgroup of finite index m in G. Let F be a field of characteristic 0 or prime to m and V an FG-module which is completely reducible as FH-module. Then V is completely reducible as FG-module.*

Proof. Let V_1 be an FG-submodule of V. We have just to prove that V_1 is a direct summand of V as FG-module. There exists an FH-submodule U of V such that $V_1 \oplus U = V$. Denote by μ the natural projection of V onto U; μ is an FH-endomorphism.

Let T be a right transversal of H to G and g an element of G. For each t in T, $tg = h_{t,g} t_g$ for some $h_{t,g}$ in H and t_g in T, and $T = \{t_g : t \in T\}$. Let

$$\phi = \frac{1}{m} \sum_{t \in T} t^{-1} \mu t \quad \text{and} \quad V_2 = V\phi.$$

$\dfrac{1}{m}$ is defined in F since $(\mathrm{char}\, F, m) = 1$. Certainly ϕ is linear so V_2 is an F-subspace.

$$t^{-1}\mu t g = t^{-1}\mu h_{t,g} t_g = t^{-1} h_{t,g}\mu t_g = g t_g^{-1}\mu t_g$$

since μ is an H-map. Hence

$$(V\phi)g = V\cdot\frac{1}{m}\sum_{t\in T} g t_g^{-1}\mu t_g = Vg\phi = V\phi.$$

Therefore V_2 is an FG-submodule of V.
$V_1\mu = \{0\}$, so $V_1\phi = \{0\}$. Also for any v in V.

$$v - v\phi = \frac{1}{m}\sum_{t\in T} v t^{-1}(1-\mu) t \in V_1$$

since $V(1-\mu) = V_1$ and V_1 is G-invariant. Thus $V = V_1 + V_2$. If $w \in V_1 \cap V_2$, $w = v\phi$ for some v in V. Then $v - v\phi \in V_1$, so $v \in V_1$. Hence $w = 0$. The theorem is now proved. □

1.6. Corollary (Maschke-Schur). *If G is a locally finite subgroup of $\mathrm{GL}(n, F)$ and either* char $F = 0$ *or G contains no elements of order* char F, *then G is completely reducible.*

Proof. Let $g_1, \ldots, g_r$ be a maximal linearly independent subset of G and put $G_1 = \langle g_1, \ldots, g_r \rangle$. By 1.5 (with $H = \{1\}$) G_1 is completely reducible. Since $F\{G\} = F\{G_1\}$, G is completely reducible. □

We now look at the converse question to that raised in 1.5. If G is a completely reducible subgroup of $\mathrm{GL}(n, F)$ which subgroups H of G are completely reducible? We make first some preliminary remarks.

Let G be a group and V an FG-module. Suppose that V contains r non-trivial subspaces $V_1, \ldots, V_r$ such that $V = V_1 \oplus \cdots \oplus V_r$ and for each i and each g in G, $V_i g = V_k$ where $k = k(i, g)$. We write $k = i\sigma$ where σ maps $\{1, \ldots, r\}$ into itself. It is easy to see that σ is a permutation and that $\pi\colon g \mapsto \sigma$ is a homomorphism of G into S_r, the symmetric group on r letters. The set $\{V_1, \ldots, V_r\}$ is called a *system of imprimitivity* of G in V. $\{V\}$ is a system of imprimitivity of G in V. If it is the only one we say that G acts *primitively* on V, otherwise we say G acts *imprimitively* on V. If G is a subgroup of $\mathrm{GL}(n, F)$ we call G *primitive* or *imprimitive* according to whether G acts primitively or imprimitively on the space of n-row vectors (by right multiplication). Clearly if G is completely reducible and primitive, then G is irreducible.

1.7. Theorem (A. H. Clifford). *Let G be a group, H a normal subgroup of G and V an irreducible FG-module of finite F-dimension. Then V is com-*

pletely reducible into irreducible FH-modules of equal F-dimension. If $W_1, \ldots, W_r$ *are the non-trivial homogeneous components of V as FH-module, then* $\{W_1, \ldots, W_r\}$ *is a system of imprimitivity of G in V.*

Proof. Let U be an irreducible FH-submodule of V and g an element of G. Then g induces an invertible linear map of U onto Ug. If $h \in H$, then $Ugh = Uh^{g^{-1}}g \subseteq Ug$ since H is normal in G, and so Ug is an FH-submodule of V. If U_1 is a proper FH-submodule of Ug then the same argument shows that $U_1 g^{-1}$ is a proper FH-submodule of U. Thus Ug is an irreducible FH-submodule of V such that $\dim_F U = \dim_F Ug$. $\sum_{g \in G} Ug$ is an FG-submodule of V. Since V is irreducible $V = \sum_{g \in G} Ug$. That is, V is generated by irreducible FH-submodules and therefore is completely reducible as FH-module.

Let U_1 and U_2 be irreducible FH-submodules of W_i. Then there exists an FH-isomorphism ϕ of U_1 onto U_2. If $g \in G$, $U_1 g$ and $U_2 g$ are irreducible FH-modules and $\phi^g = g^{-1}\phi g$ is a linear mapping of $U_1 g$ onto $U_2 g$. For any h in H,

$$\phi^g h = g^{-1}\phi g h = g^{-1}\phi h^{g^{-1}} g$$
$$= g^{-1} h^{g^{-1}} \phi g \text{ since } H \lhd G \text{ and } \phi \text{ is an } FH\text{-isomorphism,}$$
$$= h \cdot \phi^g.$$

Thus ϕ^g is an FH-isomorphism of $U_1 g$ onto $U_2 g$ and so $U_1 g + U_2 g \subseteq W_j$ for some j. Therefore $W_i g \subseteq W_j$. Since V has finite dimension $W_i g = W_j$, consequently $\{W_1, \ldots, W_r\}$ is a system of imprimitivity of G in V. ☐

1.8. Corollary. *If H is a subnormal subgroup of a completely reducible linear group, then H is completely reducible.* ☐

Let G be a group, V an FG-module and $\{V_1, \ldots, V_r\}$ a system of imprimitivity of G in V. We say G acts *transitively* on $\{V_1, \ldots, V_r\}$ if for each pair i, j there exists a g in G such that $V_i g = V_j$ [1]. Suppose that V is irreducible and that for some i and j, $V_i g \ne V_j$ for all g in G. Let W be the subspace generated by all the V_k such that $V_k g \ne V_j$ for all g in G. Then $V_i \subseteq W$ and $V_j \cap W = \{0\}$ (since $V_j 1 = V_j$). Thus W is a proper subspace of V. Let $g \in G$ and $V_k \subseteq W$. If $V_k g \nsubseteq W$, there exists an x in G such that $V_j = (V_k g) x = V_k(g x)$. This contradicts $V_k \subseteq W$ and so W is an FG-submodule of V. This contradiction of the irreducibility of V proves:

1.9. *If* $\{V_1, \ldots, V_r\}$ *is a system of imprimitivity for the group G in the irreducible FG-module V, then G acts transitively on* $\{V_1, \ldots, V_r\}$. ☐

---

[1] i.e., if $G\pi$ is a transitive subgroup of $\mathbf{S}_r$ where π is the natural map of G into $\mathbf{S}_r$. Note that $G\pi$ may or may not be a primitive subgroup of $\mathbf{S}_r$ and this notion of primitivity is not the same as the one used above.

Keeping the above notation let

$$S_i = N_G(V_i) = \{g \in G : V_i g = V_i\}$$

and put $S = \bigcap\limits_{i=1}^{r} S_i$. Then S_i is a subgroup of G and $S = \ker \pi$ is a normal subgroup of G. If V is irreducible as FG-module, the S_i are conjugate subgroups (for if $V_i g = V_j$ then $S_i^g = S_j$). The order of G/S divides $r!$, thus if V has finite F-dimension n, $(G : S)$ divides $n!$.

1.10. *Let* $\{V_1, \ldots, V_r\}$ *be a system of imprimitivity for the group G in the FG-module V and put $S_i = N_G(V_i)$. If V is irreducible as FG-module then V_i is irreducible as FS_i-module. If $\{W_1, \ldots, W_s\}$ is a system of imprimitivity for S_i in V_i and if $V_i g_j = V_j$ for $j = 1, 2, \ldots, r$ where $g_j \in G$ and $g_i = 1$, then*

$$\{W_l g_j : l = 1, 2, \ldots, s; \ j = 1, 2, \ldots, r\}$$

is a system of imprimitivity for G in V. If V has finite F-dimension there exists a system of imprimitivity $\{U_1, \ldots, U_t\}$ for G in V such that for each k, $N_G(U_k)$ acts primitively on U_k.

(Such a system is called a minimal system of imprimitivity for G in V.)

Proof. Let X be a proper FS_i-submodule of V_i and put $Y = \sum\limits_{j=1}^{r} X g_j$. If $g \in G$, $V_i g_j g = V_k$ for some k, thus $g_j g g_k^{-1} \in S_i$ and so $X g_j g \subseteq X g_k$. Therefore Y is a proper FG-submodule of V. Thus if V is irreducible as FG-module V_i is irreducible as FS_i-module.

Again, since $g_j g g_k^{-1} \in S_i$, $W_l g_j g g_k^{-1} = W_m$ for some m and so $(W_l g_j) g = W_m g_k$. The remainder of 1.10 follows. □

Let $\{V_1, \ldots, V_r\}$ be a system of imprimitivity for G in the FG-module V and suppose that $\dim_F(V_1) = \cdots = \dim_F V_r = 1$. We say that G is *monomial* on V. If v_i is a non-zero element of V_i then $v_1, \ldots, v_r$ is a basis of V. Let ϕ be the homomorphism of G into $\mathrm{GL}(r, F)$ determined by this basis. Then $g \phi$ is a generalized permutation matrix for each g in G and $g \phi$ is a diagonal matrix for each g in $\bigcap\limits_{i=1}^{r} N_G(V_i)$. If G is a subgroup of $\mathrm{GL}(n, F)$ we say that G is monomial (over F) if G is monomial on the space of n-row vectors over F. Any two diagonal matrices in $\mathrm{GL}(n, F)$ commute. Thus a monomial subgroup of $\mathrm{GL}(n, F)$ contains an abelian normal subgroup of finite index dividing $n!$.

1.11. Lemma. *Let $\mathfrak{X}$ be a class of groups, closed under taking subgroups of finite index, such that every irreducible primitive representation of finite degree of an $\mathfrak{X}$-group over the field F has degree 1. Then every completely reducible representation of finite degree of an $\mathfrak{X}$-group over F is monomial.*

Proof. Let G be an $\mathfrak{X}$-group and V a completely reducible FG-module of finite dimension. $V = V_1 \oplus \cdots \oplus V_r$ where each V_i is an irreducible FG-module. If G is monomial on each V_i then G is monomial on V. Hence we can suppose that V is irreducible. By 1.10, V contains a system of imprimitivity for G, $\{W_1, \ldots, W_s\}$ such that W_i is an irreducible primitive $N_G(W_i)$-module. Since $(G : N_G(W_i))$ is finite $N_G(W_i)$ is an $\mathfrak{X}$-group and so $\dim_F W_i = 1$. Therefore G is monomial on V. $\quad\square$

1.12. Lemma. *Let F be an algebraically closed field and G a group of linear mappings of the n-dimensional F-module V. Suppose that A is an abelian normal subgroup of G such that V is completely reducible as FA-module. Then there exists a system of imprimitivity $\{V_1, \ldots, V_r\}$ for G in V such that $C_G(A) = \bigcap\limits_{i=1}^{r} N_G(V_i)$. $G/C_G(A)$ is isomorphic to a subgroup of S_r and in particular $(G : C_G(A))$ divides $n!$*

Proof. Let $V_1, \ldots, V_r$ be the homogeneous components of V as FA-module. By 1.7, $\{V_1, \ldots, V_r\}$ is a system of imprimitivity for G in V. It just remains to show that $C_G(A) = \bigcap\limits_{i=1}^{r} N_G(A)$. If g is an element of $C_G(A)$, then g induces an FA-automorphism on V. The V_i are fully invariant FA-submodules of V and so $V_i g = V_i$ for each i. Therefore $C_G(A) \subseteq \bigcap\limits_{i=1}^{r} N_G(V_i)$.

Let $g \in \bigcap\limits_{i=1}^{r} N_G(V_i)$ and $a \in A$. V_i is a direct sum of isomorphic FA-modules and by 1.3 each of these irreducible FA-modules has dimension 1. Therefore a acts on V_i as a scalar (i.e., for some α_i in F, $va = v\alpha_i$ for all v in V_i). Hence $[g, a]$ acts as the identity mapping on each V_i and so on V. But G acts faithfully on V. Therefore $[g, a] = 1$ and the lemma is proved. $\quad\square$

1.13. Corollary (Blichfeldt). *If F is algebraically closed and G is a primitive irreducible subgroup of $\mathrm{GL}(n, F)$, then every abelian normal subgroup of G lies in the centre of F_n.*

Proof. If A is an abelian normal subgroup of G, then by 1.12, $C_G(A) = G$. By 1.2, $\zeta_1(G) \subseteq F1_n$. $\quad\square$

1.14. Theorem. *Let F be algebraically closed and G a completely reducible subgroup of $\mathrm{GL}(n, F)$. If G is abelian-by-locally-supersoluble then G is monomial.*

Proof. The class of all abelian-by-locally-supersoluble groups is both subgroup and quotient-group closed. Let G_1 be a primitive irreducible abelian-by-supersoluble subgroup of $\mathrm{GL}(n, F)$. G_1 contains a maximal abelian normal subgroup A such that G_1/A is supersoluble. Suppose that $G_1 \neq A$. Then G_1 contains a normal subgroup K such that $A \subseteq K$ and

7

K/A is a non-trivial cyclic group. By 1.13, $A \subseteq \zeta_1(G_1)$. Hence K is abelian, a contradiction of the maximality of A. Thus G_1 is an abelian group and so $n=1$ (by 1.3). We have now shown that every completely reducible abelian-by-supersoluble subgroup of $GL(n, F)$ is monomial (1.11).

Let G be as in the statement of the theorem. By 1.11 it suffices to suppose that G is primitive and irreducible and to prove that $n=1$. G contains a finitely generated subgroup H spanning G linearly. Let I be the set of all finitely generated subgroups of G, and for each K in I let S_K be the set of all abelian normal subgroups of K of finite index at most $n!$. $\langle H, K \rangle$ is irreducible and finitely generated. Hence it is abelian-by-supersoluble and so monomial by the above. Therefore S_K is non-empty. Let $K \subseteq L$ be elements of I. Define the mapping $\lambda_K^L \colon S_L \to S_K$ by $B\lambda_K^L = B \cap K$ for $B \in S_L$. Then $(S_K, \lambda_K^L; I)$ is an inverse system of finite sets and so $\varprojlim S_K$ is non-empty (see [31a] I.K.1). Let $(A_K) \in \varprojlim S_K$ where $A_K \in S_K$ for each K in I. It is easy to see that $A = \bigcup_{K \in I} A_K$ is an abelian normal subgroup of G of finite index at most $n!$.

By 1.13 every abelian normal subgroup of G is contained in $\zeta_1(G)$. Hence $(G : \zeta_1(G))$ is finite. But G is abelian-by-locally supersoluble, so $G/\zeta_1(G)$ is supersoluble. That is G is abelian-by-supersoluble and consequently $n=1$ by the first part of the proof. $\quad\square$

Exercise 1.1 (B. Huppert [25]). *F is algebraically closed and G is a finite completely reducible subgroup of* $GL(n, F)$. *G has a soluble normal subgroup N such that every Sylow subgroup of N is abelian and* G/N *is supersoluble. Prove that G is monomial.*

Exercise 1.2. F is algebraically closed and G is a completely reducible subgroup of $GL(n, F)$. *G has a periodic soluble normal subgroup N such that every Sylow subgroup of N is abelian and* G/N *is locally supersoluble. Prove that G is monomial.*

The problem of characterizing the class of groups all of whose representations are monomial is a very interesting one. For the latest state of progress on this problem see [14], [50a], [50b], [50c] and [59a].

1.7 contains only part of what is usually referred to as Clifford's Theorem. In this book we shall not use the remainder, but for completeness we close our discussion of primitivity with it. It can be, and usually is, expressed in terms of projective representations, cf. [10] 51.7. However it is used in the theory of infinite linear groups to pass, in suitable circumstances, from a given group G to two closely associated groups H and K of smaller degree, and we state it in a form that has been adapted to this use. Provided the H and K are sufficiently related to G for one to be able to transfer data from G to H and K and the conclusion from H and K

back to G, then one can induct on the degree of the group. At first sight it appears to be a powerful tool, but it turns out to be of real value only if the group is soluble (or something like it), and in such a situation one can analyse the structure of the group in much more detail anyway; see Chapter 3.

1.15. Theorem (A. H. Clifford). *Let F be an algebraically closed field, G a primitive irreducible subgroup of $\mathrm{GL}(n, F)$ and N a normal subgroup of G. Then N is completely reducible into m equivalent irreducible parts of degree r where $rm=n$, and there exist linear groups H and K such that*

a) $F^* 1_m \subseteq H \subseteq \mathrm{GL}(m, F)$ and $F^* 1_r \subseteq K \subseteq \mathrm{GL}(n, F)$,

b) *G is isomorphic to a subgroup of $(H \times K)/Z$ where Z is the central subgroup of $H \times K$ given by*

and
$$Z = \langle (\alpha 1_m, \alpha^{-1} 1_r) \colon \alpha \in F^* \rangle,$$

c) *there exist homomorphisms ϕ of G onto $H/F^* 1_m$ and ψ of G onto $K/F^* 1_r$ such that*

$$N \subseteq \ker \phi \quad and \quad N \cap \ker \psi = N \cap F^* 1_n.$$

Proof. By 1.7 there exist integers r and m with $rm=n$ and an irreducible representation ρ of N into $\mathrm{GL}(r, F)$ such that for some x in $\mathrm{GL}(n, F)$ we have
$$h^x = \mathrm{diag}(h\rho, \dots, h\rho), \quad \text{for all } h \in N.$$

For each g in G write $g^x = (g_{ij})$ where the g_{ij} are $r \times r$ matrices over F. If $h \in N$ then the equation $g\, h^g = h g$ expressed in terms of ρ and g_{ij} becomes

$$g_{ij}(h^g)\rho = (h\rho)g_{ij} \quad \text{for } i, j = 1, 2, \dots, m. \tag{1}$$

The map
$$\rho^g \colon h \mapsto (h^g)\rho,$$

where $g \in G$ and $h \in N$, is an irreducible representation of N that by 1.7 is equivalent to ρ; for if V is the n-row vector space over F made into a G-module via right multiplication and if U is an FN-submodule of V giving rise to ρ (after a suitable choice of basis), then $U g^{-1}$ is an FN-submodule giving rise to ρ^g. Hence there exists an element $g\sigma$ of $\mathrm{GL}(r, F)$ such that
$$(h^g)\rho = (h\rho)^{g\sigma} \tag{2}$$

for every h in N. Since ρ is a homomorphism we can (and do) choose $g\sigma = g\rho$ for all $g \in N$.

From (1) and (2) we obtain

$$g_{ij}(g\sigma)^{-1}(h\rho) = (h\rho)g_{ij}(g\sigma)^{-1} \quad \text{for every } h \text{ in } N,$$

and hence by Schur's Lemma $g_{ij} = g\sigma \cdot \alpha_{ij}(g)$ for some $\alpha_{ij}(g) \in F$. Write $g\tau = (\alpha_{ij}(g))$, so τ is a mapping of G into F_m. If g' is a second element of G

we obtain

$$((g\,g')\,\sigma\cdot\alpha_{ij}(g\,g'))=(g\,g')^x=g^x(g')^x$$
$$=(g\,\sigma\cdot\alpha_{ij}(g))\,(g'\,\sigma\cdot\alpha_{ij}(g'))$$
$$=\left(g\,\sigma\cdot g'\,\sigma\cdot\left(\sum_{k=1}^{m}\alpha_{ik}(g)\,\alpha_{kj}(g')\right)\right). \tag{3}$$

Clearly at least one $\alpha_{ij}(g\,g')\neq 0$, so

$$(g'\,\sigma)^{-1}(g\,\sigma)^{-1}(g\,g')\,\sigma\in F^*1_r. \tag{4}$$

Taking $g'=g^{-1}$ in (3) and (4) we obtain

$$\left(1_r\sum_{k=1}^{m}\alpha_{ik}(g)\,\alpha_{kj}(g^{-1})\right)\in F^*1_n.$$

Therefore $G\tau\subseteq\mathrm{GL}(m,F)$ and (3) and (4) yield that

$$(g'\,\tau)^{-1}(g\,\tau)^{-1}(g\,g')\,\tau\in F^*1_m. \tag{5}$$

((4) and (5) say that σ and τ are projective representations of G.)

Put $H=F^*\cdot G\tau$ and $K=F^*\cdot G\sigma$. It follows from (4) and (5) that H is a subgroup of $\mathrm{GL}(m,F)$ and K is a subgroup of $\mathrm{GL}(r,F)$. For $h\in H$ write $h(i,j)$ for its (i,j)-th entry. The map θ of $H\times K$ into $\mathrm{GL}(n,F)$ given by $(h,k)\theta=(h(i,j)\,k)$ is a homomorphism with kernel

$$Z=\{(\alpha 1_m,\alpha^{-1}1_r)\colon \alpha\in F^*\}.$$

Moreover by construction $G^x\subseteq(H\times K)\,\theta$. (After a suitable identification of bases $(H\times K)\,\theta$ is simply $H\otimes K$.)

By (4) and (5) the maps σ and τ determine homomorphisms ψ of G into K/F^*1_r and ϕ of G into H/F^*1_m, and by definition of H and K, both ϕ and ψ are onto. If $g\in N$ then $g\sigma=g\rho$ by choice and $g_{ij}=\delta_{ij}\cdot g\rho$. Hence $g\tau=1_m$ and $g\in\ker\psi$ if and only if $g\rho\in F^*1_r$. Therefore $N\subseteq\ker\phi$ and $N\cap\ker\psi=N\cap F^*1_n$, and the theorem is proved. (It is easy to see that ϕ and ψ are induced by the above embedding of G into $(H\times K)/Z$ and the obvious projections of the latter group onto H/F^*1_m and K/F^*1_r.) ☐

We give an ad-hoc proof of the following well-known theorem of Burnside. More illuminating proofs may be found in [10] and [26b].

1.16. Theorem (Burnside). *Let A be an irreducible F-subalgebra of F_n. If $C_{F_n}(A)=F1_n$, then $A=F_n$.*

An equivalent formulation of this theorem is the following. Let A be an F-algebra and V an irreducible A-module which is finite dimensional as F-module. If ϕ is the homomorphism of A into $\mathrm{End}_F(V)$ defined by the action of A on V and if $C_{\mathrm{End}_F(V)}(A\,\phi)=F$, then ϕ is onto.

Proof. Let V be the space of n-row vectors over F and let F_n operate on V by right-multiplication. If $a = (a_{ij})$ is an element of F_n denote by ϕ_i the mapping of F_n into V given by $a\phi_i = (a_{i1}, \ldots, a_{in})$. It is easy to check that ϕ_i is an F_n-module homomorphism (and so ϕ_i is an A-module homomorphism).

Suppose that A does not contain n^2 linearly independent elements (i.e., that $A \neq F_n$). Then the equations

$$\mathrm{tr}(x\,a) = 0, \quad a \in A, \quad x = (x_{ij})$$

in the n^2 indeterminates x_{ij} have a non-zero solution over F. That is

$$U = \{x \in F_n : \mathrm{tr}(x\,a) = 0 \text{ for all } a \text{ in } A\} \neq \{0\}.$$

U is an A-module of finite F-dimension and so U contains an irreducible A-submodule W. By Schur's Lemma $\phi_i : W \to V$ is either the zero map or invertible. There exists $w = (w_{ij})$ in W such that $w_{ij} \neq 0$ for some i and j. Then $w\phi_i \neq 0$. Hence $\phi_i|_W$ is an A-isomorphism of W and V. In particular $\dim_F W = n$ and $\mathrm{End}_A(W) \cong_F \mathrm{End}_A(V) \cong_F F$ as F-algebras. We can suppose that $i = 1$.

Now $\phi_j(\phi_1|_W)^{-1}$ is an endomorphism of W. Thus for each $j = 1, 2, \ldots, n$ there exists an element α_j of F such that $w\phi_j(\phi_1|_W)^{-1} = \alpha_j w$ for all w in W. If $w = (w_{ij})$ this says that $\alpha_j w_{1k} = w_{jk}$ for $j, k = 1, 2, \ldots, n$. Hence $w = (\alpha_j w_{1k})_{jk}$. But $1_n \in A$ so

$$0 = \mathrm{tr}(w\,1_n) = \mathrm{tr}(w) = \sum_{j=1}^{n} \alpha_j w_{1j}$$

and so $\dim_F W \leq n - 1$. This contradiction proves the theorem. □

1.17. Corollary (Burnside). *If F is an algebraically closed field and G is an irreducible subgroup of $\mathrm{GL}(n, F)$ then G contains n^2 linearly independent elements.*

Proof. By the proof of 1.2, $C_{F_n}(G) = F\,1_n$. Hence $F\{G\} = F_n$ by 1.16. □

Throughout this text $\bar{F}$ denotes the algebraic closure of the field F.

1.18. Corollary. *Let A be an F-algebra and V an A-module of finite F-dimension. The following are equivalent:*

   i) *V^K is irreducible as A^K-module for every extension field K of F;*
   ii) *$V^{\bar{F}}$ is irreducible as $A^{\bar{F}}$-module;*
   iii) *V is irreducible as A-module and $C_E(A) = F$ where $E = \mathrm{End}_F(V)$;*
   iv) *The natural homomorphism μ of A into $\mathrm{End}_F(V)$ is onto.*

If the A-module V satisfies any of the above V is called an *absolutely irreducible* A-module. A subset X of F_n is called absolutely irreducible if the space of n-row vectors over F is absolutely irreducible as $F\{X\}$-module. Note that in iii) $C_E(A)$ means $C_E(A\mu)$ where μ is the μ of iv).

Proof. If K is any extension field of F we have:

$$1 \otimes_F A \subseteq K \otimes_F A \xrightarrow{1 \otimes \mu} K \otimes_F E \xrightarrow{\phi} \text{End}_K(V^K)$$

where μ is the F-algebra homomorphism of iv) and ϕ is the isomorphism mentioned in the introduction to this chapter. That i) implies ii) is trivial. Suppose that ii) holds. By Schur's Lemma $C_{\bar{E}}(\bar{F} \otimes A) = \bar{F}$ where $\bar{E} = \text{End}_{\bar{F}}(V^{\bar{F}})$. Let $x \in C_E(A)$. Taking $K = \bar{F}$ in the above sequence

$$(1 \otimes x)\phi \in C_{\bar{E}}(\bar{F} \otimes A).$$

Thus $1 \otimes x \in \bar{F} \phi^{-1} = \bar{F} \otimes 1$. Therefore $x \in F$, so $C_E(A) = F$. If U were a proper A-submodule of V, then $\bar{F} \otimes U$ would be a proper $A^{\bar{F}}$-submodule of $V^{\bar{F}}$. Hence V is an irreducible A-module. iv) follows from iii) by 1.16. Suppose that iv) holds. Since $\mu: A \to E$ is onto,

$$(1 \otimes \mu)\phi: K \otimes_F A \to \text{End}_K(V^K)$$

is onto. Trivially V^K is irreducible as $\text{End}_K(V^K)$-module. The result now follows. □

The following theorem often allows one to suppose that an irreducible linear group is even absolutely irreducible. The idea is to be able to assume that the ground-field is algebraically closed. We shall have other results of this kind later.

1.19. Theorem. *Let G be an irreducible subgroup of $\text{GL}(n, F)$. Then there exists a finite extension field K of F and a subgroup H of $\text{GL}(m, K)$ isomorphic to G where $n = m(\dim_F K)$ such that H is absolutely irreducible. Moreover, if G is a primitive subgroup of $\text{GL}(n, F)$, then H is a primitive subgroup of $\text{GL}(m, K)$.*

Proof. By Schur's Lemma $D = C_{F_n}(G)$ is a division ring. Let K be any maximal subfield of D. Then $F1_n \subseteq K \subseteq D \subseteq F_n$ and $C_D(K) = K$. V the n-row F-space is naturally a vector space over K and since $F1_n \subseteq K$, $n = m(\dim_F K)$ for some integer m. Every K-linear map of V is trivially an F-linear map and so there exists a K-algebra isomorphism ϕ of $C_{F_n}(K)$ onto K_m.

Since $K \subseteq C_{F_n}(G)$, so $G \subseteq C_{F_n}(K)$. Put $H = G\phi$. Now

$$C_{F_n}(G) \cap C_{F_n}(K) = D \cap C_{F_n}(K) = K.$$

Also since $G \subseteq C_{F_n}(K)$, V is irreducible as KG-module. By 1.18, V is absolutely irreducible as KG-module.

Every K-subspace of V is an F-subspace. Thus if G is imprimitive on V as KG-module it is imprimitive on V as FG-module. We have now shown that H is an absolutely irreducible subgroup of $\mathrm{GL}(m, K)$ and that H is primitive in $\mathrm{GL}(m, K)$ whenever G is primitive in $\mathrm{GL}(n, F)$. $\square$

1.20. Lemma. *Let G be an irreducible subgroup of $\mathrm{GL}(n, \bar{F})$ and suppose that $\{\mathrm{tr}\, g : g \in G\}$ is a finite set of order m. Then G is a finite group of order at most m^{n^2}.* (tr g denotes the trace of the matrix g.)

Proof. By 1.17, G contains n^2 linearly independent elements $g_1, \ldots, g_{n^2}$. For each g in G let $g = (g(i, j))$. Then

$$\mathrm{tr}(g_k g) = \sum_{i, j} g_k(i, j)\, g(j, i).$$

For given values of the $\mathrm{tr}(g_k g)$ there is at most one element g satisfying the above n^2 equations (since $g_1, \ldots, g_{n^2}$ are linearly independent). There are at most m possible values for each $\mathrm{tr}(g_k g)$ and so $|G| \le m^{n^2}$. $\square$

$$\mathrm{Tr}(n, F) = \left\{ \text{all matrices in } \mathrm{GL}(n, F) \text{ of the form } \begin{pmatrix} a_1 & & 0 \\ & \ddots & \\ * & & a_n \end{pmatrix} \right\},$$

$$\mathrm{Tr}_1(n, F) = \left\{ \text{all matrices in } \mathrm{GL}(n, F) \text{ of the form } \begin{pmatrix} 1 & & 0 \\ & \ddots & \\ * & & 1 \end{pmatrix} \right\},$$

and $D(n, F) = \{\text{all diagonal matrices in } \mathrm{GL}(n, F)\}$. $\mathrm{Tr}(n, F)$ is called the triangular subgroup, $\mathrm{Tr}_1(n, F)$ the unitriangular subgroup and $D(n, F)$ the diagonal subgroup of $\mathrm{GL}(n, F)$.

A subgroup G of $\mathrm{GL}(n, F)$ is said to be *triangularizable over F* if there exists an element x of $\mathrm{GL}(n, F)$ such that $G^x \subseteq \mathrm{Tr}(n, F)$. We say that G is *triangularizable* if G is triangularizable over $\bar{F}$. It is easy to see that G is triangularizable if and only if G is triangularizable over some finite extension field of F. In the same way we introduce the terms *unitriangularizable* and *diagonalizable*. If V is the n-row vector space over F, the subgroup G of $\mathrm{GL}(n, F)$ is triangularizable over F if and only if the composition factors of V as FG-module have dimension 1. G is unitriangularizable over F if and only if the composition factors of V as FG-module are trivial as G-module (i.e., G acts as the identity on them).

If g is an element of $\mathrm{GL}(n, F)$ g is called *unipotent* if $(g - 1_n)^n = 0$. Trivially g is unipotent if and only if all its eigenvalues are 1. A subgroup G of $\mathrm{GL}(n, F)$ is called *unipotent* if all its elements are unipotent. Note that if ρ is a representation of G arising from a factor of the n-row vector space over F as FG-module, then $G\rho$ is unipotent whenever G is.

1.21. Corollary. *Let G be a unipotent subgroup of $\mathrm{GL}(n, F)$. Then G is unitriangularizable over F.*

The converse of this corollary is trivially true.

Proof. It suffices to suppose that G is irreducible and to prove that $n=1$. We suppose first that F is algebraically closed. Each element of G has trace $n 1_F$ and so $|G|=1$ by 1.20. Therefore $n=1$.

Suppose now that F is not algebraically closed. Let $R = \{(g-1): g \in G\}$. By what we have proved above G is unitriangularizable over $\bar{F}$. Hence R is nilpotent. That is there exists a least non-negative integer r such that

$$R^{r+1} = \{(g_0 - 1)(g_1 - 1)\ldots(g_r - 1): g_0, \ldots, g_r \in G\} = \{0\}.$$

Let V be the space of n-row vectors over F. Since $R^r \neq \{0\}$ there exists a non-zero element v in VR^r. Then $v(g-1)=0$ for all g in G and so vF is an FG-submodule of V. As G is irreducible $n=1$ and so the corollary is proved. ☐

1.22. Theorem. *Let G be a completely reducible subgroup of $\mathrm{GL}(n, F)$. Then $\mathrm{GL}(n, \bar{F})$ contains a completely reducible subgroup H such that there is an isomorphism $\phi: G \to H$ satisfying $\mathrm{tr}(g) = \mathrm{tr}(g\phi)$ for all g in G.*

Note that if G is irreducible H may be reducible.

Proof. $G \subseteq \mathrm{GL}(n, F) \subseteq \mathrm{GL}(n, \bar{F})$. There exists an x in $\mathrm{GL}(n, \bar{F})$ and irreducible representations $\rho_1, \ldots, \rho_r$, $\rho_i: G \to \mathrm{GL}(n_i, \bar{F})$, such that

$$g^x = \begin{pmatrix} g\rho_1 & & 0 \\ & \ddots & \\ * & & g\rho_r \end{pmatrix} \quad \text{for all } g \text{ in } G.$$

Let ϕ be the mapping of G into $\mathrm{GL}(n, \bar{F})$ given by

$$g\phi = \begin{pmatrix} g\rho_1 & & 0 \\ & \ddots & \\ 0 & & g\rho_r \end{pmatrix}.$$

Clearly ϕ is a completely reducible representation of G in $\mathrm{GL}(n, \bar{F})$ such that $\mathrm{tr}(g) = \mathrm{tr}(g\phi)$ for all g in G. It remains to show that $U = \ker \phi$ is the trivial group. U^x is unitriangular, so U is unipotent. Hence by 1.21, U is unitriangularizable over F. Let V be the n-row vector space over F. Then U acts trivially on each of the composition factors of V as FU-module. But by Clifford's Theorem U is completely reducible as FU-module. Hence $U = \{1\}$. ☐

1.20 can be developed somewhat. Using Exercise 1.5 and 1.22 (to reduce to the algebraically closed case) one can show that if G is a completely reducible subgroup of $\mathrm{GL}\,(n, F)$ over a field F of characteristic $<n$ such that $|\{\mathrm{tr}\,g\colon g\in G\}|\leq m$, then $|G|\leq m^{n^2}$. This is clearly false if $\mathrm{char}\,F=p>0$, for example F^*1_p is a completely reducible subgroup of $\mathrm{GL}(p, F)$ and $\{\mathrm{tr}\,x\colon x\in F^*1_p\}=\{0\}$. It is also clearly false if G is not completely reducible, even if $\mathrm{char}\,F=0$; if F is any infinite field and $n\geq2$, then $\mathrm{Tr}_1(n, F)$ is infinite and yet each of its elements has trace $n1_F$. However a weakened version of 1.20 does extend nicely.

1.23. Corollary (Burnside). *Let G be a completely reducible subgroup of $\mathrm{GL}(n, F)$ of finite exponent e. Then G is finite of order at most e^{n^3}.*

Proof. By 1.22 we may assume that F is algebraically closed. If

$$n=n_1+\cdots+n_r,$$

where each n_i is a positive integer and each e_i is a positive integer dividing e, then

$$\prod_{i=1}^{r}e_i^{n_i^3}\leq e^{n^3}.$$

Hence we may also assume that G is irreducible. If g is an element of G, each eigenvalue of g is an e-th root of 1. Thus

$$|\{\mathrm{tr}\,g\colon g\in G\}|\leq e^n,$$

and therefore $|G|\leq e^{n^3}$ by 1.20. □

If F is infinite of positive characteristic and $n\geq2$, then $\mathrm{Tr}_1(n, F)$ is infinite of finite exponent, see Exercise 1.3 below. We shall see that if $\mathrm{char}\,F=0$ and if G has finite exponent, then 1.6 implies that G must be completely reducible and hence finite by 1.23.

1.19 and 1.22 will enable us to reduce our problems to the case where F is algebraically closed. A further result of this type, which we shall not need in this book, is the following.

1.24. Theorem. *Let A be an F-algebra, U an A-module and K an extension field of F.*

i) a) *If $K\otimes_F V=V^K$ is completely reducible as A^K-module, then V is completely reducible as A-module;*

b) *If F is a perfect field, $\dim_F V$ is finite and V is completely reducible as A-module, then V^K is completely reducible as A^K-module.*

ii) a) *If V^K is irreducible as A^K-module, then V is irreducible as A-module;*

b) *If F is algebraically closed, $\dim_F V$ is finite and V is irreducible as A-module, then V^K is irreducible as A^K-module.*

The proof is not difficult and can be found in [10] or [7] for example. Part ii) follows easily from what we have proved above. Part i) a) is also very easy.

Exercise 1.3. Prove that

   i) $\mathrm{Tr}(n, F)' = \mathrm{Tr}_1(n, F)$ unless $F = GF(2)$ when $\mathrm{Tr}(n, F) = \mathrm{Tr}_1(n, F)$;

   ii) $\mathrm{Tr}(n, F)$ is the split extension of $\mathrm{Tr}_1(n, F)$ by $D(n, F)$;

   iii) $\mathrm{Tr}(n, F)$ is soluble with derived length $1 - [-\log_2 n]$ unless $F = GF(2)$ when it is soluble with derived length $-[-\log_2 n]$.

   iv) $\mathrm{Tr}_1(n, F)$ is nilpotent of class $n - 1$;

   v) If char $F = 0$, $\mathrm{Tr}_1(n, F)$ is torsion-free, and if char $F = p > 0$, $\mathrm{Tr}_1(n, F)$ has finite exponent p^m where $p^{m-1} < n \leq p^m$.

To solve iii) and iv) note that a direct computation shows that if

$$a = (a_{ij}) \quad \text{where} \quad a_{ij} = \begin{cases} 1 & \text{if } i = j \\ 0 & \text{if } i < j \text{ or if } 0 < i - j < r, \end{cases}$$

and

$$b = (b_{ij}) \quad \text{where} \quad b_{ij} = \begin{cases} 1 & \text{if } i = j \\ 0 & \text{if } i < j \text{ or if } 0 < i - j < s, \end{cases}$$

then

$$[a, b] = (c_{ij}) \quad \text{where} \quad c_{ij} = \begin{cases} 1 & \text{if } i = j \\ 0 & \text{if } i < j \text{ or if } 0 < i - j < r + s. \end{cases}$$

These parts also follow easily from the theory of stability groups, see [24 a] Corollary to 3.5, and 3.4 iii).

Exercise 1.4. If F is an infinite field and $n \geq 3$, prove that $\mathrm{Tr}_1(n, F)$ is *not* residually finite, and that $\mathrm{Tr}_1(2, F)$ is residually finite if and only if char $F \neq 0$.

Exercise 1.5 (Frobenius, Schur). F is an algebraically closed field, A is an F-algebra and $V_1, V_2, \ldots, V_r$ are a finite set of pairwise non-isomorphic, irreducible A-modules with finite F-dimension. For each k choose a basis for V_k and let $x \mapsto (\alpha_{ij}^k(x))$ be the corresponding matrix representation of A. Prove that the $r n^2$ mappings α_{ij}^k of A into F are linearly independent.
   Hint: adapt the proof of 1.16 or see [10] 27.8.

Exercise 1.6. G is a subgroup of $GL(n, F)$ that is triangularizable (resp. diagonalizable) over some extension field of F. Prove that G is triangularizable (resp. diagonalizable).

2. Some Examples of Linear Groups

In this chapter we shall have a look at what sort of group has a faithful representation of finite degree over a field. Our plan of campaign is roughly as follows. Firstly we consider abelian groups, and then soluble groups – especially those satisfying either the minimal condition or the maximal condition on subgroups. Next we prove a local theorem of Mal'cev, which states that a group has a faithful representation of degree n if and only if each of its finitely generated subgroups has such a representation. Then we prove that a free group has a faithful representation of degree 2 over most fields, and discuss similar results for certain relatively free groups. We follow this with Nisnevič's Theorem, that a free product of groups having faithful representations of degree n over fields of characteristic $p \geq 0$, has a faithful representation of degree $n+1$ over some field of characteristic p, plus some related results. Finally we consider the representability of wreath products of linear groups.

We begin with a trivial remark.

2.1. *If G is any finite group and F is any field, then G has a faithful representation over F of finite degree.* □

If G is any group $\tau(G)$ is the subgroup of G generated by all the periodic normal subgroups of G; that is $\tau(G)$ is the maximum periodic normal subgroup of G. G has *finite rank at most n* if every finite subset of G is contained in an n-generator subgroup of G. If G is abelian and periodic then G has finite rank at most n if and only if for each prime p the Sylow p-subgroup of G is a direct product of at most n cyclic and Prüfer p-groups (a Prüfer p-group is a $\mathbf{C}_{p^\infty}$-group). If π is any set of primes and G is a group with a unique maximal π-subgroup we denote this maximal π-subgroup by G_π.

2.2. Theorem (Mal'cev [34]). i) *An abelian group A has a faithful representation of degree $n \geq 1$ over some field of characteristic zero if and only if $\tau(A)$ has rank at most n.*

ii) *An abelian group A has a faithful representation of degree $n \geq 1$ over some field of characteristic $p > 0$ if and only if $\tau(A)_{p'}$ has finite rank r and*

$\tau(A)_p$ *has finite exponent* p^e *satisfying*

$$p^{e-1} + \max\{1, r\} < n + 1.$$

Proof. i) Let A be an abelian subgroup of $GL(n, F)$ where $\operatorname{char} F = 0$. We can clearly suppose that F is algebraically closed. By 1.6 and 1.3, $\tau(A)$ is diagonalizable; that is, there exists an element x of $GL(n, F)$ such that $\tau(A)^x \subseteq D(n, F)$. Now $D(n, F)$ is isomorphic to the direct product of n copies of F^* and the Sylow subgroups of F^* are Prüfer groups. Therefore $\tau(A)$ has finite rank at most n.

Suppose now that A is an abelian group such that $\tau(A)$ has rank at most n. Let F be an algebraically closed field of characteristic zero of transcendence degree greater than $\max\{\aleph_0, |A|\}$. Denote by $\mathscr{S}$ the set of all pairs (B, ϕ) where B is a subgroup of A and ϕ is a monomorphism of B into $D(n, F)$. $\mathscr{S}$ is non-empty, $(\{1\}, 1 \mapsto 1_n) \in \mathscr{S}$. Order $\mathscr{S}$ by defining $(B_1, \phi_1) \leq (B_2, \phi_2)$ whenever $B_1 \subseteq B_2$ and $\phi_2|_{B_1} = \phi_1$. $\mathscr{S}$ is inductively ordered. Let (C, ψ) be a maximal element of $\mathscr{S}$; this exists by Zorn's Lemma. If $C \neq A$ there exists an element a of $A \smallsetminus C$ such that either

1. aC has infinite order in A/C or
2. aC has order q in A/C for some prime q.

1. Suppose that $|aC| = \infty$. Then $\langle a, C \rangle = \langle a \rangle \times C$.

$$C\psi \subseteq D(n, F_1) \subseteq D(n, F)$$

for some subfield F_1 of F of cardinal $\max\{\aleph_0, |A|\}$. Hence there exists an element α of F that is algebraically independent of F_1. Define

$$\psi_1 : \langle a, C \rangle \to D(n, F) \quad \text{by} \quad (a^i c)\psi_1 = \alpha^i(c\psi).$$

ψ_1 is a monomorphism of $\langle a, C \rangle$ into $D(n, F)$ such that $\psi = \psi_1|_C$. This contradiction of the maximality of (C, ψ) proves that $C = A$.

2. Suppose that $a^q = b \in C$ where q is a prime. If a_1 is an element of A satisfying $a_1^q = b$, then $(a_1 a^{-1})^q = 1$. Since the rank of $\tau(A)$ is at most n, A contains at most q^n elements a_1 satisfying $a_1^q = b$, one of which is a. Hence C contains at most $(q^n - 1)$ elements a_1 satisfying $a_1^q = b$. $D(n, F)$ contains q^n elements x_1 satisfying $x_1^q = b\psi$ since F is algebraically closed, and $C\psi$ can contain at most $q^n - 1$ of these x_1. Therefore there exists x in $D \smallsetminus C\psi$ satisfying $x^q = b$. It is simple to check that the mapping

$$\psi_1 : \langle a, C \rangle \to D(n, F)$$

given by $(a^i c)\psi_1 = x^i(c\psi)$ is a monomorphism extending ψ. As in Case 1 this contradiction proves that $A = C$ and completes the proof of part i) of the theorem.

ii) Let A be an abelian subgroup of $GL(n, F)$ where F is an algebraically closed field of characteristic $p > 0$. If the element a of A has order p^r then

$0 = a^{p^r} - 1 = (a-1)^{p^r}$ since char $F = p$. Therefore $a - 1$ is nilpotent, so $(a-1)^n = 0$. Hence if $n \le p^t$, $0 = (a-1)^{p^t} = a^{p^t} - 1$. We have proved the following:

(∗) If $p^{t-1} < n \le p^t$, then $\tau(A)_p$ has finite exponent dividing p^t.

Let $GL(n, F)$ act on the n-row vector space V over F by right multiplication. By 1.6 and 1.3, V is completely reducible as $F\tau(A)_{p'}$-module into irreducible submodules of F-dimension 1. If $V_1, \ldots, V_s$ are the (nonzero) homogeneous components of V as $F\tau(A)_{p'}$-module then $\tau(A)_{p'}$ acts on each V_i as a group of scalars. Therefore $r = \operatorname{rank} \tau(A)_{p'} \le s$. Since $1 \le s$, so $\max\{1, r\} \le s$.

Each element of A induces an $F\tau(A)_{p'}$-automorphism on V since A is abelian and the V_i are fully invariant in V as $F\tau(A)_{p'}$-modules; thus the V_i are FA-submodules of V. Let e_i be the unique integer such that

$$p^{e_i - 1} < \dim_F V_i \le p^{e_i}.$$

If $x \in \tau(A)_p$, $x^{p^{e_i}}$ induces the identify mapping on V_i and so if p^e is the exponent of $\tau(A)_p$ then $p^e \le \max_i p^{e_i}$. Thus

$$p^{e-1} < \max_i \dim_F V_i \le n - (s-1)$$

since $V_i \ne \{0\}$ for each i. Consequently

$$p^{e-1} + \max\{1, r\} \le p^{e-1} + s < n + 1.$$

Let A be an abelian group such that $\tau(A)_{p'}$ has finite rank r and $\tau(A)_p$ has finite exponent p^e. We prove that A has a faithful representation of degree $[p^{e-1} + \max\{1, r\}]$ over some field of characteristic p. $\tau(A)_p$ is a pure (serving) subgroup of A of finite exponent and so $A = \tau(A)_p \times B$ for some subgroup B of A ([32] Vol. 1, p. 179). Just as in the proof of i) there exists a monomorphism ψ_1 of B into $D(\max\{1, r\}, F)$ where F is some algebraically closed field of characteristic p and transcendence degree greater than $\max\{\aleph_0, |A|\}$. Let ψ_2 be the monomorphism of $D(\max\{1, r\}, F)$ into $D([p^{e-1} + \max\{1, r\}], F)$ given by

$$\operatorname{diag}(d_1, d_2, \ldots) \to \operatorname{diag}(d_1, \ldots, d_1, d_2, d_3, \ldots)$$

where the d_1 is repeated $[p^{e-1} + 1]$ times. Put $\psi = \psi_1 \psi_2$.

$\tau(A)_p$ is a direct product of cyclic groups of order dividing p^e ([32] Vol. 1, p. 173). If α and β are algebraically independent elements of F let x and y be the $s \times s$ matrices

$$x = \begin{pmatrix} 1 & & & 0 \\ & \alpha & & \\ & & \ddots & \\ 0 & & & \alpha^{s-1} \end{pmatrix}, \quad y = \begin{pmatrix} 1 & & & 0 \\ & \beta & & \\ & & \ddots & \\ 0 & & & \beta^{s-1} \end{pmatrix}.$$

It is simple to compute that $xy = yx$, $\langle x, y \rangle = \langle x \rangle \times \langle y \rangle$ and

$$
x^{s-1} = \begin{pmatrix} 1 & & & & & & 0 \\ * & 1 & & & & & \\ & * & \ddots & & & & \\ & & & \ddots & & & \\ & & & & \ddots & & \\ \alpha^{s-1} & & & & & * & 1 \end{pmatrix}.
$$

Hence if $s = [p^{e-1} + 1]$ then $|x| \geq p^{e-1} p = p^e$. In fact $|x| = p^e$ by $(*)$ above since $p^e \geq s$. If $\{\alpha_i : i \in I\}$ is a transcendence basis of F then

$$
\langle x: \alpha = \alpha_i, \, i \in I \rangle
$$

is a direct product of cyclic groups of order p^e of cardinal greater than $|A|$. Therefore there exists a monomorphism ϕ_1 of $\tau(A)_p$ into $\mathrm{GL}(s, F)$. If ϕ_2 denotes the usual embedding of $\mathrm{GL}(s, F)$ into

$$
\mathrm{GL}([p^{e-1} + \max\{1, r\}], F)
$$

$\left(\text{i.e., } g\phi_2 = \mathrm{diag}(g, 1_{\max\{0, r-1\}})\right)$ put $\phi = \phi_1 \phi_2$. Then $\tau(A)_p \phi \cap B\psi = \{1\}$ and $[\tau(A)_p \phi, \, B\psi] = \{1\}$. Thus

$$
\phi \times \psi: A \to \mathrm{GL}([p^{e-1} + \max\{1, r\}], F)
$$

is a monomorphism. $\square$

2.3. Lemma. *Let H be a subgroup of finite index m in the group G and ϕ a monomorphism of H into $\mathrm{GL}(n, F)$. Then there exists a monomorphism ϕ_1 of G into $\mathrm{GL}(mn, F)$. If H is normal in G, ϕ completely reducible and either $\mathrm{char}\, F = 0$ or $\mathrm{char}\, F$ is prime to m, then ϕ_1 is also completely reducible.*

Proof. Let $1 = g_1, \ldots, g_m$ be a right transversal of H to G. If $g \in G$ then $g_i g = h_i g_{i\sigma}$ for each i, where $h_i \in H$ and σ is some element of S_m.

Let V be the n-row vector space over F regarded as FH-module via ϕ. Then $W = V \otimes_{FH} FG$ is an FG-module of dimension mn. We prove that G acts faithfully on W. Let $g \in G$ induce the identity on W. Now

$$
W = \bigoplus_{i=1}^{m} V \otimes g_i
$$

as F-modules and $V \otimes g_1$ is an FH-submodule of W isomorphic to V. Then $V \otimes g_i = (V \otimes g_i) g = V \otimes h_i g_{i\sigma} = V \otimes g_{i\sigma}$. Therefore σ is the identity permutation. Thus $g = g_1^{-1} h_1 g_1 \in H$. Now H is faithful on V and so on $V \otimes g_1$; consequently $g = 1$.

(This representation of G can be given quite explicitly in terms of matrices. We define $\phi_1: G \to \mathrm{GL}(mn, F)$ by $g\phi_1 = (g(i, j))$, $g \in G$ where

$g(i, j)$ is an $n \times n$ matrix over F such that

$$g(i, j) = \begin{cases} 0_n & \text{if } j \neq i\sigma \\ h_i \phi & \text{if } j = i\sigma. \end{cases}$$

It is an easy computation that ϕ_1 is a monomorphism of G and in fact ϕ_1 is the matrix representation of G arising from W with a particular choice of basis.)

Suppose now that H is normal in G and that ϕ is completely reducible. Then each $V \otimes g_i$ is an FH-submodule of W and $v \mapsto v \otimes g_i$ induces a lattice isomorphism between the lattices of FH-submodules of V and $V \otimes g_i$. Therefore each $V \otimes g_i$, and so W, is a completely reducible FH-module. The result now follows from 1.5. ☐

2.4. Corollary. *If p is zero or a prime and G is a soluble group with minimal condition (on subgroups) then G has a faithful representation over a field of characteristic p for all but a finite number of non-zero p.*

Proof. By Černikov's Theorem ([32] Vol. 2, p. 191) G contains a normal subgroup A of finite index that is a direct product of a finite number of Prüfer groups. By 2.2, A has a faithful representation over a field of characteristic p if and only if $p \notin \pi(A)$. The result now follows from 2.3. ☐

It follows from work of W. Feit and R. Brauer ([4]) that a group with minimal condition has a faithful representation (of finite degree) over some field if and only if it is abelian-by-finite (see Chapter 9 for further discussion of this and related results).

Hsien-Chung Wang proved for polycyclic groups (soluble groups satisfying the maximal condition on subgroups) the result corresponding to 2.4. He showed that every polycyclic group has a faithful representation over the complex numbers ([59]). One can easily show that a polycyclic group has a faithful representation over a field of positive characteristic if and only if it is abelian-by-finite, and that if this is the case then it has a faithful representation over a field of characteristic p for every possible p. H. Zassenhaus [74] claimed to show that a group with maximal condition has a faithful representation over some field if and only if it is polycyclic-by-finite. A correct proof has now been given by Tits ([57a] but see also Chapter 10).

L. Auslander [2] and R. Swan [57] independently improved Wang's result to the following:

2.5. Theorem. *Every polycyclic group has a faithful representation of finite degree over $\mathbb{Z}$.*

Conversely every soluble subgroup of $GL(n, \mathbb{Z})$ is polycyclic ([47] 2.13). The proof below of 2.5 is taken from [57]. It is divided into four parts.

1. *Let* $G = \langle g_1, \dots, g_n \rangle$ *be a finitely generated group and* $\mathfrak{a}$ *a (two-sided) ideal in* $\mathbb{Z}G$ *such that* $\mathbb{Z}G/\mathfrak{a}$ *is finitely generated as* $\mathbb{Z}$-*module. Then* $\mathfrak{a}$ *is a finitely generated ideal and for each* $r = 1, 2, \dots, \mathbb{Z}G/\mathfrak{a}^r$ *is finitely generated as* $\mathbb{Z}$-*module.*

Let $u_1, \dots, u_m$ generate $\mathbb{Z}G$ modulo $\mathfrak{a}$ as $\mathbb{Z}$-module and put

$$M = \left(\sum_1^m u_i \, \mathbb{Z} \right) + \sum_1^n (g_j \, \mathbb{Z} + g_j^{-1} \, \mathbb{Z}); \qquad \mathfrak{a} + M = \mathbb{Z}G.$$

Thus if m_1 and m_2 are elements of M, then $m_1 \, m_2 = x + m$ for some m in M and x in $\mathfrak{a} \cap (M + M^2)$. (Here M^2 denotes the $\mathbb{Z}$-submodule generated by all the products $m_1 \, m_2$.) Let $\mathfrak{b}$ be the ideal generated by $\mathfrak{a} \cap (M + M^2)$; $\mathfrak{b} \subseteq \mathfrak{a}$. Since $\mathfrak{b} + M$ is multiplicatively closed (by construction) and contains g_j and g_j^{-1} for $j = 1, 2, \dots, n$, so $\mathfrak{b} + M = \mathbb{Z}G$. Thus $\mathbb{Z}G/\mathfrak{b}$ is a finitely generated $\mathbb{Z}$-module. But $\mathfrak{b}$ is finitely generated as an ideal and $\mathfrak{b} \subseteq \mathfrak{a}$. Therefore $\mathfrak{a}$ is a finitely generated ideal of $\mathbb{Z}G$.

Suppose that $\mathbb{Z}G/\mathfrak{a}^r$ is finitely generated as $\mathbb{Z}$-module, certainly this is true if $r = 1$. $\mathfrak{a}^r/\mathfrak{a}^{r+1}$ is finitely generated as $(\mathbb{Z}G/\mathfrak{a}^r)^{\circ} \otimes_{\mathbb{Z}} (\mathbb{Z}G/\mathfrak{a}^r)$-module and so is finitely generated as $\mathbb{Z}$-module. (The $^{\circ}$ here indicates the opposite ring.) Therefore $\mathbb{Z}G/\mathfrak{a}^{r+1}$ is finitely generated as $\mathbb{Z}$-module. An inductive argument completes the proof. $\Box$

2. *Let N and H be normal subgroups of the group G and A a subgroup of the group G such that H is finitely generated and*

$$N \subseteq H, \qquad [H, G] \subseteq N, \qquad H \cap A = \{1\} \quad \text{and} \quad HA = G.$$

Suppose also that H is a subgroup of $\mathrm{GL}(n, \mathbb{Z})$ *such that N is unipotent. Then there exists an integer m and a homomorphism ϕ of G into* $\mathrm{GL}(m, \mathbb{Z})$ *such that $H \cap \ker \phi = \{1\}$ and $N \phi$ is unipotent.*

If $H = N$ and G is nilpotent then we can choose ϕ such that $G\phi$ is unipotent.

The inclusion map $H \to \mathrm{GL}(n, \mathbb{Z})$ determines a ring homomorphism $\theta \colon \mathbb{Z}H \to \mathbb{Z}_n$. Let $\mathfrak{a}$ be the ideal of $\mathbb{Z}H$ generated by the augmentation ideal of N and put $\mathfrak{k} = \ker \theta$. Since N is normal in H, $\mathfrak{a}^n$ is generated by elements of the form $(x_1 - 1)(x_2 - 1) \dots (x_n - 1)$, $x_i \in N$. By 1.21, N is unitriangularizable (over $\mathbb{Q}$) and so $\mathfrak{a}^n \subseteq \mathfrak{k}$. Now $\mathbb{Z}H/(\mathfrak{a} + \mathfrak{k})$ is a quotient of $\mathbb{Z}H/\mathfrak{k}$, which is a finitely generated $\mathbb{Z}$-module (as $\mathbb{Z}_n$ is). Hence by 1. above $V = \mathbb{Z}H/(\mathfrak{a} + \mathfrak{k})^n$ is a finitely generated $\mathbb{Z}$-module. Let T be its torsion part, T is fully invariant in V. Also since $\mathbb{Z}H/\mathfrak{k}$ is torsion-free, $T \subseteq \mathfrak{k}/(\mathfrak{a} + \mathfrak{k})^n$. Thus $U = V/T$ is a faithful H-module that is free of finite rank (m say) as $\mathbb{Z}$-module. Also N acts unipotently on U since $\mathfrak{a}^n \subseteq (\mathfrak{a} + \mathfrak{k})^n$. To complete the first part of the proof we have only to extend the action on U from H to G.

Each element g of G has a unique representation of the form $g = ha$ where $h \in H$ and $a \in A$. If $x = \sum \alpha_i h_i$ is an arbitrary element of $\mathbb{Z}H$ define $x \cdot g = (xh)^a = \sum \alpha_i (a^{-1} h_i ha)$. It is simple to check that this makes $\mathbb{Z}H$ into a right G-module in such a way that the action of $H \subseteq G$ is given by right multiplication. Let $h \in H$ and $a \in A$. Then in $\mathbb{Z}H$

$$h \cdot a = h^a = h[h, a] = h + h([h, a] - 1) \in h + \mathfrak{a},$$

since $[H, G] \subseteq N$. Therefore

$$(\mathfrak{a} + \mathfrak{f})^n \cdot a = ((\mathfrak{a} + \mathfrak{f})^n)^a = ((\mathfrak{a} + \mathfrak{f})^a)^n \subseteq (\mathfrak{a} + \mathfrak{f})^n.$$

Thus $(\mathfrak{a} + \mathfrak{f})^n$ is a G-submodule of $\mathbb{Z}H$ and so U is a $\mathbb{Z}G$-module of the required type.

Suppose now that $H = N$ and that G is nilpotent. Then N contains a series

$$N = N_1 \supseteq N_2 \supseteq \cdots \supseteq N_{s+1} = \{1\}$$

central in G. Denote by $\mathfrak{n}_i$ the ideal of $\mathbb{Z}N$ generated by the augmentation ideal of N_i. If $h \in H$ and $a \in A$ then

$$ha - 1 = (h-1)(a-1) + (h-1) + (a-1)$$

and

$$(h-1)(a-1) = (a-1)(h^a - 1) + (h^a - 1) - (h-1).$$

Hence it suffices to check that H and a act unipotently on U. Trivially $H = N$ implies that H acts unipotently on U. Let $x \in N$ and $y \in N_i$. Then

$$x \cdot (a-1) = x([x, a] - 1) \in \mathfrak{n}_1 = \mathfrak{a},$$

and

$$x(y-1) \cdot (a-1) = x^a y([y, a] - 1) + x([x, a] - 1)(y-1) \in \mathfrak{n}_{i+1} + \mathfrak{a}\mathfrak{n}_i.$$

Thus

$$\mathfrak{n}_i \cdot (a-1) \subseteq \mathfrak{n}_{i+1} + \mathfrak{a}\mathfrak{n}_i.$$

Hence

$$\mathfrak{n}_i \cdot (a-1)^2 \subseteq \mathfrak{n}_{i+1} + \mathfrak{a}(\mathfrak{n}_{i+1} + \mathfrak{a}\mathfrak{n}_i) = \mathfrak{n}_{i+1} + \mathfrak{a}^2 \mathfrak{n}_i.$$

A simple induction shows that

$$\mathfrak{n}_i \cdot (a-1)^n \subseteq \mathfrak{n}_{i+1} + \mathfrak{a}^n \mathfrak{n}_i \subseteq \mathfrak{n}_{i+1} + \mathfrak{a}^n.$$

It now follows that

$$\mathbb{Z}N \cdot (a-1)^{1+sn} \subseteq \mathfrak{n}_{s+1} + \mathfrak{a}^n = \mathfrak{a}^n$$

and so a acts unipotently on U. $\square$

3. *Let G be a finitely generated torsion-free nilpotent group. Then for some integer n, G is isomorphic to a unipotent subgroup of $GL(n, \mathbb{Z})$.*

The upper central series of G has torsion-free factors ([47] 1.63). Hence G contains a central series

$$\{1\} = G_0 \subseteq \cdots \subseteq G_r \subseteq G$$

each of whose factors are infinite cyclic. The proof is by induction on r. Suppose that G_r is a unipotent subgroup of $GL(n, \mathbb{Z})$. $G = G_r\langle a \rangle$ for some a in G and G_r is finitely generated since G satisfies the maximal condition. Thus by 2. above there exists an integer m and a homomorphism ϕ of G onto a unipotent subgroup of $GL(m, \mathbb{Z})$ such that $G_r \cap \ker \phi = \{1\}$. Then the map $\psi: G \to GL(m+2, \mathbb{Z})$ given by

$$(g a^i)\psi = \begin{pmatrix} (g a^i)\phi & \begin{matrix} 0 & 0 \\ 0 & 0 \end{matrix} \\ \hline \begin{matrix} 0 & 0 \\ 0 & 0 \end{matrix} & \begin{matrix} 1 & 0 \\ i & 1 \end{matrix} \end{pmatrix}, \quad g \in G_r$$

is a monomorphism. Clearly $G\psi$ is unipotent. $\square$

(See [24a] 7.5 for another proof of point 3.)

Proof of 2.5. Let P be a polycyclic group. By the maximal condition every factor of P is finitely generated. Also P contains a nilpotent torsion-free normal subgroup N such that P/N is abelian-by-finite ([35] or use 3.6). Hence P contains a normal subgroup G containing N and of finite index such that G/N is a finitely generated free abelian group. Thus there exists a series $N = G_0 \subseteq G_1 \subseteq \cdots \subseteq G_s = G$ such that for $i = 1, 2, \ldots, s$, G_i/G_{i-1} is infinite cyclic. The proof uses induction on s. If $s = 0$ then by 3. $G = N$ is isomorphic to a unipotent subgroup of $GL(n, \mathbb{Z})$ for some n.

Suppose that $H = G_{s-1}$ is a subgroup of $GL(n, \mathbb{Z})$ and that N is unipotent. Then by 2. there exists an integer m and a homomorphism ϕ of G into $GL(m, \mathbb{Z})$ such that $H \cap \ker \phi = \{1\}$ and $N\phi$ is unipotent. Just as in 3. above there exists a monomorphism ψ of G into $GL(m+2, \mathbb{Z})$ such that $N\psi$ is unipotent.

Since $(P:G)$ is finite P is isomorphic to a subgroup of

$$GL((m+2)(P:G), \mathbb{Z}).$$

One such isomorphism is given by ψ_1 where $x\psi_1 = (x(i, j))$,

$$x(i, j) = \begin{cases} 0_{m+2} & \text{if } j \neq i\sigma \\ g_i \phi & \text{if } j = i\sigma, \end{cases}$$

$\sigma \in S_{(P:G)}$, $x_1, \ldots, x_t$ is a transversal of G to P and for each i, $x_i x = g_i x_{i\sigma}$ (cf. the proof of 2.3). $\square$

The *Prüfer rank* of a group G is the least cardinal r such that every finitely generated subgroup of G can be generated by r elements. In an abelian group this is the same as the rank, but in a non-abelian group, even a soluble one, it can be strictly larger.

Corollary. *Let G be a torsion-free nilpotent group with finite Prüfer rank r. Then for some integer n the group G is isomorphic to a subgroup of* $\mathrm{Tr}_1(n, \mathbb{Q})$.

Proof. Let $\{1\} = Z_0 \subseteq Z_1 \subseteq \cdots \subseteq Z_s = G$ be the upper central series of G. There exist elements $h_{i1}, \ldots, h_{ir}$ of Z_i such that $Z_i / \langle Z_{i-1}, h_{i1}, \ldots, h_{ir} \rangle$ is periodic. Put

$$H = \langle h_{ij} : i = 1, 2, \ldots, r, j = 1, 2, \ldots, s \rangle.$$

We claim that some power of an element g of G lies in H. For by induction on s we may assume that $g^l = hz$ for some positive integer l, some h in H and some z in Z_1. By construction there exists a positive integer m such that $z^m \in H$. But Z_1 is central in G, so $g^{lm} = h^m z^m \in H$. So far we have not used the torsion-freeness.

H is a torsion-free, finitely generated, nilpotent group so there exists a monomorphism ϕ of H into $\mathrm{Tr}_1(n, \mathbb{Z})$ for some integer n, by Point 3 preceding the proof of 2.5. There exists a torsion-free nilpotent completion $\bar{H}$ of H containing G, and $\mathrm{Tr}_1(n, \mathbb{Q})$ is torsion-free, nilpotent and radicable. Therefore ϕ extends to a monomorphism of $\bar{H}$ into $\mathrm{Tr}(n, \mathbb{Q})$. (See [24 a] Chapter 6, or for a discussion [32] Vol. 2, pp. 254–258.) □

Exercise 2.0. Prove that a nilpotent group has a faithful representation of finite degree over the rationals if it is a subgroup of the direct product of a torsion-free group of finite Prüfer rank by a finite extension of a countable free abelian group.

The converse of this exercise is also true; to prove this use 7.14.

Recently there has been a number of developments of 2.5. V. M. Kopytov [31 b], E. M. Levič [32 b] and Ju. I. Merzljakov [36 b] have all shown that a torsion-free-by-finite soluble group with finite Prüfer rank (an $\mathfrak{S}_r$-group for short) has a faithful representation of finite degree over the rationals, thus extending the corollary above. However there exist finitely generated soluble groups of finite Prüfer rank that are not isomorphic to any linear group. (Such a group can be chosen to contain a Prüfer subgroup and we shall see – 4.2 – that no finitely generated linear group can contain a Prüfer subgroup.)

It follows from the results of the next chapter that an $\mathfrak{S}_r$-group G contains normal subgroups N and L such that N is a nilpotent subgroup of L, L/N is free abelian of finite rank and G/L is finite. It is easily seen that in the whole argument leading to 2.5 we can replace the integers by the rationals, and indeed this even makes some of the argument superfluous. (All we used about the integers was that they form a principal ideal domain.) The reason in point 2. of this proof for assuming that H is finitely generated is to be able to apply point 1. Now by the corollary above N is isomorphic to some unitriangular group over $\mathbb{Q}$. Hence once we can show that point 1. of the proof of 2.5 remains valid if we replace the assumption that G is generated by n elements by the assumption that G has Prüfer rank at most n, then it will follow that any $\mathfrak{S}_t$-group is isomorphic to some linear group over the rationals. We shall see that rank at most n will do for this part of the argument.

A glance at the proof of point 1. shows, to make only a very crude estimate, that $\mathbb{Z}G/\mathfrak{a}^r$ can be generated by

$$k = (m + 2n + 2)^{3^r}$$

elements. Now suppose that G has rank at most n. For every n-generator subgroup X of G it follows that $\mathbb{Q}X/(\mathfrak{a} \cap X)^r$ has Q-dimension at most k. Since $(\mathfrak{a} \cap X)^r \subseteq \mathfrak{a}^r$ it follows that $\mathbb{Q}G/\mathfrak{a}^r$ has finite Q-dimension at most k. This proves that every $\mathfrak{S}_t$-group has a faithful representation of finite degree over the rationals.

In [36 d] Merzljakov proves that the holomorph of a polycyclic group has a faithful representation of finite degree over the integers, which is a very striking generalization of 2.5. (The finitely generated nilpotent case had earlier been done by A. Learner.) Merzljakov [36 b] and Levič [32 b] both have some rather complex conditions for a soluble group to be isomorphic to a linear group of characteristic zero involving P. Hall's v-powered groups (see [24 a] Chapter 6).

Levič [32 c] and V.N. Remeslennikov [46 b] have shown that any finitely generated, torsion-free, metabelian group has a faithful representation over the complex numbers for example. (D. Segal has pointed out to me that $(\mathbf{C}_\infty \times \mathbf{C}_\infty) \wr (\mathbf{C}_\infty \times \mathbf{C}_\infty)$ is a finitely generated, torsion-free, metabelian group whose automorphism group is not isomorphic to any linear group.) We shall prove in Chapter 10, see 10.22, that $\mathbf{C}_\infty \wr \mathbf{C}_\infty$ is not isomorphic to any linear group, so the finite generation here is essential. Another example of such a group is

$$\langle a_p, h: [a_p, a_q] = 1, \, a_p^h = a_p^p, \text{ for all primes } p \text{ and } q \rangle.$$

The torsion-freeness cannot be deleted either; if E is a non-cyclic free group then $E/(E')^6 E''$ is not isomorphic to any linear group since $E'/(E')^6 E''$ is not by 2.2. Kopytov [31 c] describes precisely which finitely generated soluble groups have faithful representations of finite degree over algebraic number fields.

We collect together the results that we have proved so far about linear p-groups.

2.6. *Let G be a locally finite p-subgroup of* $\mathrm{GL}(n, F)$. *If* char $F \neq p$ *then G satisfies the minimal condition (on subgroups). Conversely if K is the algebraic closure of any prime field of characteristic other than p, then every locally finite p-group satisfying the minimal condition has a faithful representation of finite degree over K. If* char $F = p$, *then G is nilpotent of class at most* $n - 1$ *and has finite exponent dividing* p^e *where* $p^{e-1} < n \leq p^e$.

We shall prove later (4.9) that every linear p-group is locally finite.

Proof. Let $p \neq$ char F. We can suppose that F is algebraically closed. G is completely reducible by 1.6 and monomial by 1.14. Hence G contains an abelian normal subgroup A of finite index. A satisfies the minimal condition by 2.2 and so G does also. The converse follows from 2.4.

Let $p =$ char F. If $x \in \mathrm{GL}(n, F)$ then x is a p-element if and only if $(x - 1)^{p^r} = 0$ for some r, which happens if and only if $(x - 1)^n = 0$. Therefore the p-subgroups of $\mathrm{GL}(n, F)$ are exactly the unipotent subgroups. Thus G is unitriangularizable by 1.21 and the result now follows from Exercise 1.3. ☐

Question 1. *Can one characterise the class of p-groups having faithful representations of finite degree over a field of characteristic p by abstract-group properties?*

For example we know (2.6) that such group must be nilpotent and of finite exponent, but 6.1 will show that this is not sufficient to characterise the above class. What else is required?

2.7. Theorem (Mal'cev [34]). *A group G has a faithful representation of degree n over some field if and only if every finitely generated subgroup of G has a faithful representation of degree n over some field. A group G has a faithful representation of degree n over a field of characteristic p* (≥ 0) *if and only if every finitely generated subgroup of G has a faithful representation of degree n over a field of characteristic p.*

We shall need the following concepts during the proof of 2.7.

Let S be a non-empty set. $\mathscr{B}(S)$ denotes the set of all subsets of S. A *filter* on S is a non-empty subset $\mathscr{F}$ of $\mathscr{B}(S)$ satisfying:

    i) If $A, B \in \mathscr{F}$ then $A \cap B \in \mathscr{F}$.

    ii) If $A \in \mathscr{F}$ and $A \subseteq B \subseteq S$ then $B \in \mathscr{F}$.

    iii) The empty set is not a member of $\mathscr{F}$.

A subset $\mathscr{C}$ of $\mathscr{B}(S)$ is said to have the *finite intersection property* (f.i.p.) if for every finite non-empty subset $\mathscr{T}$ of $\mathscr{C}$, $\bigcap_{T \in \mathscr{T}} T \neq \emptyset$. A subset $\mathscr{C}$ of $\mathscr{B}(S)$ has the f.i.p. if and only if $\mathscr{C}$ is a subset of some filter on S. For if $\mathscr{C} \subseteq \mathscr{F}$, a filter on S, then $\mathscr{C}$ has the f.i.p. since $\mathscr{F}$ does (by i) and iii) above). Conversely suppose that $\mathscr{C}$ has the f.i.p. Put $\mathscr{F}$ equal to the set of all

$B \in \mathscr{B}(S)$ such that there exists a finite non-empty subset $\mathscr{T}$ of $\mathscr{C}$ such that $\left(\bigcap_{T \in \mathscr{T}} T\right) \subseteq B$. It is trivial to check that $\mathscr{F}$ is a filter.

A maximal filter is called an ultrafilter. If S is a non-empty set and $\mathscr{F}$ is a filter on S, then $\mathscr{F}$ is contained in an ultrafilter on S. This follows at once from Zorn's Lemma; it is easily seen that the set of filters on S is inductively ordered by inclusion (as subsets of $\mathscr{B}(S)$). If $\mathscr{D}$ is an ultrafilter on S and A is a subset of S then either A or $\tilde{A} = \{x \in S : x \notin A\}$ belongs to $\mathscr{D}$. For suppose that A is not a member of $\mathscr{D}$. Then by the maximality of $\mathscr{D}$ and the remarks above $\mathscr{C} = \{A\} \cup \mathscr{D}$ does not have the f.i.p. But the intersection of the elements of a finite subset of $\mathscr{D}$ is in $\mathscr{D}$. Hence there exists some D in $\mathscr{D}$ such that $A \cap D = \emptyset$. Thus $D \subseteq \tilde{A}$ proving that $\tilde{A}$ belongs to $\mathscr{D}$.

For each x in the non-empty set S let F_x be a field and suppose that $\mathscr{D}$ is an ultrafilter on S. If (a_x), (b_x) are elements of $\prod_S F_x$ define $\sim$ by

$$(a_x) \sim (b_x) \iff \{x \in S : a_x = b_x\} \in \mathscr{D}.$$

$\sim$ is an equivalence relation. Denote the set of all equivalence classes by $(\prod F_x)/\mathscr{D}$. We claim that $F = (\prod F_x)/\mathscr{D}$ can be made into field in a natural way. For $\prod F_x$ is a commutative ring under pointwise addition and multiplication and it is trivial to check that $\sim$ respect these laws; that is F is a commutative ring. We have only to show the existence of inverses. Let $(a_x) \in \prod F_x$ and suppose that $(a_x) \neq (0)$. If $A = \{x \in S : a_x = 0\}$ then $A \notin \mathscr{D}$, so $\tilde{A} \in \mathscr{D}$. Define $(b_x) \in \prod F_x$ by

$$b_x = \begin{cases} a_x^{-1} & \text{if } x \in \tilde{A} \\ 0 & \text{if } x \in A. \end{cases}$$

Then $a_x b_x = 1$ for all x in $\tilde{A} \in \mathscr{D}$ and thus $(a_x)(b_x) \sim (1)$. Trivially F has characteristic $p > 0$ if each of the fields F_x does. Suppose that for each x in S, F_x has characteristic zero. For each prime p, $(p) \neq (0)$ and so F also has characteristic zero. Note also that the natural projection ψ of $\prod F_x$ onto F is a ring homomorphism, so in particular the map ψ_n of $(\prod F_x)_n$ onto F_n induced by ψ is a ring homomorphism.

We come now to the proof of the theorem. The "only if" parts are trivial. Let G be a group and suppose that every finitely generated subgroup of G has a faithful representation of degree n over some field (depending on the subgroup).

Let P be the subset of $\mathscr{B}(G)$ consisting of all $\{g, h, k\}$ for $g, h, k \in G$ with $ghk = 1$ and all $\{g, h\}$ for $g, h \in G$ with $g \neq h$. Let S denote the set of all finite non-empty subsets of P. Define

$$\mathscr{C} = \{A \in \mathscr{B}(S) : \exists x \in S \text{ with } A = \{u \in S : x \subseteq u\}\}.$$

$\mathscr{C}$ has the f.i.p.; for if $A_i = \{u \in S: x_i \subseteq u\}$, then

$$A_1 \cap \cdots \cap A_r = \{u \in S: x_1 \cup \cdots \cup x_r \subseteq u\},$$

which is non-empty. Therefore there exists an ultrafilter $\mathscr{D}$ on S containing $\mathscr{C}$.

If $x \in S$ then $x \subseteq \mathscr{B}(H_x)$ for some finite subset H_x of G. There exists a field F_x and a monomorphism ρ_x of $\langle H_x \rangle$ into $\mathrm{GL}(n, F_x)$. Define

$$\bar{\rho}_x \colon G \to \mathrm{GL}(n, F_x)$$

by

$$g\bar{\rho}_x = \begin{cases} g\rho_x & \text{if } g \in \langle H_x \rangle, \\ 1 & \text{otherwise}. \end{cases}$$

Let F be the field $(\prod_x F_x)/\mathscr{D}$ and ψ_n the natural mapping of $(\prod_x F_x)_n$ onto F_n. Denote by θ the mapping of G into $\prod_x (F_x)_n$ determined by the $\bar{\rho}_x$ and by ϕ the natural isomorphism of $\prod_x (F_x)_n$ onto $(\prod_x F_x)_n$.

$$G \xrightarrow{\ \theta\ } \prod_x (F_x)_n \xrightarrow{\ \phi\ } (\prod_x F_x)_n \xrightarrow{\ \psi_n\ } F_n.$$

We claim that $\rho = \theta \phi \psi_n$ is a monomorphism of G into $\mathrm{GL}(n, F)$.

Suppose that g, h and k are elements of G with $ghk = 1$. Put

$$y = \{\{g, h, k\}\} \in S \quad \text{and} \quad A = \{x \in S: y \subseteq x\}.$$

If $x \in A$ then $\{g, h, k\} \subseteq H_x$. Hence $g\rho_x$, $h\rho_x$ and $k\rho_x$ are defined and $g\rho_x \cdot h\rho_x \cdot k\rho_x = 1_n$ since ρ_x is a homomorphism. Therefore

$$g\bar{\rho}_x \cdot h\bar{\rho}_x \cdot k\bar{\rho}_x = 1_n$$

for every x in A. Now $A \in \mathscr{D}$ and thus

$$(g\bar{\rho}_x \cdot h\bar{\rho}_x \cdot k\bar{\rho}_x)\phi\psi_n = 1_n \quad \text{in } F_n.$$

But

$$g\rho \cdot h\rho \cdot k\rho = (g\theta \cdot h\theta \cdot k\theta)\phi\psi_n$$

$$\text{since } \phi \text{ and } \psi_n \text{ are homomorphisms,}$$

$$= ((g\bar{\rho}_x) \cdot (h\bar{\rho}_x) \cdot (k\bar{\rho}_x))\phi\psi_n,$$

$$= (g\bar{\rho}_x \cdot h\bar{\rho}_x \cdot k\bar{\rho}_x)\phi\psi_n = 1_n,$$

since multiplication in the cartesian product is pointwise.

Let g and h be any two elements of G and set $k = (gh)^{-1}$. Then the above yields $g\rho \cdot h\rho = ((gh)^{-1}\rho)^{-1}$. In this equation take gh for g and 1 for h. Clearly $1\rho = 1$, so $(gh)\rho = ((gh)^{-1}\rho)^{-1}$. It now follows that ρ is a homomorphism of G into $\mathrm{GL}(n, F)$.

Let g and h be distinct elements of G. Put

$$z=\{\{g,h\}\}\in S \quad \text{and} \quad B=\{x\in S:\, z\subseteq x\}.$$

Now if $x\in B$ then $\{g,h\}\subseteq H_x$ and

$$g\,\bar\rho_x = g\,\rho_x \neq h\,\rho_x = h\,\bar\rho_x.$$

But $g\rho=(g\bar\rho_x)\phi\psi_n$ and $h\rho=(h\bar\rho_x)\phi\psi_n$. Hence if $g\rho=h\rho$ then there exists an element D of $\mathscr{D}$ such that $g\,\bar\rho_x=h\,\bar\rho_x$ for every x in D. Consequently $B\cap D=\varnothing$ which contradicts the third filter axiom. Therefore $g\rho\neq h\rho$ and so ρ is a monomorphism. The theorem is now proved. $\quad\square$

A more general but less explicit proof of 2.7, also using ultraproducts, is given in the appendix to § 1.L of [31a]. Further, this result and Exercise 2.1 below are both very special cases of a general local theorem, see [9] Theorem VI.2.1. A quite different proof of 2.7 may be based on 5.13 of [47]; see also Mal'cev's original paper [34] for yet another variation.

Let $\mathfrak{X}$ be a class of groups. If G is a group we say that G is super-residually $\mathfrak{X}$ if for every finite set of elements X of G there exists a normal subgroup N of G such that $G/N\in\mathfrak{X}$ and $xN\neq yN$ for all x and y in X with $x\neq y$. If $\mathfrak{X}$ is R_0-closed then G is super-residually $\mathfrak{X}$ if and only if G is residually $\mathfrak{X}$.

Let $\mathscr{L}(n)$ be the class of all groups having a faithful representation of degree n over some field and $\mathscr{L}(n,p)$ be the class of all groups having a faithful representation of degree n over some field of characteristic p.

Exercise 2.1 (Mal'cev [34]). Prove that a group G lies in $\mathscr{L}(n)$ (resp. $\mathscr{L}(n,p)$) if G is locally super-residually $\mathscr{L}(n)$ (resp. $\mathscr{L}(n,p)$).

The proof of this is identical to the proof of 2.7. We did not really use that ρ_x was monic, merely that it was monic on a certain finite subset of $\langle H_x\rangle$. Note that "super-residually" cannot be replaced by "residually" in Exercise 2.1. If A is a direct product of an infinite number of cyclic groups of order 6, then A has no faithful representations of finite degree over any field whatever by 2.2. But A is residually cyclic of order 6, so $A\in R\,\mathscr{L}(1)$.

2.8. Lemma. *Let* $a=\begin{pmatrix}1&0\\x&1\end{pmatrix}$ *and* $b=\begin{pmatrix}1&x\\0&1\end{pmatrix}$ *lie in* $\mathrm{GL}(n,F)$ *where* x *is transcendental over the prime field of* F. *Then*

$$\langle a,b\rangle = \langle a\rangle * \langle b\rangle.$$

Proof. A simple inductive argument shows that for all integers $m_1, n_1, \ldots, m_r, n_r$,

$$a^{m_1} b^{n_1} a^{m_2} b^{n_2}\ldots a^{m_r} b^{n_r} = \begin{pmatrix} s & t \\ u & v+m_1\,n_1\ldots m_r\,n_r\,x^{2r} \end{pmatrix}$$

where s, t, u and v are polynomials in x over the prime field of F of degree at most $2r-1$. Hence for all non-zero integers $m_1, n_1, \ldots, m_r, n_r$, prime to char F if char F is positive, $a^{m_1} b^{n_1} \ldots a^{m_r} b^{n_r} \neq 1$.

$$a^{m_1} b^{n_1} \ldots a^{m_r} b^{n_r} a^{m_r+1} = \begin{pmatrix} e & f \\ g + m_1 n_1 \ldots m_r n_r m_{r+1} x^{2r+1} & h \end{pmatrix}$$

where e, f, g and h are polynomials in x of degree at most $2r$. Thus for all non-zero integers $m_1, n_1, \ldots, m_r, n_r, m_{r+1}$, prime to char F if char F is positive, $a^{m_1} b^{n_1} \ldots a^{m_r} b^{n_r} a^{m_r+1} \neq 1$. In the same way we deal with words of the form $b^{n_0} a^{m_1} b^{n_1} \ldots a^{m_r} (b^{n_r})$. If char $F = p > 0$ then a and b have order p, see 2.6. The result now follows. ☐

If char $F = 0$ then $\langle a, b \rangle$ is free of rank 2. If char $F \geq 3$ then $\langle a, b \rangle$ is not free but does contain non-cyclic free subgroups. If char $F = 2$, $\langle a, b \rangle$ is of course soluble; however, GL(2, F) does contain non-cyclic free subgroups (by 10.16 for example and the simplicity of SL(2, F) for this F, [50] 10.8.4). If F has transcendence degree at least two an example of a free subgroup of rank two is given in the following exercise.

Exercise 2.2. Let x and y be algebraically independent elements (over the prime subfield) of the field F. Put

$$a = \begin{pmatrix} 1 & 0 \\ x & 1 \end{pmatrix}, \qquad b = \begin{pmatrix} 1 & y \\ 0 & 1 \end{pmatrix},$$

$$c = \begin{pmatrix} y & 0 \\ x & y^{-1} \end{pmatrix}, \qquad d = \begin{pmatrix} y & x \\ 0 & y^{-1} \end{pmatrix}.$$

Prove that
$$\langle a, b \rangle = \langle a \rangle * \langle b \rangle,$$
$$\langle c, d \rangle = \langle c \rangle * \langle d \rangle$$

and that $\langle c, d \rangle$ is free of rank 2.

(The first part is a simple consequence of 2.8. The second part can be proved in a manner analogue to 2.8, see [45] for details.)

Exercise 2.3. Show that

$$\left\langle \begin{pmatrix} 1 & 0 \\ 1 & 1 \end{pmatrix}, \begin{pmatrix} 1 & 1 \\ 0 & 1 \end{pmatrix} \right\rangle = \mathrm{SL}(2, \mathbb{Z}) \cong \langle x, y : x^4, x^2 = y^3 \rangle$$

is not free but that $\left\langle \begin{pmatrix} 1 & 0 \\ 2 & 1 \end{pmatrix}, \begin{pmatrix} 1 & 2 \\ 0 & 1 \end{pmatrix} \right\rangle$ is free. Show further that

$$\mathrm{PSL}(2, \mathbb{Z}) \cong \langle x, y : x^2, y^3 \rangle.$$

(See [33] pp. 44–47 for the details of this.)

Exercise 2.4 (Brenner [5]). If x is a complex number such that $|x| \geq 2$ prove that $\left\langle \begin{pmatrix} 1 & 0 \\ x & 1 \end{pmatrix}, \begin{pmatrix} 1 & x \\ 0 & 1 \end{pmatrix} \right\rangle$ is free of rank 2.

There are a number of known conditions of this type for the complex numbers, the most general that I know of being contained in a paper by R.C. Lyndon and J.L. Ullman ([32d]).

2.9. Theorem (Nisnevič [38]). *If p is zero or a prime, every free group has a faithful representation of degree 2 over a field of characteristic p.*

Proof. By 2.8 (Exercise 2.3 if $p=2$) a free group of rank 2 has a faithful representation of degree 2 over a field of characteristic p. Hence the free group of countable rank does, and so by 2.7 every free group does. □

We shall give Nisnevič's proof of 2.9 when we come to free products. Now we look briefly at relatively free groups. First we give a preliminary lemma.

2.10. Lemma. *Let G be a group and $1 \to R \to E \to G \to 1$ a free presentation of G (i.e. the sequence is exact and E is a free group). Suppose that there exists a ring monomorphism ϕ of $\mathbb{Z}G$ into F_n. Then E/R' has a faithful representation of degree $2n$ over some purely transcendental extension of F.*

Proof. Let $Y = \{y_i : i \in I\}$ be a family of independent indeterminates over F where $|I| = \operatorname{rank} E$, and set $K = F(Y)$. Put $S = (\mathbb{Z}G)\phi$ and $A = \sum_{i \in I} y_i S$. Then A is a free $\mathbb{Z}G$-submodule of K_n of rank $|I|$, the action being given by

$$\left(\sum y_i \sigma_i \right) g = \sum y_i \sigma_i(g\phi), \quad \text{where } \sigma_i \in S.$$

A theorem of Magnus (an easy consequence of [20a] §9.1 Theorem 1; use [20a] §3.6, §3.2 Proposition 3 and §3.1 Theorem 2 ii) states that E/R' is isomorphic to a subgroup of the multiplicative group

$$\left\{ \begin{pmatrix} g & 0 \\ a & 1 \end{pmatrix} : g \in G, \ a \in A \right\}.$$

This latter group has a faithful representation of degree $2n$ over K given by

$$\begin{pmatrix} g & 0 \\ a & 1 \end{pmatrix} \mapsto \begin{pmatrix} g\phi & 0 \\ a & 1 \end{pmatrix}. \quad □$$

If I freely generates E, the faithful representation of E/R' into $GL(2n, K)$ guaranteed by 2.10 is given by

$$iR' \mapsto \begin{pmatrix} (iR)\phi & 0 \\ y_i 1_n & 1_n \end{pmatrix}.$$

2.11. Theorem (W. Magnus, 1939). *A free metabelian group has a faithful representation of degree 2 over some purely transcendental extension of the rationals (depending on the rank of the group).*

Proof. Let E be the free group on the set I and $\{x_i, y_i: i \in I\}$ a family of independent indeterminates over the rationals $\mathbb{Q}$. Put

$$F = \mathbb{Q}(x_i: i \in I) \quad \text{and} \quad K = F(y_i: i \in I).$$

Then E/E' is free abelian of rank $|I|$ and $\mathbb{Z}(E/E')$ is isomorphic to the subring $\mathbb{Z}[x_i, x_i^{-1}: i \in I]$ of F. Hence by 2.10 the free metabelian group E/E'' is isomorphic to a subgroup of $\mathrm{GL}(2, K)$. □

2.12. Theorem (W. Magnus, 1937). *A free nilpotent-group-of-class-c has a faithful representation of degree $c+1$ over some polynomial ring over the integers (depending on the rank of the group).*

Proof. Suppose that G is a free nilpotent-group-of-class-c on the set $X = \{x_i: i \in I\}$. Choose a family $\{\xi_{ij}: i \in I, j = 1, 2, \ldots, c\}$ of independent indeterminates over $\mathbb{Z}$ and put $R = \mathbb{Z}[\xi_{ij}: i \in I, j = 1, 2, \ldots, c]$.

Denote the free associative ring (with identity) on X by S and the ideal of S generated by X by $\mathfrak{s}$. The map of X into R_{c+1} given by

$$x_i \mapsto y_i = \begin{pmatrix} 0 & & & & 0 \\ & \ddots & & & \\ \xi_{i1} & & \ddots & & \\ \vdots & & & \ddots & 0 \\ 0 & & \cdots & & \xi_{ic} \end{pmatrix}$$

determines a ring homomorphism θ of S into R_{c+1} and clearly $\mathfrak{s}^{c+1} \leq \ker \theta$. Now for $1 \leq r \leq c$, a simple calculation shows that

$$y_{i_1} y_{i_2} \cdots y_{i_r} = \begin{pmatrix} 0 & & & & & 0 \\ \vdots & \ddots & & & & \\ 0 & & \ddots & & & \\ \omega_1 & & & \ddots & & \\ & \ddots & & & & \\ 0 & & \omega_{c-r+1} & 0 & \cdots & 0 \end{pmatrix} \quad r \text{ rows}$$

where $\omega_j = \xi_{i_1, r+j-1} \xi_{i_2, r+j-2} \cdots \xi_{i_r j}$.

Thus the elements $y_{i_1} y_{i_2} \cdots y_{i_r}$, for $1 \leq r \leq c$ and all possible choices of the $i_j \in I$, are linearly independent over $\mathbb{Z}$. Therefore $\ker \theta = \mathfrak{s}^{c+1}$.

Now the group of units of the ring $S/\mathfrak{s}^{c+1}$ contains an isomorphic copy of G, see [24a] 5.6, [21] Chapter 11, especially p. 175, or [33] §§ 5.6, 5.7. Specifically the map of X into R_{c+1} given by $x_i \mapsto 1 + y_i$ extends to a monomorphism of G into $\mathrm{GL}(c+1, R)$. □

2.10 cannot be used to extend 2.12 to free $\mathfrak{A}\mathfrak{N}_c$-groups, where $\mathfrak{N}_c$ denotes the variety of groups of class at most c: if the group ring $\mathbb{Z}G$ of a group G is isomorphic to a subring of F_n then G is abelian-by-finite. This latter result, due to I.M. Isaacs and D.S. Passman, follows from [26b] §X.5 and 10.23, 9.1, 6.4, 5.9 and 5.11.

Exercise 2.5. Let ξ be a single indeterminate over $\mathbb{Z}$ and for $k=1, 2, \ldots$ let $x_k=(x_k(i, j))\in T_{r_1}(c+1, \mathbb{Z}[\xi])$ where

$$x_k(i, j)=0 \quad \text{if } i-j\geq 2 \text{ or if } j-i\geq 1,$$
$$=1 \quad \text{if } i=j$$
$$=\xi^{2ck+i-2} \quad \text{if } i-1=j.$$

Set $X=\{x_k: k=1, 2, \ldots\}$. Prove that $G=\langle X\rangle$ is a free nilpotent-group-of-class-c on X. (I am indebted to N.D. Gupta for showing me this representation of G. He credits the original idea to P. Hall.)

In [20d] C.K. Gupta and N.D. Gupta prove that a free $\mathfrak{N}_c\mathfrak{A}$-group has a faithful representation of degree $c+1$ over any field of characteristic zero of sufficiently large transcendence degree[2]. Further in [20c] C.K. Gupta shows that a free $(\mathfrak{N}_2\mathfrak{A}\cap\mathfrak{A}\mathfrak{N}_2)$-group is isomorphic to a linear group of degree 4 over a similar such field.

Question 2. *For which varieties $\mathfrak{B}$ do the free $\mathfrak{B}$-groups have faithful representation of finite degree over fields?*

It follows from theorems of Mal'cev and Platonov (3.6 and 10.15) that a linear group either generates the variety of all groups or is nilpotent-by-abelian-by-finite (which again shows why 2.10 cannot be used to extend 2.12, see comment above). Thus Question 2 really relates only to sub-varieties of $\mathfrak{N}_c\mathfrak{A}, c=1, 2, \ldots$. By 2.2 the free abelian-groups-of-exponent-6 with infinite rank are not isomorphic to linear groups. For varieties in which the free groups are torsion-free I know of no such example.

Of course it is quite possible for the finitely generated free groups of a variety to be isomorphic to linear groups without them all being so. If G is a finitely generated free group of a nilpotent variety then G is isomorphic to a subgroup of $GL(n, \mathbb{Z})$ for some n by 2.5 and the variety of abelian groups of exponent 6 again shows that the finite generation of G is essential. Similarly a torsion-free finitely generated free group of a metabelian variety is isomorphic to a linear group by results of Levič and Remeslennikov, see comments before 2.6. C.K. Gupta proves in [20b] that a free centre-by-metabelian group on three generators has a faithful representation of degree three over a purely transcendental extension of $\mathbb{Q}$.

---

[2] Jacques Lewin has also proved this.

Question 3. *Which finitely generated relatively free groups are isomorphic to linear groups?*

We now come to free products. The following result is a variation of the main theorem of Nisnevič's paper [38].

2.13. Theorem. *Let $\{G_\alpha : \alpha \in \Lambda\}$ be a family of subgroups of $GL(n, F)$ and put $Z_\alpha = G_\alpha \cap F1_n$. Suppose that $\{x_\alpha : \alpha \in \Lambda\}$ is a set of $n \times n$ matrices such that the family of entries of all the x_α forms an algebraically independent set over F. If $G = \langle G_\alpha^{x_\alpha} : \alpha \in \Lambda \rangle$ and $Z = G \cap F1_n$, then*

$$Z = \langle Z_\alpha : \alpha \in \Lambda \rangle \quad and \quad G/Z \simeq \mathop{\text{\Large $*$}}_{\alpha \in \Lambda} G_\alpha/Z_\alpha,$$

the free product of the G_α/Z_α.

2.14. Corollary (Nisnevič [38]). *For each $\alpha \in \Lambda$ suppose that G_α is a group having a faithful representation of degree n over some field of characteristic $p \geq 0$. Then the free product of the G_α has a faithful representation of degree $n + 1$ over some field of characteristic p.*

2.13 implies 2.14 as follows. Any set of fields of characteristic p can be embedded into a single field. Hence there exists a field F of characteristic p and for each $\alpha \in \Lambda$, a monomorphism ρ_α of G_α into $GL(n, F)$. Let $\bar{\rho}_\alpha$ be the monomorphism of G_α into $GL(n+1, F)$ given by

$$g\bar{\rho}_\alpha = \begin{pmatrix} g\rho_\alpha & 0 \\ 0 & 1 \end{pmatrix}, \quad g \in G_\alpha.$$

Then $G_\alpha \bar{\rho}_\alpha$ contains no non-trivial scalar matrices and the result follows from 2.13. ∎

Notice that 2.9 is an immediate consequence of 2.14; for a free group is simply the free product of infinite cyclic groups, and clearly the infinite cyclic group has a faithful representation of degree 1 over a field of characteristic p. Incidentally this shows that the $n+1$ of 2.14 cannot be replaced by n (see also Exercise 2.6 below). In contrast to this of course 2.13 implies that the free product of groups with faithful projective representations of degree n and characteristic p does have a faithful projective representation of degree n and characteristic p.

We cannot allow the p to vary in 2.14. For example, if A_r is an infinite elementary abelian r-group for $r = 2, 3$ then A_r has a faithful representation of degree 2 over any infinite field of characteristic r but $A_2 * A_3$ is not isomorphic to any linear group at all by 2.2.

Exercise 2.6. If p is a field characteristic and q and r are distinct primes distinct from p, let A (resp. B) be an elementary abelian group of order q^n (resp. r^n), where $n \geq 1$. Prove that A and B have faithful representations of degree n over some field of characteristic p while $A * B$ is not isomorphic to any linear group of degree n. (This exercise shows that the $n+1$ of 2.14 is the best possible for every n and p.)

We break the proof of 2.13 into three parts. In what follows all matrices are $n \times n$, and if x is a matrix $x(i, j)$ denotes its (i, j)-th entry.

1. *Let a be a non-scalar matrix of* $\mathrm{GL}(n, F)$ *and x a matrix whose n^2 entries are algebraically independent over F. Then every entry of a^x is non-zero.*

Since a is not a scalar there exist linearly independent row vectors u and v over F such that $ua = u + v$. Choose a basis for the n-row vector space over F containing u and v as its k-th and l-th members (for some $k \neq l$), and let b be the element of $\mathrm{GL}(n, F)$ whose rows are the elements of this basis in order. Then $a^{b^{-1}}(k, k) = a^{b^{-1}}(k, l) = 1$. The map

$$x(i, j) \mapsto b^{-1}(i, j), \quad 1 \leq i, j \leq n,$$

determines an F-algebra homomorphism ϕ of

$$F[(\det x)^{-1}, x(i, j): i, j = 1, 2, \dots, n]_n$$

into F_n such that $a^x \phi = a^{b^{-1}}$. Therefore the (k, k)-th and (k, l)-th entries of a^x are non-zero and 1. follows. ☐

2. *Suppose that $a_1, a_2, \dots, a_r$ are non-scalar matrices in $\mathrm{GL}(n, F)$ and that $x_1, x_2, \dots, x_s$ are matrices whose sn^2 entries are algebraically independent over F. Let $b = a_1^{y_1} a_2^{y_2} \dots a_r^{y_r}$ where each y_t is equal to one of the x_i and where $y_t \neq y_{t+1}$ for $1 \leq t < r$. Then b is not a scalar matrix.*

The proof of 2. uses induction on s; the case $s = 1$ being trivial. Then by induction we may assume that $b = c_1^x b_1 \, c_2^x b_2 \dots c_h^x b_h$, where $x = y_1$, for each t, c_t is a non-scalar matrix in $\mathrm{GL}(n, F)$ (in fact one of the a's) and b_t is an invertible non-scalar matrix over

$$E = F\big(x_k(i, j): 1 \leq i, j \leq n, 1 \leq k \leq s, x_k \neq x\big),$$

except for the possibility that $b_h = 1_n$.

Let u, v, w and z be matrices whose $4n^2$ entries are algebraically independent over E. There exists an E-algebra homomorphism of

$$R = E[(\det x)^{-1}, x(i, j): i, j = 1, 2, \dots, n]$$

that induces an E-algebra homomorphism θ of R_n such that

$$x\theta = uvw^{-1}z^{-1};$$

so

$$b\theta = \prod_{t=1}^{h} c_t^{uvw^{-1}z^{-1}} b_t = z \left(\prod_{t=1}^{h} c_t^{uvw^{-1}} b_t^z \right) z^{-1}.$$

By 1. every entry of $e_t = c_t^u$ and every entry of $d_t = b_t^z$ is non-zero, with the one exception that possibly $d_h = 1_n$.

Consider $g = (\det v)^h (\det w)^h (b\theta)^z = \prod\limits_{t=1}^{h} f_t$ where

$$f_t = w \cdot \text{adj}\, v \cdot e_t \cdot v \cdot \text{adj}\, w \cdot d_t.$$

If K denotes the field generated by E and the entries of u and z, and if S is the K-algebra generated by the entries of v and w, then there exists a K-algebra homomorphism of S onto K that induces a K-algebra homomorphism ϕ of S_n onto K_n satisfying.

$$v\phi = \text{diag}(0, 1, 1, \ldots, 1) \quad \text{and} \quad w\phi = \text{diag}(1, 1, \ldots, 1, 0).$$

Then $(\text{adj}\, v)\phi = \text{diag}(1, 0, \ldots, 0)$, $(\text{adj}\, w)\phi = \text{diag}(0, \ldots, 0, 1)$ and a simple calculation shows that

$$f_t \phi(i, j) = \begin{cases} e_t(1, n)\, d_t(n, j) & \text{if } i = 1 \\ 0 & \text{otherwise.} \end{cases}$$

Hence

$$g\phi(1, n) = \left(\prod\limits_{t=1}^{h-1} e_t(1, n)\, d_t(n, 1) \right) e_h(1, n)\, d_h(n, n) \neq 0.$$

Since ϕ is given by a map of the ground ring it follows that $g(1, n) \neq 0$. Consequently $b\theta$ is not a scalar matrix and therefore neither is b. □

Proof of 2.13. Clearly $Y = \langle Z_\alpha : \alpha \in \Lambda \rangle \subseteq Z$. If $a_i \in G_{\alpha_i} \setminus Z_{\alpha_i}$ for $i = 1, 2, \ldots, m$, where $\alpha_i \neq \alpha_{i+1}$ for each $i < m$ then

$$b = a_1^{x_{\alpha_1}} \cdot a_2^{x_{\alpha_2}} \ldots a_m^{x_{\alpha_m}} \notin Z$$

by 2. above. Thus by the normal form theorem for free products ([21] §17.1 or [33] §4.1) it follows that

$$G/Z \cong \underset{\alpha \in \Lambda}{\bigstar} (G_\alpha^{x_\alpha}/Z_\alpha) \cong \underset{\alpha \in \Lambda}{\bigstar} (G_\alpha/Z_\alpha).$$

Finally since every element of $G \setminus Y$ lies in a coset of the form bY with b as above, we obtain $Y = Z$. □

For a discussion of generalized free products of linear groups and a generalization of 2.13 see [69 f].

We now come to our final topic of this chapter, (restricted) wreath products. In Chapter 10 we shall show that every wreath and every complete wreath product isomorphic to a linear group can be constructed from 2.3 and the two results below.

2.15. Lemma. *If A is a torsion-free abelian group and H is a group with a torsion-free abelian subgroup H_1 of finite index k, then $G = A \wr H$ has a faithful representation of degree $2k$ over some field of characteristic zero.*

Proof. Let B denote the base group of G. It suffices by 2.3 to prove that BH_1 has a faithful representation of degree 2 over some field of characteristic zero. Further by the theorem of Mal'cev (2.7) we need only prove this statement for the finitely generated subgroups of BH_1. Hence we may confine our attention to BK where K is a finitely generated subgroup of H_1. K is a free abelian group; let r denote its rank.

A can be embedded in a direct sum of copies of $\mathbb{Q}^+$ the additive group of the rationals, and if F is any field of characteristic zero, then F^+ is a direct sum of copies of $\mathbb{Q}^+$. Hence there exists a field F of characteristic zero such that F^+ contains a subgroup A_1 isomorphic to A. As K-module B is isomorphic to a direct sum of (say γ) copies of the base group of $A \wr K$. Let $\{x_1, x_2, \ldots, x_r\} \cup Y$ be an algebraically independent set over F, where $|Y| = \gamma$, and set

$$E = F(x_1, x_2, \ldots, x_r, Y)$$

and

Put

$$K_1 = \langle x_1, x_2, \ldots, x_r \rangle \subseteq E^*.$$

$$W = \left\langle \begin{pmatrix} 1 & 0 \\ ya & 1 \end{pmatrix}, \begin{pmatrix} k & 0 \\ 0 & 1 \end{pmatrix} : y \in Y, a \in A_1, k \in K_1 \right\rangle \subseteq \mathrm{GL}(2, E).$$

If U denotes the maximal unipotent subgroup of W, then W is isomorphic to the split extension of U by the group $\mathrm{diag}(K_1, 1)$ isomorphic to K_1. For $y \in Y$ and $k \in K_1$, let

$$_yU_k = \left\{ \begin{pmatrix} 1 & 0 \\ yak & 1 \end{pmatrix} : a \in A_1 \right\}$$

and

$$_yU = \langle _yU_k : k \in K_1 \rangle.$$

Each $_yU_k$ is a subgroup of U isomorphic to A and $U = \langle _yU : y \in Y \rangle$. Since $\{yK_1 : y \in Y\}$ is a linearly independent subset of E over F, it follows that U is the direct product of the $_yU_k$. Clearly therefore, each $_yU$ is isomorphic as K_1-module to the base group of $A_1 \wr K_1$ and W is isomorphic to BK. □

2.16. Lemma. *If A is an abelian p-group of finite exponent p^e, here p is some prime, and H is a group with a torsion-free abelian subgroup H_1 of finite index k, then $G = A \wr H$ has a faithful representation of degree $[p^{e-1}+1]k$ over some field of characteristic p.*

Proof. A can be embedded into a direct product of cyclic groups of order p^e (see [32] Vol. 1, p. 173) and we may assume that A is such a direct product. Let B denote the base group of G and K a finitely generated subgroup of H_1 of rank r say. As in the proof of the preceding lemma it suffices to prove that BK has a faithful representation of degree

$$n = [p^{e-1}+1]$$

over some field of characteristic p.

If Z is a set of independent indeterminates over $\mathbb{F}_p$ the field of p elements of sufficiently large cardinal and if $F = \mathbb{F}_p(Z)$, then the group

$$A_1 = \langle u(z) \colon z \in Z \rangle \subseteq GL(n, F)$$

where

$$u(t) = \begin{pmatrix} 1 & & & & 0 \\ t & 1 & & & \\ & \ddots & 1 & & \\ & & & \ddots & \\ 0 & & & t & 1 \end{pmatrix},$$

is isomorphic to A (see proof of 2.2). As K-module B is isomorphic to the direct product of (say γ) copies of the base group of $A \wr K$. Let

$$\{x_1, x_2, \ldots, x_r\} \cup Y$$

be an algebraically independent set over F where $|Y| = \gamma$ and set

$$E = F(x_1, x_2, \ldots, x_r, Y)$$

and

$$K_1 = \langle x_1, x_2, \ldots, x_r \rangle \subseteq E^*.$$

Let $W = \langle u(yz), \operatorname{diag}(k^{n-1}, k^{n-2}, \ldots, k, 1) \colon y \in Y, z \in Z, k \in K_1 \rangle$, here

$$W \subseteq GL(n, E).$$

If U denotes the maximal unipotent subgroup of W, then W is isomorphic to the split extension of U by a group isomorphic to K_1 (consisting of the exhibited diagonal matrices in the above definition of W). For $y \in Y$ and $k \in K$, put

$$_yU_k = \langle u(yzk) \colon z \in Z \rangle$$

and

$$_yU = \langle {_yU_k} \colon k \in K_1 \rangle.$$

It is not difficult to check that for any $y \in Y$, $k \in K_1$ and $t \in E$

$$\operatorname{diag}(k^{n-1}, \ldots, k, 1)^{-1} u(t) \operatorname{diag}(k^{n-1}, \ldots, k, 1) = u(tk),$$

$$_yU_k \cong A \quad \text{and} \quad U = \underset{y \in Y}{\times} {_yU}.$$

Once we have shown that $_yU = \underset{k \in K_1}{\times} {_yU_k}$ it will be clear that $_yU$ is isomorphic as K_1-module to the base group $A_1 \wr K_1$ and that W is isomorphic to BK.

Since $\{x_1, x_2, \ldots, x_r\} \cup Y \cup Z$ is an algebraically independent set over $\mathbb{F}_p$ it suffices to prove that $\{u(k) \colon k \in K_1\}$ freely generates an abelian group of exponent p^e. If $e = 0$ there is nothing here to prove so assume that $e > 0$. Suppose that

$$u(k_1)^{e_1} u(k_2)^{e_2} \ldots u(k_s)^{e_s} = 1$$

where the k_i are s distinct elements of K_1 and $0 < e_i < p^e$ for each i. By induction on e we may assume that p^{e-1} divides e_i for each i, say $e_i = f_i p^{e-1}$.

An easy computation shows that

$$u(t)^m = \begin{pmatrix} 1 & 0 & 0 & \cdots & & 0 \\ u_1 & 1 & 0 & & & \\ u_2 & u_1 & 1 & & & \\ \vdots & & & \ddots & & \\ u_m & & & & & \\ 0 & u_m & \cdots & u_2 & u_1 & 1 \end{pmatrix}$$

where $u_i = {}^m C_i t^i$ and $m \le p^{e-1}$. (${}^m C_i$ is the binomial coefficient.) It follows easily that

$$\prod_{i=1}^{s} u(k_i)^{e_i} = \begin{pmatrix} 1 & & & & & 0 \\ 0 & & & & & \\ \vdots & & & \ddots & & \\ 0 & & & & & \\ & & & & & \\ \prod_{i=1}^{s} f_i k_i^{p^{e-1}} & & 0 & \cdots & 0 & 1 \end{pmatrix}.$$

Hence $\sum f_i k_i^{p^{e-1}} = 0$. Now each f_i lies strictly between 0 and p, and the elements of K_1 are independent over $\mathbb{F}_p$. Hence for some distinct i and i' we have $k_i^{p^{e-1}} = k_{i'}^{p^{e-1}}$, and therefore $k_i = k_{i'}$. This contradicts the choice of $k_1, k_2, \ldots, k_s$ and completes the proof of the lemma. □

Added in Proof

D. Segal has shown that the holomorph of a nilpotent $\mathfrak{S}_f$-group has a faithful representation of finite degree over the rationals, cf. p. 26.

D. I. Edelkind (Faithful representations of relatively free groups (Russian). Uspehi Mat. Nauk **26**, 253–254 (1971)) has announced some interesting results relating to the discussion on pp. 33–35.

3. Soluble Linear Groups

Direct computation shows that the triangular group $\text{Tr}(n, F)$ is soluble of derived length at most $(1 - [-\log_2 n])$ (see Exercise 1.3). By 2.3 every extension of a subgroup of $\text{Tr}(n, F)$ by a finite soluble group is soluble and linear. The principal object of this chapter is to prove that every soluble linear group essentially has this form.

3.1. Lemma. *Let H and L be subgroups of $\text{GL}(n, F)$ such that $[H, L] \subseteq F^* 1_n$ and put $K = C_H(L)$. Then K is normal in H, $(H:K) \leq n^2$, $H^n \subseteq K$ and the elements of any transversal of K to H are linearly independent over F.*

Proof. Since $[H, L] \subseteq F^* 1_m$, we have

$$K \subseteq C_H(L \cdot [H, L]) = C_H(L^H) \subseteq C_H(L).$$

Hence $K = C_H(L^H)$ and so K is normal in H.

Let $a_1, \ldots, a_t$ be a finite part of a transversal of K to H and suppose that $a_1, \ldots, a_t$ are linearly dependent over F. By remembering if necessary we can suppose that $0 = \lambda_1 a_1 + \cdots + \lambda_s a_s$ is a non-trivial linear relation between the a_i containing the least number of terms. Thus $\lambda_i \neq 0$ for each i and $s \geq 2$. Suppose that for each x in L, $[a_1, x] = [a_2, x]$. Then

$$[a_1 a_2^{-1}, x] = [a_1, x][a_2, x]^{-1} = 1$$

for all x in L and so $a_1 a_2^{-1} \in K$. This contradiction proves that for some y in L, $[a_1, y] \neq [a_2, y]$. Now for each i, $[a_i, y] = \alpha_i 1_n$ where $\alpha_i \in F^*$ and $\alpha_1 \neq \alpha_2$. Thus

$$0 = \alpha_1 \left(\sum_1^s \lambda_i a_i \right) - \left(\sum_1^s \lambda_i a_i \right)^y = \sum_2^s (\alpha_1 - \alpha_i) \lambda_i a_i.$$

Since $(\alpha_1 - \alpha_2) \lambda_2 \neq 0$ this contradicts the minimality of s. Hence $a_1, \ldots, a_t$ are linearly independent over F and so $t \leq n^2$. Therefore $(H:K) \leq n^2$.

If $x \in H$ and $y \in L$, $xy = \lambda y x$ for some λ in F^*. Then $\det x \cdot \det y = \lambda^n \det y \cdot \det x$ and so $\lambda^n = 1$. Hence $x^n y = \lambda^n y x^n = y x^n$ and this is for all y in L. Thus $x^n \in K$ and therefore $H^n \subseteq K$. $\square$

3.2. Corollary. *Let F be algebraically closed and A an irreducible subgroup of $\text{GL}(n, F)$ which is nilpotent of class 2. Then $(A : \zeta_1(A)) = n^2$.*

Proof. $\zeta_1(A) \subseteq F^* 1_n$ by Schur's Lemma. By 3.1 (with $A = H = L$ and $\zeta_1(A) = K$) $(A : \zeta_1(A)) \leq n^2$. Let $a_1, \ldots, a_t$ be a transversal of $\zeta_1(A)$ to A. Then $a_1, \ldots, a_t$ are linearly independent over F by 3.1. If $a \in A$, $a = a_i z$ for some i and some z in $\zeta_1(A)$. Then $z = \alpha 1_n$ where $\alpha \in F^*$ and so

$$A \subseteq F a_1 + \cdots + F a_t.$$

A is irreducible and F is algebraically closed. Thus Burnside's Theorem (1.17) implies that $n^2 \leq t$. $\square$

3.3. Theorem (Suprunuenko). *Let F be an algebraically closed field, n an integer greater than one, G a primitive irreducible subgroup of $\mathrm{GL}(n, F)$ and H a soluble normal subgroup of G. Then G contains normal subgroups*

$$\{1\} \subseteq Z \subseteq A \subseteq H \subseteq G$$

such that

i) $Z = H \cap \zeta_1(G)$.

ii) *A/Z is abelian, every Sylow subgroup of A/Z is elementary abelian and $|A/Z| = r^2$ for some integer r dividing n.*

iii) *H/A is isomorphic to a subgroup of*

$$\prod_{i=1}^{t} \mathrm{Sp}(2l_i, p_i) \quad \text{where} \quad r = p_1^{l_1} p_2^{l_2} \ldots p_t^{l_t} \quad \text{and} \quad p_1, \ldots, p_t$$

are distinct primes.

In this work we shall not use the full force of part iii) but merely that $|H/A| \leq (r^2)!$. This follows from point 4 of the proof below.

Proof. By 1.13, $\zeta_1(G)$ is the unique maximal abelian normal subgroup of G. Let $Z = H \cap \zeta_1(G)$ and let A be a subgroup of G maximal subject to the restrictions,

$$A \subseteq H, \quad A \triangleleft G, \quad A/Z \text{ abelian}.$$

1. $C_H(A) = Z$.

Trivially $Z \subseteq C_H(A)$. Also $C_H(A)$ is normal in G and so $A \cap C_H(A)$ is an abelian normal subgroup of G. Hence $Z = A \cap C_H(A)$. $C_H(A)$ is soluble, so if $Z \neq C_H(A)$ then G contains a normal subgroup K such that

$$Z \subsetneq K \subseteq C_H(A)$$

and K/Z is abelian. But $A \subseteq KA \subseteq H$ and KA/Z is an abelian normal subgroup of G/Z. Hence $K \subseteq A$. But then

$$Z \subsetneq K \subseteq A \cap C_H(A) = Z.$$

Consequently $C_H(A) = Z$.

42

2. $(A:Z)=r^2$ where r divides n.

By Clifford's Theorem A is completely reducible into m equivalent irreducible representations of degree r, where $rm=n$. Since these representations are equivalent they are faithful. By 3.2 and 1.

$$r^2 = (A:\zeta_1(A)) = (A:Z).$$

3. *Every Sylow subgroup of A/Z is elementary abelian.*

Let A_p/Z be the Sylow p-subgroup of A/Z and suppose that A_p/Z has exponent p^e where $e \geq 2$. Put $B = (A_p)^{p^{e-1}}$; B is a normal subgroup of G. If x and y are elements of A_p, since A is nilpotent of class 2 and $2e - 2 \geq e$, $[x,y]^{p^e} = [x^{p^e}, y] = 1$ and

$$[x^{p^{e-1}}, y^{p^{e-1}}] = [x,y]^{p^{2e-2}} = 1.$$

Hence B is abelian and thus is contained in Z. Therefore A_p/Z has exponent p^{e-1}. This contradiction proves 3.

4. $C_H(A/Z) = A$.

If c is an element of $C_H(A/Z)$, the mapping ϕ_c of A/Z into Z given by $\phi_c: xZ \mapsto [x,c]$ is a well-defined homomorphism and the mapping $c \mapsto \phi_c$ is a homomorphism of $C_H(A/Z)$ into the additive group $\text{Hom}(A/Z, Z)$. ϕ_c is the identity map if and only if $[A,c] = \{1\}$, i.e. if and only if $c \in Z$ (by 1.). Since every finite subgroup of Z is cyclic and A/Z is finite

$$|\text{Hom}(A/Z, Z)| \leq |A/Z|$$

(in fact they are equal). Therefore

$$|C_H(A/Z)/Z| \leq |A/Z|.$$

But trivially $A \subseteq C_H(A/Z)$, so this proves 4.

It follows from 4. that H/A is isomorphic to a subgroup of the automorphism group of A/Z. Hence $|H/A| \leq (r^2)!$.

5. $\zeta_1(A_p) = Z$ for any prime p.

Trivially $Z \subseteq \zeta_1(A_p)$. But $\zeta_1(A_p)$ is an abelian normal subgroup of G so $\zeta_1(A_p) \subseteq Z$.

By 3., A_p/Z is a finite elementary abelian p-group. If we regard it as a vector space over $\mathbb{F}_p$ the field of p-elements, then H/A induces a group of linear mappings on A_p/Z (by conjugation). The map $\rho: A_p/Z \times A_p/Z \to \mathbb{F}_p$ given by $(xZ, yZ)\rho = u$ where $[x,y] = \eta^u 1_n$, η is a primitive p-th root of 1 in F and $0 \leq u < p$, is a non-degenerate alternating form[3]. For ρ is bilinear since A is nilpotent of class 2. ρ is alternating since $[x,x] = 1$ implies that

---
[3] $A_p = Z$ if $p = \text{char } F$ by 5., since ϕ_c is a homomorphism.

$(xZ, xZ)\rho = 0$ for all x in A_p. If ρ were degenerate there would exist an x in $A_p \setminus Z$ such that $(xZ, A_p/Z)\rho = \{0\}$, i.e., that $[x, A_p] = \{1\}$. This is impossible by 5. Therefore A_p/Z is a symplectic space. If $g \in G$ then

$$[x^g, y^g] = [x, y]^g = [x, y] \qquad \text{for all } x \text{ and } y \text{ in } A.$$

Hence $G/C_G(A_p/Z)$ is isomorphic to a subgroup of $\mathrm{Sp}(2l, p)$ where $(A_p : Z) = p^{2l}$.

Let $r = p_1^{l_1} p_2^{l_2} \ldots p_t^{l_t}$ be the prime decomposition of r. By 4.

$$A = C_H(A/Z) = \bigcap_{i=1}^{t} C_H(A_{p_i}/Z).$$

Therefore H/A is isomorphic to a subgroup of $\prod_{i=1}^{t} \mathrm{Sp}(2l_i, p_i)$. $\square$

3.4. Corollary (Zassenhaus [73]). *Let F be an algebraically closed field and G a soluble primitive irreducible subgroup of $\mathrm{GL}(n, F)$. Then the centre of G has finite index in G.*

In the notation of 3.3 (taking $H = G$) $(G:A) \le (n^2)!$ and $(A:Z) \le n^2$. Thus $(G : \zeta_1(G)) \le n^2 (n^2!)$. Notice that the bound depends only on n. It is not of course the best possible bound. $\square$

3.5. Lemma (Mal'cev [35]). *There exists an integer-valued function $\mu(n)$ of n only such that:*

i) *If $n_1, \ldots, n_t$ are positive integers satisfying $n_1 + \cdots + n_t = n$ then*

$$\mu(n) \ge \prod_{i=1}^{t} \mu(n_i).$$

ii) *Every completely reducible soluble linear group of degree n contains an abelian normal subgroup of finite index at most $\mu(n)$.*

Proof. Let G be an irreducible soluble subgroup of $\mathrm{GL}(n, F)$ where F is algebraically closed and denote by V the space of n-row vectors over F (regarded as FG-module). Suppose that

$$V = V_1 \oplus \cdots \oplus V_r \qquad \text{where } \{V_1, \ldots, V_r\}$$

is a minimal system of imprimitivity for G. Put

$$H_i = N_G(V_i) \quad \text{and} \quad \bar{H}_i = H_i/C_{H_i}(V_i).$$

$\bar{H}_i$ has a faithful primitive irreducible representation on V_i and so its centre has finite index at most $m^2(m^2!)$ by 3.4, where $mr = n$.

Let $A_i/C_{H_i}(V_i)=\zeta_i(\bar{H}_i)$,

$$K=\bigcap_{i=1}^{r}H_i \quad \text{and} \quad A=\bigcap_{i=1}^{r}A_i.$$

Then K is normal in G and $(G:K)\leq r!$. Also

$$[K,A]\subseteq\bigcap_{i=1}^{r}C_G(V_i)=\{1\};$$

thus $A\subseteq\zeta_1(K)$. But $(K:A)\leq(m^2(m^2!))^r$ and so $\zeta_1(K)$ is an abelian normal subgroup of G such that

$$(G:\zeta_1(K))\leq(m^2(m^2!))^r\cdot r!\leq(n^2(n^2!))^n\,n!=s(n)$$

say. Put

$$\mu(n)=\max_{\substack{n_i\geq 1\\ \Sigma n_i=n}}\left(\prod_{i=1}^{t}s(n_i)\right).$$

Trivially μ satisfies i). Let G be a completely reducible soluble subgroup of $GL(n,F)$. By 1.22, we can suppose that F is algebraically closed. By construction if G is irreducible G contains an abelian normal subgroup of finite index at most $\mu(n)$. Hence by i), G does in general. $\square$

3.6. Theorem (Mal'cev [35]). *A soluble linear group of degree n contains a triangularizable normal subgroup of finite index at most $\mu(n)$. In particular such a group is nilpotent-by-abelian-by-finite.*

There is a slightly weaker variant of this theorem due to Lie and Kolchin. This will be discussed in Chapter 5 where a second proof will be given.

Proof. Let G be a soluble subgroup of $GL(n,F)$. We can clearly suppose that F is algebraically closed. There exist integers $r, n_1, \ldots, n_r, x$ in $GL(n,F)$ and irreducible representations ρ_i of G into $GL(n_i,F)$, $i=1,2,\ldots,r$ satisfying

$$n=n_1+\cdots+n_r \quad \text{and} \quad g^x=\begin{pmatrix} g\rho_1 & & 0 \\ & \ddots & \\ * & & g\rho_r \end{pmatrix} \quad \text{for all } g \text{ in } G.$$

By 3.5, $G\rho_i$ contains an abelian normal subgroup A_i of finite index at most $\mu(n_i)$. Put $T=\bigcap_{i=1}^{r}A_i\rho_i^{-1}$. T is a normal subgroup of finite index at most $\mu(n)$ in G by 3.5i). By Clifford's Theorem and Schur's Lemma A_i is a diagonalizable subgroup of $GL(n_i,F)$. Therefore T is a triangularizable subgroup of $GL(n,F)$. $\square$

3.7. Theorem (Zassenhaus [73]). *There exists an integer-valued function $f(n)$ of n only such that every soluble linear group of degree n has derived length at most $f(n)$.*

Proof. If G is a soluble subgroup of $\mathrm{GL}(n, F)$ then G contains a triangularizable normal subgroup T such that $(G:T) \leq \mu(n)$. Direct computation shows that $\mathrm{Tr}(n, F)$ is soluble of derived length at most $1 - [-\log_2 n]$ (Exercise 1.3). Hence G has derived length at most

$$\mu(n) + 1 - [-\log_2 n]. \quad \square$$

3.8. Corollary (Zassenhaus [73]). *A locally soluble linear group is soluble. Every linear group contains a unique maximal soluble normal subgroup.* $\quad \square$

The bounds in 3.5, 3.6 and 3.7 can be drastically reduced by making a study of the symplectic groups arising in 3.3. This was first done by B. Huppert [26], improved by J.D. Dixon [12], [13a] and essentially completed by M.F. Newman [37a] and M. Frick & Newman [14b]. We state their results without proof.

3.9. Theorem (Dixon [12]). *Let $a = 2 \cdot 3^{\frac{1}{3}}$ and $b = 2 \cdot 3^{\frac{1}{3}}$. If G is a completely reducible soluble linear group of degree n, then*

i) $(G: \eta_1(G)) \leq a^{-1} b^n$ *and this bound is attained whenever*

$$n = 2 \cdot 4^k, \quad \text{for } k = 0, 1, 2, \dots.$$

ii) *There is a constant c such that G contains a subnormal abelian subgroup A satisfying $(G:A) \leq (2a)^{-1} c^n$. The best value of c lies between $b\sqrt{2}$ and $2b$.*

If G is any group $\eta_1(G)$ denotes the Fitting subgroup of G, that is $\eta_1(G)$ is the product of all the nilpotent normal subgroups of G. We shall prove later (8.2) that $\eta_1(G)$ is nilpotent whenever G is linear. (In [12] there appears to be some confusion between the Fitting subgroup and the Hirsch-Plotkin radical.)

Let $\rho(n)$ (resp. $\sigma(n)$) denote the maximal derived length of any soluble (resp. soluble completely reducible) linear group of degree n. In his paper [37a] Newman gives a recipe that enables one to compute $\rho(n)$ and $\sigma(n)$ for any given n. It is relatively simple but messy to state. He appends a table of values for $n \leq 74$, and then proves the following.

3.10. Theorem (Newman [37a]). *Let $D = \frac{1}{2} \cdot 17 - 15 (\log 2)(2 \log 3)^{-1}$.*

i) *For $n \geq 66$,*

$$5 \log_9 (n-1) + D \leq \rho(n) \leq 5 \log_9 (n-2) + D + \tfrac{3}{2}$$

with the lower bound being attained whenever $n = 24 \cdot 9^k + 1$, and the upper whenever $n = 8 \cdot 9^k + 2$, for $k = 0, 1, 2, \dots.$

ii) *For* $n \geq 64$,

$$5 \log_9 (n+1) + D - 2 \leq \sigma(n) \leq 5 \log_9 n + D - \tfrac{1}{2}$$

with the lower bound being attained whenever $n = 24 \cdot 9^k - 1$ *and the upper whenever* $n = 8 \cdot 9^k$, *for* $k = 0, 1, 2, \ldots$.

The Fitting length of a soluble group is the minimal length of a series running from $\{1\}$ to G with nilpotent factors.

3.11. Theorem (Frick & Newman [14 b]). *Let G be a soluble linear group of degree n. Then the Fitting length of G is at most*

$$4 + 2r(n) + [(2n-1)/8 \cdot 3^{r(n)}]$$

where

$$r(n) = [\log_3 ((2n-1)/4)],$$

and for any every $n \geq 1$ this bound is attained. If G is also completely reducible then G has Fitting length at most

$$3 + 2t(n) + [n/4 \cdot 3^{t(n)}]$$

where

$$t(n) = [\log_3 (n/2)],$$

and for every $n \geq 1$ this bound is attained.

We conclude this chapter by looking briefly at a couple of special types of soluble linear groups. The following result is a consequence of 1.14, cf. the proof of 3.6.

3.12. Theorem. *If G is an abelian-by-locally-supersoluble subgroup of $GL(n, F)$ then G contains a triangularizable normal subgroup of finite index dividing $n!$.* □

Our next result is similar to 3.4. It is, in fact, a special case of a theorem giving bounds on the orders of various upper central factors of an arbitrary linear group. This result we shall prove in Chapter 8, but for completeness we give a proof of the special case now.

3.13. Theorem (Suprunenko [54])[4]. *Let G be an irreducible subgroup of $GL(n, F)$ which is nilpotent of class l. Then $G/\zeta_1(G)$ has finite order dividing $n! \, n^{(l-1)(n-1)}$ and finite exponent dividing n^{l-1}.*

Proof. We can suppose that G is algebraically closed (1.19) and that $n \geq 2$. G is monomial by 1.14 and so contains a diagonalizable normal

---

[4] See J.D. Dixon [11] for another proof of 3.13 (yielding a different bound).

subgroup A such that $(G:A)$ divides $n!$. By Schur's Lemma $\zeta_1(G) \subseteq F^* 1_n$. Applying 3.1 with $L=G$, $H=\zeta_2(G)$ and $K=\zeta_1(G)$ we see that $\zeta_2(G)/\zeta_1(G)$ has finite exponent dividing n. Hence $\zeta_{i+1}(G)^n \subseteq \zeta_i(G)$ for $i=1, 2, \ldots$ ([47], Lemma 1.62) and in particular $G/\zeta_1(G) = \zeta_l(G)/\zeta_1(G)$ has finite exponent dividing n^{l-1}. A is diagonalizable; thus $A/A \cap \zeta_1(G)$ is isomorphic to a periodic subgroup of $D(nF)/F^* 1_n$ and so has rank at most $n-1$. Therefore the order of $A/A \cap \zeta_1(G)$ divides $n^{(l-1)(n-1)}$ and so the order of $G/\zeta_1(G)$ divides $n! \, n^{(l-1)(n-1)}$. $\quad \square$

Exercise 3.1 (Gruenberg [20]). Let G be a linear group of degree n and H a soluble normal subgroup of G. Prove that H contains a triangularizable subgroup of finite index at most $\mu(n)$ which is normal in G.

This may be proved in the same manner as 3.6.

Exercise 3.2 (Kolchin-Kaplansky-Suprunenko, see [28] and [55]). S is a subsemigroup of $GL(n, F)$ such that every element of S has the form $\alpha 1_n + x$ where $\alpha \in F$ and x is nilpotent. Prove that S is triangularizable over F.

Hint: use the Burnside trace argument, cf. 1.21.

Exercise 3.3. F is an algebraically closed field of characteristic $p>0$ and G is an irreducible subgroup of $GL(n, F)$ that is nilpotent of class 2. Prove that p does not divide n.

Use 3.2. This exercise follows immediately from 7.7, but there is an easy direct proof.

Exercise 3.4. Derive 1.14 from 3.4 and 3.8.

Exercise 3.5 (J. D. Dixon). If G is an irreducible nilpotent linear group of degree n and class l prove that the order of G' is finite and divides

$$n! \, n^{1 + (l-1)(n-1)}.$$

Exercise 3.6. G is an irreducible primitive subgroup of $GL(n, F)$ where F is algebraically closed, and $Z = \zeta_1(G)$. Prove the following.

a) If $A = \eta_1(G)$ then A is nilpotent of class at most 2 and $A = \eta(G)$, the Hirsch-Plotkin radical of G.

b) A/Z is the maximal abelian normal subgroup of G, $|A/Z| = r^2$ for some integer r dividing n and prime to $c = \operatorname{char} F$ if the latter is non-zero, and every Sylow subgroup of A/Z is elementary abelian.

c) If $r = p_1^{l_1} p_2^{l_2} \ldots p_t^{l_t}$ where $p_1, p_2, \ldots, p_t$ are distinct primes, then $G/C_G(A/Z)$ is isomorphic to a subgroup of the direct product

$$D = \prod_{i=1}^{t} \operatorname{Sp}(2l_i, p_i).$$

d) If H is any soluble normal subgroup of G, then

$$A \cap H = C_H(A/Z) \quad \text{and} \quad Z \cap H = C_H(A).$$

In particular $H/(A \cap H)$ is isomorphic to a subgroup of D.

e) $C_G(A/Z)/C_G(A)$ is isomorphic to a subgroup of A/Z.

f) $C_G(A)$ is isomorphic to a subgroup of $GL(n/r, F)$.

Hints: To prove a) use P. Hall's formula $[\gamma^r G, \gamma^s G] \subseteq \gamma^{r+s} G$ for the first part, and either Exercise 3.1 or a refinement of 1.14 for the second part. The proofs of b), c) and d) are similar to the corresponding parts of the proof of 3.3 (in fact they are easier if you make use of a). For e) use stability groups and the duality $\text{Hom}(X, \mathbb{C}^*) \cong X$ for a finite abelian group X. Part f) is a consequence of Clifford's Theorem.

We halt our discussion of soluble linear groups as such at this point. For results on soluble d-groups see Chapter 7, for locally nilpotent linear groups see the second half of Chapter 7 and Chapter 8, and for (locally) supersoluble linear groups see Chapter 11.

4. Finitely Generated Linear Groups

Let G be a finitely generated subgroup of $\mathrm{GL}(n, F)$. Then G is generated as a semi-group by a finite subset say $\{g_1, \ldots, g_r\}$. If

$$g_k = (g_k(i, j)) \quad \text{for } k = 1, 2, \ldots, r,$$

denote by R the subring of F generated by all the $g_k(i, j)$, i.e.

$$R = P[g_k(i, j): k = 1, 2, \ldots, r; \ i, j = 1, 2, \ldots, n]$$

where P is the subring of F generated by 1_F. Then $G \subseteq \mathrm{GL}(n, R) \subseteq \mathrm{GL}(n, F)$. For most of this chapter we shall be somewhat more general than the title indicates and study subgroups of the groups $\mathrm{GL}(n, R)$ for R a finitely generated integral domain. We shall study the properties of these groups in the following order: their residual properties (especially finite ones), their Frattini properties, their centrality properties, and finally their chief factors and maximal subgroups.

If R is any commutative ring $J(R)$ denotes the intersection of the maximal ideals of R and $N(R)$ denotes the set of all nilpotent elements of R. We need the following well-known result.

4.1. *If R is a finitely generated commutative ring, then:*
  i) $J(R) = N(R)$.
  ii) *If $\mathfrak{m}$ is a maximal ideal of R then $R/\mathfrak{m}$ is a finite field.*

We briefly indicate a proof modulo results in Kaplansky's book [29]. If R is a commutative ring an ideal $\mathfrak{a}$ of R is called a *G-ideal* if either it is maximal or if it is prime and the intersection of all the non-zero prime ideals of $R/\mathfrak{a}$ is non-zero. R is called a *Hilbert ring* if every G-ideal of R is maximal. Clearly a homomorphic image of a Hilbert ring is a Hilbert ring.

  i) Trivially $\mathbb{Z}$ is a Hilbert ring. Hence $\mathbb{Z}[X_1, \ldots, X_n]$ is a Hilbert ring by [29] Theorem 31. Therefore R is a Hilbert ring. But now [29] Theorem 30 gives that $J(R) = N(R)$.

  ii) We can clearly suppose that $R = \mathbb{Z}[X_1, \ldots, X_n]$. The proof of [29] Theorem 32 shows that $R/\mathfrak{m}$ is a finite field. $\square$

Suppose that R is a finitely generated integral domain and that $a_1, \ldots, a_r$ are distinct elements of R_n. Let $a_k = (a_k(i, j))$. For each pair of distinct integers $h, k, 1 \leq h, k \leq r$, there exists a pair i and j such that

$$a_h(i, j) \neq a_k(i, j).$$

Denote such an (i, j) by $(h, k)\sigma$ and put

$$a = \prod_{\substack{1 \leq h, k \leq r \\ h \neq k \\ (i, j) = (h, k)\sigma}} (a_h(i, j) - a_k(i, j))$$

a is a non-zero element of R. Since R is an integral domain, 4.1 implies that $J(R) = \{0\}$. Hence there exists a maximal ideal $\mathfrak{m}$ of R such that $a \notin \mathfrak{m}$. The natural map $\chi: R \to R/\mathfrak{m} = K$ induces a ring homomorphism χ_n of R_n onto K_n. Now $a\chi \neq 0$ and so if $h \neq k$ and $(i, j) = (h, k)\sigma$ then $a_h(i, j)\chi \neq a_k(i, j)\chi$. Therefore $a_1 \chi_n, \ldots, a_r \chi_n$ are distinct elements of K_n. In particular, R_n is residually finite. If g is a unipotent element of $\mathrm{GL}(n, R)$ then $g\chi_n$ is a unipotent element of K_n, for

$$0 = (g-1)^n \chi_n = (g\chi_n - 1)^n.$$

Note also that if char $R \neq 0$ then char $R = $ char K. We have now proved the following theorem.

4.2. Theorem (Mal'cev [34]). *Let R be a finitely generated integral domain. For every finite set $g_1, \ldots, g_r$ of distinct elements of $\mathrm{GL}(n, R)$ there exists a finite field K and a homomorphism ϕ of $\mathrm{GL}(n, R)$ into $\mathrm{GL}(n, K)$ such that $g_1 \phi, \ldots, g_r \phi$ are all distinct. If g is a unipotent element of $\mathrm{GL}(n, R)$ then $g\phi$ is a unipotent element of $\mathrm{GL}(n, K)$. If char $R \neq 0$ then char $K = $ char R. In particular, $\mathrm{GL}(n, R)$ is a residually-finite group.* $\square$

4.3. Corollary (Gruenberg [19]). *Let $\mathfrak{X}$ be a class of groups such that $Q\mathfrak{X} = \mathfrak{X}$, $\mathfrak{X} \subseteq L(\mathfrak{G} \cap \mathfrak{X})$, $\mathfrak{X} \cap \mathfrak{F} \subseteq \mathfrak{S}$. Then linear $\mathfrak{X}$-groups are soluble.*

$\mathfrak{G}$ is the class of finitely generated groups, $\mathfrak{F}$ the class of finite groups and $\mathfrak{S}$ the class of soluble groups. Q and L are respectively the quotient and local closure operators. The condition $\mathfrak{X} \subseteq L(\mathfrak{G} \cap \mathfrak{X})$ is trivially satisfied if $\mathfrak{X}$ is subgroup closed.

Proof. Let G be an $\mathfrak{X}$-subgroup of $\mathrm{GL}(n, F)$ and suppose that G is not soluble. Put $r = 2^{f(n)}$ where f is the function of 3.7. G contains elements $a_1, \ldots, a_r$ such that if w is the r-th derived word then $a = w(a_1, \ldots, a_r) \neq 1$. There exists a finitely generated $\mathfrak{X}$-subgroup H of G such that

$$\langle a_1, \ldots, a_r \rangle \subseteq H.$$

By 4.2 there exists a normal subgroup N of H such that $a\notin N$ and such that H/N is isomorphic to a finite linear group of degree n. Since $Q\mathfrak{X}=\mathfrak{X}$ and $\mathfrak{X}\cap\mathfrak{F}\subseteq\mathfrak{S}$, H/N is soluble. Therefore by 3.7, $w(a_1,\dots,a_r)\in N$. This contradiction completes the proof. $\square$

By 3.8 locally soluble linear groups are soluble, so in particular hypercentral linear groups and locally nilpotent linear groups are soluble. It follows from 4.3 that if G is a linear group lying in any one of the following classes then G is soluble; SI^*, SN^*, N, Engel groups (see [32] Vol. 2, pp. 182 and 218 for the definitions). Since free groups are linear the corresponding statements for the classes ZD, Z, SD, SI and SN are false. They are also false for the classes $\bar{Z}$ and $\overline{SI}$; P. Hall in [24] gives an example of a $\bar{Z}$-group that is linear of degree 2 over the p-adic integers and contains non-abelian free subgroups, and $\bar{Z}\subseteq\overline{SI}$. This leaves the class $\overline{SN}$.

Question 4. *Does there exist a non-soluble linear $\overline{SN}$-group?*

It is still an open question whether a subgroup of an $\overline{SN}$-group is necessarily an $\overline{SN}$-group. If there is a group G giving a positive answer to Question 4 then by 4.3, G will contain a subgroup that is not an $\overline{SN}$-group.

In a soluble group the four characteristic subsets of Engel elements (left, right, bounded left, bounded right) are in fact subgroups (Gruenberg [17]). It follows from this result and 4.3 that the four characteristic subsets of Engel elements in a linear group are necessarily subgroups. See [19] §1.3 or the end of Chapter 8 for the details.

4.4. Corollary (Mal'cev [34]). *Let G be a finitely generated linear group. Then*

i) *G is Hopfian (i.e. every epi-endomorphism is an isomorphism), and*

ii) *if G is simple, G is finite.*

Any finitely generated residually finite group is Hopfian ([37] 41.44). Hence both parts of the corollary follow from the residual finiteness of G (4.2). $\square$

Let t be a non-zero element of the finitely generated integral domain R. Then there exists a maximal ideal $\mathfrak{m}$ of R not containing t. Put $K=R/\mathfrak{m}$, $\chi: R\to K$ and $\chi_n: R_n\to K_n$ as before and let $\mathfrak{a}=\ker\chi_n$. If $a\in R_n$,

$$\mathrm{tr}(a)\chi=\mathrm{tr}(a\chi_n).$$

Hence $\mathrm{tr}(a)\neq t$ for all a in $\mathfrak{a}$.

4.5. Lemma (Greenberg). *Let $t_1,\dots,t_m$ be elements of the field F such that $t_i\neq n1_F$ for each i. If G is a subgroup of $\mathrm{GL}(n, R)$ where R is a finitely generated subring of F, then G contains a normal subgroup N of finite index such that no element of N has trace t_i for any $i=1,2,\dots,m$.*

Trivially we cannot do this for $n1_F$ since this is the trace of the identity element.

Proof. If t_i does not lie in R put $N_i = G$. Suppose now that $t_i \in R$ and let χ, χ_n and $\mathfrak{a}$ be as above with $t = t_i - n1_F$. χ_n induces a (group) homomorphism ϕ of $GL(n, R)$ into $GL(n, K)$. Put $N_i = \ker \phi$; N_i is a normal subgroup of finite index in G. Now $N_i = G \cap (1_n + \mathfrak{a})$ and so if $x \in N$, then $\operatorname{tr} x \neq n1_F + t = t_i$. The subgroup $N = N_1 \cap \cdots \cap N_m$ has all the required properties. $\Box$

If in 4.5 we were also told that the set $\{\operatorname{tr} x : x \in G\}$ is finite then we could find a normal subgroup M of finite index in G such that every element of M has trace $n1_F$. This is also a consequence of Burnside's trace argument, at least in characteristic zero, see comments before 1.23.

Question 5. *What is the relationship between 4.5 and 1.20?*

In 1.20 we are given a bound in terms of n and m. It is not possible to bound the index of N in G (of 4.5) just by a function of m and n only.

Example. *Let F be a field containing an element α of infinite order (multiplicatively). Then $GL(n, F)$ contains an infinite cyclic subgroup G such that for every positive integer s there exists an element t of F not equal to $n1_F$ so that every subgroup N of G none of whose elements have trace t, has index at least s in G.*

Put

$$a = \begin{pmatrix} \alpha & 0 \\ 0 & 1_{n-1} \end{pmatrix}, \quad G = \langle a \rangle \quad \text{and} \quad t = \alpha^{s!} + (n-1)1_F.$$

Since α has infinite order $t \neq n1_F$. Let N be a subgroup of G such that $(G:N) \leq s$. Then $a^{s!} \in N$ and $\operatorname{tr}(a^{s!}) = t$. $\Box$

Let G be a finitely generated subgroup of $GL(n, F)$, $t \in F \smallsetminus \{n1_n\}$ and put $H = \langle x \in G : \operatorname{tr} x \neq t \rangle$. It follows from 4.5 that H has finite index in G.

Question 6. *Does there exist a function $\rho(n)$ of n only such that $(G:H) \leq \rho(n)$ for all possible G and t?*

If x is an element of $GL(n, F)$, then $\operatorname{tr} x$ is a certain coefficient of the characteristic polynomial of x. 4.5 guarantees the existence of a large subgroup N of G such that the possible values that this coefficient can take for elements of N is restricted. It is possible to find similar large subgroups N of G such that other coefficients of the characteristic polynomial (or even the characteristic polynomial itself) of the elements of N are restricted. The following two exercises indicate results along these lines which further suggest a connection between 4.5 and 1.20.

Exercise 4.1. Prove the following. Let R be a finitely generated subring of the field F and G a subgroup of $\mathrm{GL}(n, R)$. Suppose that $f_1(X), \dots, f_m(X)$ are polynomials over F of degree n such that for each i we have

$$f_i(X) \neq (X - 1_F)^n.$$

Then there exists a normal subgroup N of finite index in G such that for each x in N for each $i = 1, 2, \dots, m$

$$\det(X 1_n - x) \neq f_i(X).$$

Exercise 4.2. Use Exercise 4.1 to prove the following special case of Burnside's Exponent Theorem (1.23). A finitely generated linear group of finite exponent is finite.

(It follows from 1.23 that such a group has a unipotent normal subgroup U of finite index. U is also finitely generated and of finite exponent. Also U is nilpotent and hence U is finite.)

4.6. Lemma. *Let R be a ring, $\mathfrak{p}$ an ideal of R, A an R-module satisfying*

$$\bigcap_{i=1}^{\infty} A \mathfrak{p}^i = \{0\}$$

and m a positive integer such that $m 1_R \in \mathfrak{p}$. If G is a group of R-automorphisms of A satisfying $A(G-1) \subseteq A\mathfrak{p}$, and $K_i = C_G(A/A\mathfrak{p}^{i+1})$, then $G^{m^i} \subseteq K_i$ and $\bigcap_{i=1}^{\infty} K_i = \{1\}$.

Proof. For every positive integer j

$$A \mathfrak{p}^j (G - 1) = A(G-1)\mathfrak{p}^j \subseteq A\mathfrak{p}^{j+1}.$$

(So G/K_i stabilizes the series $A\mathfrak{p}^{i+1} \subseteq \cdots \subseteq A\mathfrak{p} \subseteq A$!) Now $G^{m^i} \subseteq K_i$ if and only if $A(G^{m^i} - 1) \subseteq A\mathfrak{p}^{i+1}$. We prove the latter statement by induction on i. If $i = 0$ the result is given. Suppose that

$$A(G^{m^{i-1}} - 1) \subseteq A\mathfrak{p}^i.$$

If $a \in A$, the mapping of $G^{m^{i-1}}$ into $A\mathfrak{p}^i / A\mathfrak{p}^{i+1}$ given by

$$g \mapsto a(g - 1) + A\mathfrak{p}^{i+1}$$

is a (group) homomorphism. Hence modulo $A\mathfrak{p}^{i+1}$,

$$a(g^m - 1) \equiv a m(g - 1) \in A\mathfrak{p}^{i+1}$$

and so

$$A(G^{m^i} - 1) \subseteq A\big((G^{m^{i-1}})^m - 1\big) \subseteq A\mathfrak{p}^{i+1}.$$

Also if $g \in \bigcap_{i=1}^{\infty} K_i$, then $A(g - 1) \subseteq \bigcap_{i=1}^{\infty} A\mathfrak{p}^i$, $= \{0\}$. Since G acts faithfully on A, $g = 1$. ∎

The following result is Proposition 1.6 of [69]. Part i) has also been proved by Ju. Merzljakov [36] and M.I.Kargarpolov [30a]. See also V.P.Platonov [45c] for the full result.

4.7. Theorem. *Let R be a finitely generated integral domain and*

$$G = \mathrm{GL}(n, R).$$

Then:

i) *If* char $R = 0$ *there exists a finite set S of primes such that for each prime p not in S, G contains a normal subgroup of finite index which is residually a finite p-group.*

ii) *If* char $R = p > 0$, *G contains a normal subgroup of finite index which is residually a finite p-group.*

We require the following result of Krull ([29] Theorem 77, or [72] Vol. 1, p. 216, Theorem 12). If R is a Noetherian integral domain and $\mathfrak{a}$ is an ideal of R (which implies that $\mathfrak{a} \neq R$ as always in this text), then

$$\bigcap_{i=1}^{\infty} \mathfrak{a}^i = \{0\}.$$

We shall also use the following: the group of units of a finitely generated integral domain is a finitely generated abelian group. For a short proof see [48] Théorème 1. In the proof immediately below, with a little extra argument, we could get by with a special (and much simpler) case of this theorem. However since we shall repeatedly need the full force of [48] Théorème 1 in the remainder of this chapter, we may as well quote it in full now.

Proof. If $\mathfrak{p}$ is a maximal ideal of R, then $R/\mathfrak{p}$ is a finite field of characteristic p say (4.1). Also $\bigcap_{i=1}^{\infty} \mathfrak{p}^i = \{0\}$ by Krull's Theorem. We can regard G as a group of R-automorphisms of a free R-module A of rank n. Trivially $\bigcap_{i=1}^{\infty} A\mathfrak{p}^i = \{0\}$. Let $H = C_G(A/A\mathfrak{p})$. Then G/H is isomorphic to a subgroup of $\mathrm{GL}(n, R/\mathfrak{p})$ and so is a finite group. Put $K_i = C_H(A/A\mathfrak{p}^{i+1})$. K_i is a normal subgroup of H (even of G) and H/K_i is a p-group (4.6). $A/A\mathfrak{p}^{i+1}$ is a free $R/\mathfrak{p}^{i+1}$ module of rank n. Hence H/K_i is a finite group if $R/\mathfrak{p}^{i+1}$ is a finite ring. $R/\mathfrak{p}$ is finite. For each j, $\mathfrak{p}^j$ is finitely generated as R-module. Therefore $\mathfrak{p}^j/\mathfrak{p}^{j+1}$ is finitely generated as $R/\mathfrak{p}$-module and thus is finite. We have now shown that H is residually a finite p-group.

If char $R > 0$, clearly $p = \mathrm{char}\, R/\mathfrak{p} = \mathrm{char}\, R$ and part ii) is proved. Let char $R = 0$. If q is a prime integer such that $q 1_R$ is not a unit of R then $q 1_R$ is contained in a maximal ideal of R and G contains a normal subgroup of finite index which is residually a finite q-group. Suppose that q_i, $i = 1, 2, \ldots$ are distinct integer primes such that $q_i 1_R$ is a unit of R for each i. Then $\langle q_i 1_R : i = 1, 2, ... \rangle$ is a (multiplicative) free abelian group of infinite rank (since $\mathbb{Z}[(q_i 1_R)^{-1} : i = 1, 2, ...] \subseteq R \cap \mathbb{Q} 1_R$). But the group of units of a finitely generated integral domain is finitely generated. Therefore for at most a finite number of primes q, is $q 1_R$ a unit of R. The theorem is now proved. $\square$

Exercise 4.3. G is a soluble linear group over a field of positive characteristic. If G is residually a finite π-group and residually a finite π'-group for some set of primes π, prove that G is abelian.

It is very easy to give examples for which the set S of 4.7 i) is non-empty.

Example. *There exists a 2-generator metabelian linear group over the rationals none of whose subgroups of finite index is residually a finite 2-group. The groups also contains an abelian subgroup which cannot be finitely generated.*

Let $G = \left\langle a = \begin{pmatrix} 1 & 0 \\ 1 & 1 \end{pmatrix}, b = \begin{pmatrix} 2 & 0 \\ 0 & 1 \end{pmatrix} \right\rangle \subseteq GL(2, \mathbb{Q})$. G is triangular and so is metabelian. It is easy to check that $a^{b^i} = \begin{pmatrix} 1 & 0 \\ 2^i & 1 \end{pmatrix}$ and that $\langle a^G \rangle$ is isomorphic to the additive group of dyadic rationals (i.e. to $\{r/2^s : r, s \in \mathbb{Z}\}^+$). This group is 2-divisible and so no subgroup of finite index in $\langle a^G \rangle$ can be residually a finite 2-group. It is easy to see that $\langle a^G \rangle$ cannot be finitely generated. $\square$

4.8. Corollary [69]. *Let R be a finitely generated integral domain. Then $G = GL(n, R)$ contains a normal subgroup T of finite index such that all the elements of finite order in T are unipotent (so if char $R = 0$, T is torsion-free).*

Proof. If char $R = 0$ then there exists a pair of distinct primes p_1 and p_2 such that for each i, G contains a normal subgroup H_i of finite index so that H_i is residually a p_i-group. Then every periodic subgroup of H_i is a p_i-group and $T = H_1 \cap H_2$ is torsion-free.

If char $R = p > 0$ then G contains a normal subgroup T of finite index such that T is residually a finite p-group (4.7 ii)). Hence an element of T of finite order is a p-element and consequently is unipotent. $\square$

The char $R = 0$ case of 4.8 was proved by A. Selberg [51] and independently by M.I. Kargapolov [30]. Selberg's method of proof can be extended to give a proof of the whole of 4.8. In this approach the subgroup T arises as the kernel of a certain representation and probably yields a "larger" subgroup. See [69] Proposition 3.2 where both proofs are given.

By 4.8 if char $R = 0$ then $GL(n, R)$ is torsion-free-by-finite. This is definitely false if char $R \neq 0$.

Example. *If F is the field of rational functions in a single indeterminate x over the field of p elements* $\mathbb{F}_p$, *then* $\mathrm{GL}(2, F)$ *contains a 2-generator metabelian subgroup which is not torsion-free-by-finite and which contains an infinitely generated abelian subgroup. If* $R = \mathbb{F}_p[x, x^{-1}]$ *then* $\mathrm{GL}(2, R)$ *is not torsion-free-by-finite.*

Let $G = \left\langle a = \begin{pmatrix} 1 & 0 \\ 1 & 1 \end{pmatrix}, b = \begin{pmatrix} x & 0 \\ 0 & 1 \end{pmatrix} \right\rangle$. Clearly G is metabelian. Also

$$a^{b^i} = \begin{pmatrix} 1 & 0 \\ x^i & 1 \end{pmatrix}$$

so $\langle a^G \rangle$ is an infinite elementary abelian p-group. Trivially $G \subseteq \mathrm{GL}(2, R)$. The result now follows. ☐

4.9. Corollary (Schur [49]). *A periodic linear group is locally finite.*

Proof. Let G be a finitely generated periodic subgroup of $\mathrm{GL}(n, F)$. We have just to prove that G is finite. By 4.8, G contains a normal subgroup T of finite index such that T is unipotent. Then T is nilpotent and periodic, and hence locally finite ([32] Vol. 2, p. 190). Therefore G is locally finite and so finite ([32] Vol. 2, p. 153). ☐

We wish now to consider the Frattini and centrality properties of finitely generated linear groups. To do this we need a series of lemmas. We give them in a somewhat more general form than we use here since we shall need their full force in Chapter 8.

4.10. Lemma ([67]). *Let R be a finitely generated subring of the field F and T a triangularizable subgroup of* $\mathrm{GL}(n, F)$ *contained in* $\mathrm{GL}(n, R)$. *If U is the maximal unipotent subgroup of T then* T/U *is a finitely generated abelian group.*

Proof. We can clearly suppose that F is algebraically closed. Then there exists an x in $\mathrm{GL}(n, F)$ such that $T^x \subseteq \mathrm{Tr}(n, F)$. Let $x = (x_{ij})$, $x^{-1} = (y_{ij})$ and put $S = R[x_{ij}, y_{ij}: i, j = 1, 2, \ldots, n] \subseteq F$. S is a finitely generated integral domain and its group of units is finitely generated ([48] Théorème 1). The map θ of T^x into $D(n, S)$ given by $(a_{ij})\theta = \mathrm{diag}(a_{11}, \ldots, a_{nn})$ is a homomorphism with kernel U^x. $D(n, S)$ is isomorphic to a direct product of n copies of the group of units of S and so is finitely generated. Hence $T/U \cong T^x/U^x$ is a finitely generated abelian group. ☐

4.11. Corollary. *Let G be a finitely generated subgroup of* $\mathrm{GL}(n, F)$. *Then every completely reducible soluble subgroup of G is finitely generated.*

Proof. Let A be a completely reducible soluble subgroup of G. By Mal'cev's Theorem (3.6) A contains a triangularizable normal subgroup T. The maximal unipotent subgroup U of T is completely reducible over F

(by Clifford's Theorem) and is unitriangularizable over F by 1.21. Therefore $U = \{1\}$. By 4.10, T is finitely generated. Since $(A : T)$ is finite, so A is finitely generated. □

For every characteristic 4.11 becomes false if the words "completely reducible" are removed. The two examples above afford appropriate counter examples.

4.12. Lemma (Gruenberg [19] Prop. 5). *Let G be a group, V a finite-F-dimensional faithful FG-module, W an FG-submodule of V and H a normal subgroup of G stabilizing the series $\{0\} \subseteq W \subseteq V$. Then $H \mathbf{e} G$ implies that $H \subseteq \zeta_t(G)$ where $t \le \dim_F W(\dim_F V - \dim_F W)$.*

$\zeta_i(G)$ denotes the i-th term of the upper central series of G and we write $H \mathbf{e} G$ if for each h in H and g in G there exists a positive integer $s = s(h, g)$ such that $[h, {}_s g] = 1$. Note that if $H \subseteq \zeta(G)$, then $H \mathbf{e} G$.

Proof. Let $n = \dim_F V$ and $d = \dim_F W$. If h is any element of H then

$$h' : a + W \mapsto a(h - 1)$$

is a linear mapping of V/W into W, that is $h' \in \mathrm{Hom}_F(V/W, W)$. The mapping $\phi : h \mapsto h'$ is a group homomorphism of H into the additive group $\mathrm{Hom}_F(V/W, W)^+$. If h' is the zero map then $ah = a$ for all a in V. Hence ϕ is one-to-one. In particular H is abelian.

H is a G-module via conjugation and $\mathrm{Hom}_F(V/W)$ is an FG-module where the action of G is given by $\lambda^g = g^{-1} \lambda g$, $\lambda \in \mathrm{Hom}_F(V/W, W)$. Now

$$h^g \phi : a + W \mapsto a(h^g - 1) \quad \text{and} \quad (h\phi)^g : a + W \mapsto a g^{-1}(h - 1) g.$$

Thus ϕ is a G-homomorphism and hence

$$[h, {}_r g] \phi = (h^{(g-1)^r}) \phi = h'^{(g-1)^r}.$$

Let M denote the F-submodule of $\mathrm{Hom}_F(V/W, W)$ generated by $H\phi$. Since $H \mathbf{e} G$ and $\dim_F M = t$ is finite, each g in G acts unipotently on M. By 1.21, G acts unitriangularly relative to some basis of M. Hence if $g_1, g_2, \dots, g_t$ are any t elements of G

$$[h, g_1, \dots, g_t] \phi = h'^{(g_1 - 1) \cdots (g_t - 1)} = 0.$$

But ϕ is one-to-one, so $[h, g_1, \dots, g_t] = 1$. This says that $h \in \zeta_t(G)$. It remains only to point out that

$$t = \dim_F M \le \dim_F \mathrm{Hom}_F(V/W, W) = d(n - d). \quad □$$

4.13. Corollary (Gruenberg [20]). *Let G be a subgroup of* GL(n, F) *and U a unipotent normal subgroup of G. If* $U \in G$ *then* $U \subseteq \zeta_t(G)$ *where*

$$t \le \tfrac{1}{2} n(n-1).$$

Proof. Let V be the n-row vector space over F regarded as FG-module. Since U is unipotent U is unitriangularizable over F (1.21). Thus there exists a non-zero element w of V such that $wx=w$ for all x in U. If $g \in G$, then $gxg^{-1} \in U$ and so $wgx = wx^g{}^{-1}g = wg$. Let W be the FG-submodule of V generated by w. Then $vx = v$ for all x in U and v in W. Denote by ρ the representation of G induced by V/W. Suppose we know that

$$U\rho \subseteq \zeta_k(G\rho),$$

i.e. that $[U, {}_kG] \subseteq K = U \cap \ker \rho$. By 4.12, $K \subseteq \zeta_{d_1(n-d_1)}(G)$, where

$$d_1 = \dim_F W.$$

Therefore $U \subseteq \zeta_t(G)$ where $t = k + d_1(n-d_1)$.

By induction on n there exist positive integers $r, d_1, d_2, \dots, d_r$ such that $d_1 + d_2 + \cdots + d_r = n$ and $U \subseteq \zeta_t(G)$ where

$$t = \sum_{i=1}^{r} d_i(n - d_1 - d_2 - \cdots - d_i).$$

The following lemma completes the proof.

Lemma. *Let* $n, r, d_1, \dots, d_r$ *be positive integers such that*

$$d_1 + d_2 + \cdots + d_r = n.$$

Then

$$y = \sum_{i=1}^{r} d_i(n - d_1 - d_2 - \cdots - d_i)$$

is at most $\tfrac{1}{2} n(n-1)$, *and y attains this value if* $r = n$ *and* $d_1 = d_2 = \cdots = d_r = 1$.

Proof. It is immediate that y attains the given value at the given point. If $r = 1$ and $d_1 = n$ then $y = 0 \le \tfrac{1}{2} n(n-1)$. Suppose now that $r > 1$. By induction we may suppose that

$$y \le d_1(n-d_1) + \tfrac{1}{2}(n-d_1)(n-d_1-1) = \tfrac{1}{2}(n^2 - d_1^2 - n - d_1) = z \quad \text{say.}$$

A trivial calculation shows that z takes its maximum value in the range $[1, n]$ at the point $d_1 = 1$. Thus $z \le \tfrac{1}{2} n(n-1)$ and the lemma is proved. ☐

4.14. Lemma (Gruenberg [20]). *Let A be an abelian normal subgroup of the group G such that* $A \in G$. *If* $C_G(A)$ *has finite index k in G, then* $[A, G]$ *is a k-group (i.e. every prime in* $\pi([A, G])$ *divides k).*

Proof. Let $a \in A$ and $g \in G$. Since $[A, G]$ is abelian we need only show that the order of $[a, g]$ divides some power of k. There exists an integer $t \ge 0$ such that $[a, {}_{t+1}g] = 1$. Put $S = \langle g \, C_G(A) \rangle \subseteq G/C_G(A)$ and let $\mathfrak{s}$ denote the augmentation ideal of S over $\mathbb{Z}$. If $x \in S$ then $x^k = 1$. Hence

$$0 = x^k - 1 = (1 + (x-1))^k - 1 \equiv k(x-1) \bmod \mathfrak{s}^2.$$

Therefore $k\mathfrak{s} \subseteq \mathfrak{s}^2$ and so by induction $k^t \mathfrak{s} \subseteq \mathfrak{s}^{t+1}$. Consequently

$$[a, g]^{k^t} \in a^{k^t \mathfrak{s}} \subseteq a^{\mathfrak{s}^{t+1}} = 1$$

and the result follows. □

4.15. Lemma. *If N is a nilpotent normal subgroup of the group G then $N' \subseteq \phi(G)$ the Frattini subgroup of G.*

This lemma is a very special case of a result of P. Hall ([23] Lemma 3).

Proof. Let M be a maximal subgroup of G and suppose that $M \cap N \ne N$. If $\zeta_i(N)$ is the first term of the upper central series of N not contained in $M \cap N$, then $G = \zeta_i(N) \cdot M$. Also $[\zeta_i(N), M \cap N] \subseteq M \cap N$, so $\zeta_i(N)$ normalizes $M \cap N$. Hence $M \cap N$ is normal in $\zeta_i(N) \cdot M = G$.

If K is any normal subgroup of G satisfying $M \cap N \subset K \subseteq N$, then $KM = G$ and
$$K = K(M \cap K) \supseteq K(M \cap N) = N.$$

Hence $N/M \cap N$ is a chief factor of G. But N is nilpotent, so $N/M \cap N$ is abelian. This is for all maximal subgroups of G; therefore $N' \subseteq \phi(G)$. □

Suppose that F is an algebraically closed field, that G is a subgroup of $GL(n, F)$ and that K is a soluble normal subgroup of G. By Mal'cev's Theorem (3.6) K contains a triangularizable normal subgroup T of finite index. Although T may not be normal in G, K does contain a triangularizable subgroup of finite index that is normal in G. (See Exercise 3.1.) In Chapter 5 we shall give a second and quite different proof of this. There we shall define a subgroup K^0 of K such that $(K : K^0)$ is finite, K^0 is normal in G and whenever K is soluble K^0 is triangularizable. We shall use these remarks in the proof of 4.16.

4.16. Lemma ([67], [43]). *Let R be a finitely generated integral domain, G a subgroup of $GL(n, R)$ and K a normal subgroup of G such that whenever a subgroup L of finite index in K is normal in G, K/L is nilpotent. Then K is nilpotent.*

We give here the proof from [67]. Platonov's approach will be given in Chapter 10 (see especially 10.5). A special case of 4.16 is that if R is a finitely generated integral domain and G is a subgroup of $GL(n, R)$ such that G/N is nilpotent for every normal subgroup N of G of finite index, then G is nilpotent. Normal in this context cannot be replaced by characteristic (which would simplify the statement of 4.16), for G could be free of infinite rank, but if G is finitely generated then this substitution can be made since such a group contains only a finite number of subgroups of a given finite index ([50] 7.1.6 for example).

Proof. By 4.2 there exists a set $\mathscr{H}$ of normal subgroups of G such that $\bigcap_{H \in \mathscr{H}} H = \{1\}$, and such that for each H in $\mathscr{H}$, G/H is isomorphic to a finite linear group of degree n. Moreover, we can choose $\mathscr{H}$ such that the image of any unipotent subgroup of G under any of these representations is also unipotent. If H lies in $\mathscr{H}$, than $K/K \cap H$ is nilpotent. Thus $K/K \cap H$ is isomorphic to a soluble linear group of degree n and so has bounded derived length (3.7). Since $\bigcap_{\mathscr{H}} H = \{1\}$, K is soluble.

K contains a triangularizable subgroup K_1 of finite index in K such that K_1 is normal in G (see above remarks). If U is the maximal unipotent subgroup of K_1, U is also normal in G. Thus UH/H is a unipotent normal subgroup of the nilpotent linear group KH/H of degree n. By 4.13, $[U, {}_k K] \subseteq H$ where $k = \frac{1}{2} n(n-1)$.[4a] Therefore $[U, {}_k K] = \{1\}$.

Let $J = K/U$. K_1/U is a finitely generated abelian group (4.10) and so contains a free abelian characteristic subgroup A of finite index. A is normal in G/U, so J/A is nilpotent. We prove that A lies in the centre of J. This will complete the proof that K is nilpotent.

$C_J(A)$ has finite index in J (since A does). Let p be any prime greater than $(J : C_J(A))$ and i any positive integer. A^{p^i} is normal in G/U and J/A^{p^i} is finite. Therefore J/A^{p^i} is nilpotent. Hence by 4.14 (applied to J/A^{p^i} and A/A^{p^i}) it follows that $[A, J] A^{p^i}/A^{p^i}$ is a $(J : C_J(A))$-group. Since $p > (J : C_J(A))$, $[A, J] \subseteq A^{p^i}$. Now A is free abelian so

$$\bigcap_{i=1}^{\infty} A^{p^i} = \{1\}.$$

Therefore $[A, J] = \{1\}$ and the lemma is proved. $\quad \square$

4.17. Theorem ([67][5]). *Let R be any finitely generated integral domain and G a subgroup of $GL(n, R)$. If K is a normal subgroup of G such that $\phi(G) \subseteq K$ and $K/\phi(G)$ is locally nilpotent then K is nilpotent. In particular*

i) $\phi(G)$ *is nilpotent;*
ii) $\eta(G) = \eta_1(G)$ *is nilpotent;*
iii) $\eta(G)/\phi(G) = \eta(G/\phi(G))$;
iv) G *is nilpotent if and only if $G' \subseteq \phi(G)$.*

[4a] Actually $n-1$ would suffice here.
[5] But see also [43] for part i).

Proof. Let H be a subgroup of finite index in K that is normal in G. Then $\phi(G)H/H \subseteq K/H \cap \phi(G/H)$ and $(K/H)/(\phi(G)H/H)$ is a finite nilpotent group. Let P be any Sylow subgroup of K/H and g any element of G/H. Then $P \cdot (\phi(G)H/H) = P^g(\phi(G)H/H)$. Since $P \cdot (\phi(G)H/H)$ is a finite group there exists an element x of $\phi(G)H/H$ such that $P = P^{gx}$. Therefore

$$G/H = N_{G/H}(P) \cdot (\phi(G)H/H).$$

Now $\phi(G)H/H$ is a finite subgroup of the Frattini subgroup of G/H and so $G/H = N_{G/H}(P)$. Thus every Sylow subgroup of K/H is normal in K/H; consequently K/H is nilpotent. By 4.16, K is nilpotent.

Taking $K = \phi(G)$ trivially gives i). Thus $\phi(G) \subseteq \eta(G)$. Since $\eta(G)/\phi(G)$ is locally nilpotent, so $\eta(G)$ is nilpotent. This implies that $\eta(G) = \eta_1(G)$. Trivially $\eta(G)/\phi(G) \subseteq \eta(G/\phi(G))$. Let K be that subgroup of G such that $K/\phi(G) = \eta(G/\phi(G))$. Then K is nilpotent and iii) follows. If $G' \subseteq \phi(G)$, $G/\phi(G)$ is locally nilpotent, so G is nilpotent. If G is nilpotent it is well known (and very easy to prove) that $G' \subseteq \phi(G)$. ⬜

4.18. Example (P. Hall, [23]). *If F is any algebraically closed field there exists a metabelian subgroup of* $GL(4, F)$ *satisfying the minimal condition on subgroups such that $\phi(G)$ is not locally nilpotent and $G' \subseteq \phi(G)$.*

Let p and q be primes such that q^2 divides $(p-1)$. Then $GF(p)$ contains a primitive q^2-root of unity. By Hensel's Lemma the ring of p-adic integers $\mathbb{Z}_p$ contains a primitive q^2-root of unity.

Let A be a $\mathbf{C}_{p^\infty}$-group. End $A \cong \mathbb{Z}_p$. Hence A has an automorphism b of order q^2. Put $B = \langle b \rangle$, $C = \langle b^q \rangle$ and $G = AB$.

$\eta(G) = A$: Trivially $A \subseteq \eta(G)$. The only subgroups of G containing A are A, AC and G. Now A is a p-group and C is a q-group. Thus if AC is locally nilpotent $AC = A \times C$. But then b^q has trivial action on A, a contradiction. Therefore $\eta(G) = A$.

$\phi(G) = AC$: Let M be a maximal subgroup of G. If A is not a subgroup of M, $A \cap M$ is finite. But then M is finite and yet G is an infinite locally finite group. Therefore $A \subseteq M$. It is trivial to confirm that $\phi(B) = C$ and so $\phi(G) = AC$.

$G' \subseteq \phi(G)$: Trivially $G' \subseteq A \subseteq \phi(G)$. (In fact $G' = A$.)

If $q = 2$ then we can take $p = 5$ or 13 for example. Thus if char $F \neq 5$, let $p = 5$, $q = 2$ and G be as above and if char $F = 5$ let $p = 13$, $q = 2$ and G be as above. By 2.3, G is isomorphic to a subgroup of $GL(4, F)$. ⬜

The following example of a *torsion-free linear group whose Frattini subgroup is not nilpotent* is due to O.H. Kegel, and answers a question raised in my original notes. *The group is also metabelian and residually finite.*

If F is any sufficiently large field (F uncountable is more than enough) then F^* contains a subgroup A that is isomorphic to the additive group of rational numbers with odd denominators (2.2). There exists a homomorphism of A onto a cyclic group of order 4. Let ϕ be any homomorphism of A onto the group

$$\left\langle \begin{pmatrix} 0 & 1 & 0 & 0 \\ 0 & 0 & 1 & 0 \\ 0 & 0 & 0 & 1 \\ 1 & 0 & 0 & 0 \end{pmatrix} \right\rangle \subseteq GL(4, F).$$

Put

$$G = \left\langle \begin{pmatrix} \alpha_1 & 0 & 0 & 0 & 0 \\ 0 & \alpha_2 & 0 & 0 & 0 \\ 0 & 0 & \alpha_3 & 0 & 0 \\ 0 & 0 & 0 & \alpha_4 & 0 \\ 0 & 0 & 0 & 0 & 1 \end{pmatrix}, \begin{pmatrix} & & & & 0 \\ & \alpha_5\phi & & & 0 \\ & & & & 0 \\ & & & & 0 \\ \hline 0 & 0 & 0 & 0 & \alpha_5 \end{pmatrix} : \alpha_i \in A \right\rangle.$$

G is isomorphic to the split extension of four copies of A by a further copy of A that acts by permuting the first four copies cyclically via ϕ. By [47] 6.12 and 6.25, G is residually finite and every maximal subgroup of G has finite index. (This is not difficult to prove directly.) Moreover every finite homomorphic image of G is a 2-group.

Let M be any maximal subgroup of G. Then $(G:M)$ is finite, so

$$N = \bigcap_{g \in G} M^g$$

has finite index in G. Thus G/N is a finite 2-group and so $G^2 \subseteq M$. Therefore $G^2 \subseteq \phi(G)$. It is easy to see that G^2 contains a factor isomorphic to $C_\infty \wr C_2$ and consequently is not locally nilpotent. By construction $G \subseteq GL(5, F)$. It can also be shown that G is isomorphic to a subgroup of $GL(10, \mathbb{Q})$. ∎

We shall prove later that every locally nilpotent linear group is hypercentral. Hence if G is any locally nilpotent linear group $G' \subseteq \phi(G)$, extending one part of 4.17 iv). However, 4.18 shows that the converse is false.

If G is any group $\phi_1(G)$ denotes the intersection of all the maximal subgroups of finite index in G, or G itself if none such exist.

Exercise 4.4 (V. P. Platonov [43]). If R is a finitely generated integral domain and G is a subgroup of $GL(n, R)$ prove that $\phi_1(G)$ is nilpotent. (May be proved in a similar manner to 4.16 and 4.17: it also follows easily from 10.5.)

If G is any group

$\delta(G)$ denotes the intersection of the non-normal maximal subgroups of G, or G itself if none such exist;

$\psi(G)$ denotes the intersection of the centralizers in G of the chief factors of G;

$\psi_1(G)$ denotes the intersection of the centralizers in G of the finite chief factors of G, or G itself if non such exist.

Always

$$\delta(G)/\phi(G) = \zeta_1(G/\phi(G)) = \zeta(G/\phi(G)).$$

For suppose that H is any group satisfying $\phi(H) = \{1\}$. Every normal maximal subgroup of H contains H'. Hence

$$[\delta(H), H] \subseteq \delta(H) \cap H' \subseteq \phi(H) = \{1\}.$$

Therefore $\delta(H) \subseteq \zeta_1(H)$. If M is any maximal subgroup of H not containing $\zeta_1(H)$ then $M \cdot \zeta_1(H) = H$. Hence M is normal in H and so $\delta(H) = \zeta_1(H)$.

Taking $G/\phi(G)$ for H we get

$$\delta(G)/\phi(G) = \delta(G/\phi(G)) = \zeta_1(G/\phi(G)).$$

Now let $H = G/\delta(G)$. Then

$$\zeta_1(G/\delta(G)) = \delta(G/\delta(G)) = \{1\}.$$

Therefore

$$\zeta_1(G/\phi(G)) = \zeta(G/\phi(G)).$$

4.19. Corollary ([67]). *Let R be any finitely generated integral domain and G a subgroup of $GL(n, R)$. Then*

i) $\eta(G) = \eta_1(G) = \psi(G) = \psi_1(G)$.

ii) $\delta(G)$ *is nilpotent and* $\delta(G)/\phi(G)$ *is the centre of* $G/\phi(G)$.

iii) $\eta(G)$ *is nilpotent and* $\eta(G)/\phi(G)$ *is the unique maximal abelian normal subgroup of* G.

Proof. i) From 4.17, $\eta(G) = \eta_1(G)$ and clearly $\eta_1(G) \subseteq \psi(G) \subseteq \psi_1(G)$. Thus we have only to show that $\psi_1(G)$ is nilpotent. If H is a subgroup of finite index in $\psi_1(G)$ that is normal in G, then $\psi_1(G)$ centralizes every chief factor of G between H and $\psi_1(G)$. Hence $\psi_1(G)/H$ is nilpotent. By 4.16, $\psi_1(G)$ is nilpotent.

ii) In any group $\delta(G)/\phi(G) = \zeta_1(G/\phi(G))$. Since $\delta(G)/\phi(G)$ is abelian, $\delta(G)$ is nilpotent (4.17).

iii) $\eta(G)$ is nilpotent by 4.17. If $A/\phi(G)$ is an abelian normal subgroup of $G/\phi(G)$ then A is nilpotent (4.17). Therefore $A \subseteq \eta(G)$. Also by 4.15, $\eta(G)' \subseteq \phi(G)$. The result now follows. □

4.20. *If G is a finitely generated soluble linear group then $\eta(G)$ is finitely G-generated.*

Proof. G contains a triangularizable normal subgroup T of finite index (3.6). Let U be the maximal unipotent subgroup of T. Then U is normal in G and G/U is finitely generated and abelian-by-finite. Thus G/U is polycyclic and so finitely presented ([47], Corollary to 2.24). By [47] 2.24, U is finitely G-generated. U is nilpotent so $U \subseteq \eta(G)$, and since G/U is polycyclic $\eta(G)/U$ is finitely generated. Therefore $\eta(G)$ is finitely G-generated. $\square$

The hypercentre of G need not be finitely G-generated.

4.21. Example. *If $p \geq 0$ is any characteristic there exists a 3-generator 3-step soluble linear group G of degree 3 over a field of characteristic p such that G has central height 1 and $\zeta_1(G)$ is not finitely generated.*

Note that $\zeta(G) = \zeta_1(G)$ and that $\zeta_1(G)$ is finitely G-generated if and only if $\zeta_1(G)$ is finitely generated.

Proof. If $p > 0$ let $R = \mathbb{F}_p[x, x^{-1}]$, where $\mathbb{F}_p$ is the field of p elements and x is an indeterminate. If $p = 0$ let $R = \mathbb{Z}[x] \subseteq \mathbb{Q}$ where $x = \frac{1}{2}$. (For $p = 0$ it would suffice to take $R = \mathbb{Z}[x, x^{-1}]$ with x an indeterminate, but we prefer to make R as "small" as possible.) Put $G = \langle a, b, t \rangle \subseteq \mathrm{Tr}(3, R)$ where

$$a = \begin{pmatrix} 1 & 0 & 0 \\ 1 & 1 & 0 \\ 0 & 0 & 1 \end{pmatrix}, \quad b = \begin{pmatrix} 1 & 0 & 0 \\ 0 & 1 & 0 \\ 0 & 1 & 1 \end{pmatrix} \quad \text{and} \quad t = \begin{pmatrix} 1 & 0 & 0 \\ 0 & x & 0 \\ 0 & 0 & 1 \end{pmatrix}.$$

Let U denote the maximal unipotent subgroup of G and let W be the subgroup of G consisting of all the elements (of G) of the form

$$\begin{pmatrix} 1 & 0 & 0 \\ 0 & 1 & 0 \\ r & 0 & 1 \end{pmatrix}.$$

A trivial calculation shows that W is a central subgroup of G. Suppose that $\zeta_1(G/W) \neq \{1\}$, say $ut^i W \in \zeta_1(G/W) \smallsetminus \{1\}$, where $u \in U$. Then $[ut^i, t] \in W$, so $[u, t] \in W$. But U/W is an abelian group and thus $u W \in \zeta_1(G/W)$. Consequently $t^i W \in \zeta_1(G/W)$. This implies that $[a, t^i] \in W$ and a direct calculation shows that this is not so. Therefore $W = \zeta_1(G) = \zeta(G)$.

Direct computation shows that

$$[b^{mt^i}, a] = \begin{pmatrix} 1 & 0 & 0 \\ 0 & 1 & 0 \\ mx^i & 0 & 1 \end{pmatrix}$$

for any integer m. Hence

$$\prod_{i=j}^{k} [b^{m_i t^i}, a] = \begin{pmatrix} 1 & 0 & 0 \\ 0 & 1 & 0 \\ \sum_{i=j}^{k} m_i x^i & 0 & 1 \end{pmatrix},$$

and therefore W is isomorphic to the additive group of R. Consequently W is not finitely generated. ☐

4.22. Example. *For each characteristic $p \geq 0$ there exists a 3-generator metabelian linear group over a field of characteristic p that is not finitely presentable.*

Proof. Let G be the group constructed in 4.21 and put $H = G/W$. Certainly H is 3-generator and metabelian. If H were finitely presented then W would be finitely generated ([47] 2.24), thus H is not finitely presentable. It only remains to show that H has a faithful representation over some field of characteristic p. It is not difficult to see that H is isomorphic to

$$\left\langle \begin{pmatrix} 1 & 0 & 0 & 0 \\ 1 & 1 & 0 & 0 \\ 0 & 0 & 1 & 0 \\ 0 & 0 & 0 & 1 \end{pmatrix}, \begin{pmatrix} 1 & 0 & 0 & 0 \\ 0 & 1 & 0 & 0 \\ 0 & 0 & 1 & 0 \\ 0 & 0 & 1 & 1 \end{pmatrix}, \begin{pmatrix} 1 & 0 & 0 & 0 \\ 0 & x & 0 & 0 \\ 0 & 0 & x & 0 \\ 0 & 0 & 0 & 1 \end{pmatrix} \right\rangle.$$

(The representability of H also follows for more general reasons, see 6.4.) ☐

The next result is a special case of a theorem to be proved in Chapter 8, but it seems worthwhile to record it here as well.

4.23. Theorem (Gruenberg [20]). *Let R be a finitely generated integral domain and G a subgroup of $GL(n, R)$. Then G has finite central height and nilpotent Hirsch-Plotkin radical.*

Proof. We have already pointed out that the nilpotence of $\eta(G)$ is a special case of 4.17. By Exercise 3.1 (or by the results of Chapter 5, see comments before 4.16) $\zeta(G)$ contains a triangularizable subgroup T of finite index that is normal in G. If U denotes the maximal unipotent subgroup of T, then $U \subseteq \zeta_{\frac{1}{2}n(n-1)}(G)$ by 4.13 and $\zeta(G)/U$ is a finite extension of a finitely generated abelian group by 4.10. Hence $\zeta(G)/U$ satisfies the maximal condition on subgroups and thus G has finite central height. ☐

It is not possible to give a bound for the central heights of such groups G in terms of n only, but one can in terms of n and R; see 1.2 of [69 b]. Curiously one does get a bound in terms of n for torsion-free groups.

This further result of Gruenberg we give in Chapter 8, where the whole subject of central height in linear groups is discussed in detail.

In [32b] Lennox and Rosenblade introduce the concept of central eremiticity. A group G is centrally eremitic of eccentricity e (a positive integer) if for any subset X of G, if some positive power of an element g of G centralizes X then so does g^e, and if e is the least such positive integer. For example G is centrally eremitic of eccentricity 1 if and only if every centralizer in G is isolated. Clearly every subgroup of a centrally eremitic group is centrally eremitic of not larger eccentricity.

For the next result, a special case of Theorem 2.8 of [69b] see Exercise 6.1, we need the following simple remark. *Suppose that G is a group, $\{H_i: i \in I\}$ a set of normal subgroups of G such that $\bigcap_i H_i = \{1\}$ and X any subset of G. Then $C_G(X) = \bigcap_i H_i \cdot C_G(X)$. For if $x \in X$ and $y \in \bigcap_i H_i \cdot C_G(X)$, then $[x, y] \in \bigcap_i H_i = \{1\}$. Hence $y \in C_G(X)$. The containment the other way round is trivial.*

4.24. Theorem ([69b]). *If R is a finitely generated integral domain of characteristic $p \geq 0$, then $G = \mathrm{GL}(n, R)$ is centrally eremitic and contains a normal subgroup of finite index with eccentricity dividing 1 if $p = 0$ and $p^{-[-\log_p n^2]}$ if $p > 0$.*

Proof. By 4.7 the group G contains a normal subgroup T of finite index m say such that T is residually a finite p-group if $p > 0$ and is residually a finite q-group and residually a finite q'-group for some prime q if $p = 0$.

Let X be any subset of G, g an element of G and suppose that $g^r \in C_G(X)$ for some positive integer r. Put $C = C_G C_G(g^r) \subseteq C_G(X)$ and $N = N_G(C)$. By the remark above $(N \cap T) C/C$ is residually a finite q-group and residually a finite q'-group if $p = 0$. Thus in this case $(N \cap T) C/C$ is torsion-free. But g lies in N, so $g^m \in N \cap T$. Also $g^r \in C$, so $g^m \in C \subseteq C_G(X)$. Thus G is centrally eremitic with eccentricity dividing m. If $g \in T$ then we actually get $g \in C$ and so T has eccentricity 1.

Now suppose that $p > 0$. Here $(N \cap T) C/C$ is only residually a finite p-group. It follows from 6.2[6] that it is also isomorphic to a linear group of degree n^2 over the quotient field of R, so its p-subgroups have exponent dividing $p^{-[-\log_p n^2]}$. As before we have $g^m \in N \cap T$ and $g^r \in C$. Hence the $m p^{-[-\log_p n^2]}$-th power of g lies in $C \subseteq C_G(X)$ and the result follows as in the case where $p = 0$. $\square$

Clearly a periodic group G is centrally eremitic if and only if $G/\zeta_1(G)$ has finite exponent. We also have the following example.

---

[6] This result and its proof may be read immediately.

4.25. Example. $\mathrm{GL}(2, \mathbb{C})$ *contains a countable torsion-free metabelian group that is not centrally eremitic.*

Proof. Let $\{\alpha_i : i = 1, 2, \ldots\}$ be a basis of a free abelian subgroup of $\mathbb{C}^*$ and for $i = 1, 2, \ldots$ let ζ_i be a primitive i-th root of unity. Put

$$G = \left\langle \begin{pmatrix} 1 & 0 \\ 1 & 1 \end{pmatrix}, \begin{pmatrix} \alpha_i \zeta_i & 0 \\ 0 & \alpha_i \end{pmatrix} : i = 1, 2, \ldots \right\rangle.$$

Clearly G is countable, torsion-free and metabelian. Also the centre Z of G contains only scalar matrices and by construction G/Z contains elements of every finite order. □

Exercise 4.5. For any characteristic $p \geq 0$, construct a torsion-free linear group over a field of characteristic p that is not centrally eremitic. (For example use 2.10.)

Exercise 4.6. If G is a torsion-free subgroup of $\mathrm{SL}(2, F)$ each of whose elements is diagonalizable, prove that G is centrally eremitic with eccentricity 1.

The final topic of this chapter is chief factors of finitely generated linear groups. Since a finitely generated linear group can be free of any (finite) rank, the chief factors of finitely generated linear groups are no simpler than those of finitely generated groups in general. The following result, therefore, is perhaps slightly surprising. Another way of looking at it is that it says that although the structure of the unipotent subgroups of a finitely generated linear group is not usually much simpler than unipotent groups in general (in contrast to the diagonalizable subgroups of a finitely generated linear group, see 4.11) the way they are embedded into the finitely generated linear group is very restricted.

4.26. Theorem ([69 d]). *Let G be a subgroup of $\mathrm{GL}(n, F)$, U a unipotent normal subgroup of G such that G/U is finitely generated and X/Y a chief factor of G covered by U (i.e. $X \leq YU$). Then X/Y is finite and $\mathbb{Z}G/\mathrm{Ann}_{\mathbb{Z}G}(X/Y)$ is isomorphic to a matrix ring of finite degree at most $[(n^4 + 16)/32]$ over a finite field.*

Here $\mathrm{Ann}_{\mathbb{Z}G}(X/Y) = \{g \in \mathbb{Z}G : X^g \subseteq Y\}$ is the annihilator of X/Y as (multiplicative) $\mathbb{Z}G$-module. (X/Y is abelian since U is nilpotent.) Using non-trivial results of Amitsur and Levitzki the bound can be reduced to $n^2/4$ (see Theorem 4 of [69 d]). Notice that although the theorem can be applied to finitely generated linear groups it does not cover the more general case of linear groups over finitely generated integral domains.

Question 7. *R is a finitely generated integral domain, G is a subgroup of $\mathrm{GL}(n, R)$ and U is a unipotent normal subgroup of G. Is every chief factor of G covered by U finite?*

Proof. We induct on n; if $n=1$ then G is abelian and everything is trivial. Let V be the n-row vector space and put

$$W = C_V(U) = \{v \in V: vx = v \text{ for all } x \in U\}.$$

By 1.21 we have $d = \dim_F W \geq 1$. Now $X/Y \cong X \cap U/Y \cap U$ as G-modules, so replacing X by $X \cap U$ we may suppose that $X \subseteq U$. If $C_G(V/W)$ does not cover X/Y then X/Y is G-isomorphic to a chief factor of $G/C_G(V/W)$ covered by $U C_G(V/W)/C_G(V/W)$ and this latter group acts unipotently on V/W. Since $\dim_F(V/W) < n$ the induction hypothesis gives the desired conclusion.

Therefore we are left with the case where $C_G(V/W)$ covers X/Y. Again we may replace X by $X \cap C_G(V/W)$, that is we may assume that

$$X \subseteq C_G(V/W) \cap C_G(W) = K$$

say. Now $H = \operatorname{Hom}_F(V/W, W)$ is an FG-module ($\lambda^g = g^{-1}\lambda g$ for $g \in G$ and $\lambda \in H$), and if for $k \in K$ we denote by k' the map $(x+W) \mapsto x(k-1)$, then as in the proof of 4.12 the group K is abelian and the map $\phi: k \mapsto k'$ is a monomorphism of G-modules. Thus X/Y is G-isomorphic to the irreducible G-factor $\overline{X} = X\phi/Y\phi$ of H.

Let $S = \mathbb{Z}G/\operatorname{Ann}_{\mathbb{Z}G}(X/Y) = \mathbb{Z}G/\operatorname{Ann}_{\mathbb{Z}G}(\overline{X})$. Then S is isomorphic to a quotient of a subring of $\operatorname{End}_F H$, which is turn is isomorphic to the matrix algebra over F of degree $d(n-d)$. Hence S satisfies the standard polynomial identity of degree $d^2(n-d)^2 + 1$, see [26b] §10.6, especially Proposition 1. (It is here that the theorem of Amitsur and Levitzki can be used to reduce the bound.) But S is a primitive ring, so the arguments of [26b] §10.5 imply that S is isomorphic to a matrix ring over a division ring and that S has dimension

$$m^2 \leq [(d^2(n-d)^2 + 1)/2]^2$$

over its centre E.

Now U is nilpotent and covers X/Y. Hence U centralizes X/Y. It now follows from the finite generation of G/U that S is a finitely generated ring. Let $x_1, x_2, \ldots, x_r$ generate S as a ring and let $1 = a_1, a_2, \ldots, a_s$ be a basis for S over its centre E. For $i = 1, 2, \ldots, r$ and $j, k = 1, 2, \ldots, s$ there exist elements α_{ijk} of E such that

$$x_i a_j = \sum_{k=1}^{s} a_k \alpha_{ijk}.$$

Let E_1 denote the subring of E generated by 1_E and the rs^2 elements α_{ijk}. If $\alpha \in E$ then $\alpha = f(x_1, x_2, \ldots, x_r) 1_E$ for some polynomial f with integer coefficients and hence

$$\alpha = \alpha\, a_1 = f(x_1, \ldots, x_r) a_1 \in \left(\sum_k a_k E_1\right) \cap E = E_1.$$

Hence $E_1 = E$ and E is finitely generated as a ring. Thus E is a finite field by 4.1, and Wedderburn's Theorem implies that $S \cong E_m$. Finally

$$m \leq [(d^2(n-d)^2 + 1)/2] \leq [(n^4 + 16)/32]$$

since $1 \leq d \leq n$. ☐

The following result follows at once from 4.10, 4.26 and Exercise 3.1.

4.27. Corollary. *If G is a finitely generated linear group and N is any soluble-by-finite normal subgroup of G, then every chief factor of G covered by N is finite.* ☐

4.28. Corollary ([69 d]). *Let R be a finitely generated integral domain and G a soluble-by-finite subgroup of $\mathrm{GL}(n, R)$. Then every chief factor of G is finite and every maximal subgroup of G has finite index.*

Proof. By 3.6 and 4.8 the group G contains a unipotent normal subgroup U such that G/U is a finite extension of a finitely generated abelian group. Clearly every chief factor of G/U is finite and by 4.26 every chief factor of G covered by U is finite. Hence every chief factor of G is finite.

Let M be any maximal subgroup of G and put $K = \bigcap_{x \in G} M^x$. There exists a normal subgroup L of G such that L/K is non-trivial and either abelian or finite. Clearly $G = LM$ and $(G:M) \leq (L:K)$, so it suffices to prove that L/K is finite. Suppose that L/K is abelian. Then $L \cap M$ is normal in both L and M and hence also in $LM = G$. Therefore $L \cap M = K$. Let L_1/K be any non-trivial G-submodule of L/K. Then $L_1 M = G$ and

$$L = L_1(L \cap M) = L_1.$$

Thus L/K is a chief factor of G and consequently is finite by the first paragraph. ☐

Exercise 4.7. Prove 4.28 using 4.1 instead of 4.26.

Exercise 4.8. If G is a finitely generated linear group use 4.26 and Exercise 4.3 to prove that $\phi(G) = \phi_1(G)$. (See [69 d] for details.)

A trivial corollary of 4.28 is that whenever G is a soluble-by-finite subgroup of $\mathrm{GL}(n, R)$, where R is some finitely generated integral domain, then $\phi(G) = \phi_1(G)$. Further such conditions are given in [69 d] and they prompt the following question.

Question 8. *Does $\phi(G) = \phi_1(G)$ for all subgroups G of $\mathrm{GL}(n, R)$ and all finitely generated integral domains R?*

Exercise 4.9. R is a finitely generated integral domain, G is a subgroup of $\mathrm{GL}(n, R)$ and M is a maximal subgroup of G.

i) If M is abelian prove that G is soluble of derived length at most 3.

ii) If M is locally nilpotent, char $R = 2$, and $n \leq 3$, prove that G is soluble.

We shall prove similar results for periodic linear groups in Chapter 12 with less restrictions.

A group is called *computable* if there exists a one-to-one mapping ϕ of G into the natural numbers $\mathbb{N}$ such that

i) $G\phi$ is a recursive subset of $\mathbb{N}$,

ii) the mapping ψ of $G\phi \times G\phi$ into $G\phi$ given by

$$(g\phi, h\phi) = (gh)\phi, \quad g, h \in G, \text{ is recursive.}$$

M.O. Rabin has announced the following two results.

4.29. Theorem (Rabin). *Every finitely generated linear group is computable.*

4.30. Corollary (Rabin). *Every finitely generated linear group has soluble word problem.*

One of the most important results on finitely generated linear groups is the following very recent theorem of Tits [57a]. *A finitely generated linear group either is soluble-by-finite or contains a non-cyclic free subgroup.* This result along with further results on finitely generated linear groups is discussed in Chapter 10. The first part of Chapter 11 studies finitely generated linear groups whose finite images are all supersoluble.

5. *CZ-Groups and the Zariski Topology*

Let U be the space of n-row vectors over the field F and $R = F[X_1, \ldots, X_n]$, the polynomial ring over F in n indeterminates. A subset A of U is said to be *closed* in U if there exists a subset S of R such that A is the set of zeros of S, that is if

$$A = \{(a_1, \ldots, a_n) \in U : f(a_1, \ldots, a_n) = 0 \text{ for all } f \text{ in } S\}.$$

If S is any subset of R let $V(S)$ denote the set of zeros of S (in U). Note that

$$V(S) = V(\text{ideal generated by } S),$$

and

$$\bigcap_\alpha V(S_\alpha) = V(\bigcup_\alpha S_\alpha).$$

Suppose that $A = V(\mathfrak{a})$ and $B = V(\mathfrak{b})$ where $\mathfrak{a}$ and $\mathfrak{b}$ are ideals. We claim that $A \cup B = V(\mathfrak{ab})$. For $\mathfrak{ab} \subseteq \mathfrak{a}$, so $A = V(\mathfrak{a}) \subseteq V(\mathfrak{ab})$, and ditto for B. Conversely let $x \in V(\mathfrak{ab}) \smallsetminus A$. Since $A = V(\mathfrak{a})$ there exists an element f of $\mathfrak{a}$ such that $f(x) \neq 0$. If $g \in \mathfrak{b}$, then $fg \in \mathfrak{ab}$, so $0 = fg(x) = f(x)g(x)$. Hence $g(x) = 0$ for all g in $\mathfrak{b}$ and so $x \in V(\mathfrak{b}) = B$.

Thus the closed subsets of U define a topology on U, the *Zariski topology*; note that $U = V(\{0\})$ and $\emptyset = V(\{1\})$. If $(a_1, \ldots, a_n) \in U$ then

$$\{(a_1, \ldots, a_n)\} = V(\{X_1 - a_1, X_2 - a_2, \ldots, X_n - a_n\}).$$

Hence every one element subset of U is closed (i.e. the topology is a T_1-topology). If $A \subseteq B$ are closed sets there exist ideals $\mathfrak{a}$ and $\mathfrak{b}$ of R such that $A = V(\mathfrak{a})$, $B = V(\mathfrak{b})$ and $\mathfrak{b} \subseteq \mathfrak{a}$. For if $A = V(\mathfrak{a}')$ and $B = V(\mathfrak{b})$, put $\mathfrak{a} = \mathfrak{a}' + \mathfrak{b}$. Thus the Hilbert Basis Theorem implies that the Zariski topology satisfies the descending chain condition on closed sets, or equivalently, the ascending chain condition on open sets.

5.1. *Let U and W be m- and n-dimensional row vector spaces over F taken with the Zariski topology. Suppose that $r_1, \ldots, r_n$ are rational functions over F in m indeterminates $X_1, \ldots, X_m$. Let S be the subset of U where any of the denominators of the r_i vanish and put $T = U \smallsetminus S$. Then the mapping*

$\phi: T \rightarrow W$ *defined by*

$$(a_1, \ldots, a_m)\, \phi = \big(r_1(a_1, \ldots, a_m), \ldots, r_n(a_1, \ldots, a_m)\big)$$

is continuous.

Proof. The r_i can be put into the form $r_i = p_i/r$ where p_i and r are polynomials and $S = V(r)$. We have to show that the inverse image of a closed set is closed. Let A be a non-empty closed subset of W; $A = V(g_1, \ldots, g_t)$ for some $g_1, \ldots, g_t$ in $F[Y_1, \ldots, Y_n]$. Then

$$A\phi^{-1} = \{x \in T: g_i(r_1(x), \ldots, r_n(x)) = 0, \ i = 1, 2, \ldots, t\}$$
$$= \{x \in T: r(x)^k\, g_i(r_1(x), \ldots, r_n(x)) = 0, \ i = 1, 2, \ldots, t\},$$

when k is some integer, since $r(x) \neq 0$ for all x in T. For sufficiently large k all the $r^k g_i(r_1, \ldots, r_n)$ are polynomials. Therefore $A\phi^{-1}$ is a closed subset of T. $\quad\square$

If V is an arbitrary finite-dimensional vector space over F, then V can be endowed with a Zariski topology after a suitable choice of basis. A consequence of 5.1 is that different choices of basis of V yield the same topology. If X is a subspace of V, then X carries its own Zariski topology as well as the topology induced on X by the Zariski topology of V. It is easy to see that these two topologies are the same.

If K is an extension field of F then $V^K = K \otimes_F V$ carries the Zariski topology as K-space and this induces a topology on $1 \otimes V$ and thus on V. This is in fact the same as the Zariski topology on V as F-space. For we may regard V as the space of n-row vectors over F, and V^K as the space of n-row vectors over K. If A is an F-closed subset of V there exists a subset S of $F[X_1, \ldots, X_n]$ such that for v in V, we have $v \in A$ if and only if $f(v) = 0$ for every f in S. But $S \subseteq K[X_1, \ldots, X_n]$ and so A is also a K-closed subset of V.

Now suppose that A is a K-closed subset of V, that is $A = B \cap V$ for some closed subset B of V^K. There exists a subset T of $K[X_1, \ldots, X_n]$ such that if $w \in V^K$ then $w \in B$ if and only if $g(w) = 0$ for every g in T. Each element g of T can be expressed in the form

$$g = \alpha_1 g_1 + \cdots + \alpha_r g_r,$$

for some elements $g_1, \ldots, g_r$ of $F[X_1, \ldots, X_n]$ and some elements $\alpha_1, \ldots, \alpha_r$ of K, where the α_i are linearly independent over F and the r depends on g. Then for v in V, we have $g(v) = 0$ if and only if

$$g_1(v) = \cdots = g_r(v) = 0.$$

Denote by S the set of all the g_i as g ranges over T. Then for v in V, we have $v \in A$ if and only if $f(v) = 0$ for every f in S, and consequently A is an F-closed subset of V.

Definition. A *Z-space* is a topological space in which every one element set is closed (i.e. is a T_1-space) and which satisfies the descending chain condition on closed sets.

Every subspace of a Z-space is a Z-space. Also a non-empty Z-space is uniquely the union of a finite number of non-empty, open, closed, connected sets. (A topological space X is connected if $\varnothing$ and X are its only open, closed subsets.) For if X_0 is a non-empty Z-space, by the descending chain condition X_0 contains a minimal, non-empty, open, closed subset C_1. Put $X_1 = X_0 \smallsetminus C_1$. If X_1 is non-empty it also contains a minimal non-empty open closed subset C_2. Put $X_2 = X_1 \smallsetminus C_2$. Note that $X_0 \supset X_1 \supset X_2$. Continue in this way. Again by the descending chain condition there exists an integer k such that $X_k = \varnothing$ and $X_{k-1} \neq \varnothing$. Then

$$X_0 = C_1 \cup C_2 \cup \cdots \cup C_k.$$

If C is any open closed subset of X_0 and $C_i \cap C \neq \varnothing$, then $C_i \cap C$ is a non-empty open closed subset of C_i. Therefore $C_i \subseteq C$ and so the above decomposition is unique. The C_i are called the connected components of X_0.

Definition. A *CZ-group* is a Z-space G whose underlying set carries a group structure such that for every a in G the four mappings given by

$$x \mapsto a x \qquad x \mapsto x^{-1}$$

$$x \mapsto x a \qquad x \mapsto x^{-1} a x, \qquad x \in G$$

are continuous. This notion of a CZ-group is taken from Kaplansky [27].

Every linear group is a CZ-group. For let G be a subgroup of $\mathrm{GL}(n, F) \subseteq F_n$. F_n is a vector space over F of dimension n^2 and so carries the Zariski topology. The induced topology on G makes G into a CZ-group. The four mappings above are continuous by 5.1. Since the topology induced on a given linear group is unaffected by ground-field extension (see above), we can unambiguously speak of its closed subsets without specifying the ground field.

A closed subgroup of $\mathrm{GL}(n, F)$ is called a *linear algebraic group*. If S is any subset of $\mathrm{GL}(n, F)$, then $\mathscr{A}_F(S)$ denotes the intersection of all the closed subgroups of $\mathrm{GL}(n, F)$ containing S. It is of course the unique minimal closed subgroup containing S. If K is any extension field of F then $S \subseteq \mathrm{GL}(n, F) \subseteq \mathrm{GL}(n, K)$ and $\mathscr{A}_F(S)$ and $\mathscr{A}_K(S)$ are both defined. In general $\mathscr{A}_F(S) \neq \mathscr{A}_K(S)$ but clearly we have

$$S \subseteq \mathscr{A}_F(S) = \mathscr{A}_K(S) \cap \mathrm{GL}(n, F) \subseteq \mathscr{A}_K(S).$$

Every abelian group is a CZ-group. If A is an abelian group, call the subset S of A closed if $S = \varnothing$, or if $S = A$, or if S is finite. It is simple to check that this makes A into a CZ-group. Incidentally, this shows that not every CZ-group is linear (by 2.2).

Exercise 5.1. The centre of the group G has finite index in G. Prove that G is a CZ-group.

Question 9. *Is every finite extension of a CZ-group a CZ-group?* The motivation behind this question comes from 2.13.

5.2. Lemma. *If G is a CZ-group and C is the connected component of G containing 1, then C is a normal subgroup of finite index in G and the connected components of G are exactly the cosets of C in G.*

Proof. The continuous image of a connected set is connected. Thus C^{-1}, xC and $x^{-1}Cx$ are connected sets for each x in G. Now $1 \in C^{-1} \cap C$, so $C^{-1} \cap C$ is a non-empty open closed subset of C^{-1}. Therefore $C^{-1} \subseteq C$. If $x \in C$, the $x \in x C \cap C$ and so $xC \subseteq C$. Hence C is a subgroup of G. Since for all x in G, the fact $1 \in C \cap x^{-1}Cx$ implies that $x^{-1}Cx \subseteq C$, so C is normal in G. xC is the inverse image of C under the map given by $y \mapsto x^{-1}y$ and so is open and closed. Thus xC is a connected component of G. Both the cosets of C in G and the connected components of G cover G and are pairwise disjoint. Therefore the connected components of G are exactly the cosets of C in G. In particular $(G:C)$ is finite. $\square$

If G is a CZ-group we denote the connected component of G containing 1 by G^0.

5.3. Lemma. *H is a closed subgroup of the CZ-group G. The following are equivalent.*

a) $G^0 \subseteq H$;
b) $(G:H)$ is finite;
c) H is open.

Proof. Trivially if $G^0 \subseteq H$, then $(G:H)$ is finite. Suppose that $(G:H)$ is finite. If $G = H$, then H is open. If not there exists a finite set $x_1, \ldots, x_r$ of elements of G such that $H \cap (x_1 H \cup x_2 H \cup \cdots \cup x_r H) = \emptyset$ and

$$G = H \cup (x_1 H \cup \cdots \cup x_r H).$$

H is closed so each $x_i H$ is closed. Therefore $(x_1 H \cup \cdots \cup x_r H)$ is closed and H is open. Suppose that H is open. Since $1 \in G^0 \cap H$, so $G^0 \cap H$ is a non-empty open closed subset of G^0. Hence $G^0 \subseteq H$. $\square$

5.4. Lemma. *If G is a CZ-group then the normalizer of a closed subset of G is closed and the centralizer of any subset of G is closed.*

Proof. Let S be any closed subset of G and a any element of S. Denote by $S(a)$ the inverse image of S in G under the continuous mapping given by

$x \mapsto x^{-1} a x$. Then $N_1 = \bigcap_{a \in S} S(a) = \{x \in G : x^{-1} S x \subseteq S\}$ is closed in G. If
$(a) S$ denotes the inverse image of S in G under the continuous mapping given by $x \mapsto x a x^{-1} = (x^{-1})^{-1} a x^{-1}$, then

$$N_2 = \bigcap_{a \in S} (a) S = \{x \in G : x S x^{-1} \subseteq S\}$$

is closed. Therefore $N_G(S) = N_1 \cap N_2$ is closed in G.

Let T be any subset of G.

$$C_G(T) = \bigcap_{t \in T} C_G(t) = \bigcap_{t \in T} N_G(t),$$

and $N_G(t)$ is closed by the above. Consequently $C_G(T)$ is closed in G. □

5.5. Lemma. *If G is a CZ-group then every finite conjugacy class of elements of G contains at most $k = (G:G^0)$ elements, and centralizes G^0.*

Proof. Suppose that x is an element of G having exactly r conjugates in G. Then $C_G(x)$ is a closed subgroup of finite index r by 5.4. Hence by 5.3, $G^0 \subseteq C_G(x)$ and so r divides k. It also follows that G^0 centralizes every conjugate of x. □

5.6. Corollary. *A linear FC-group is centre-by-finite.*

An FC-*group* is a group each of whose conjugacy classes is finite. If G is a linear FC-group and x is an element of G then $G^0 \subseteq C_G(x)$ by 5.5. Hence $G^0 \subseteq \zeta_1(G)$. □

In fact, this argument and Exercise 5.1 show that a group is both an FC-group and a CZ-group if and only if it is centre-by-finite. Using the structure of FC-groups (see [50] §15.1 for example) it is not difficult to give a direct proof of 5.6.

5.7. Lemma. *If G is a connected CZ-group then G' is also connected.*

Proof. Let D_k denote the set of all elements of G of the form

$$[a_1, b_1] [a_2, b_2] \ldots [a_k, b_k], \qquad a_i, b_i \in G.$$

Then $\{1\} = D_0 \subseteq D_1 \subseteq \cdots \subseteq D_k \subseteq \cdots$ and $G' = \bigcup_k D_k$ (since $[x, y]^{-1} = [y, x]$).
It suffices to prove that D_k is connected. This we do inductively. Trivially D_0 is connected, suppose that D_{k-1} is connected. If $D_k(a_2, \ldots, a_k, b_1, \ldots, b_k)$ is the image of G under the continuous mapping of G into itself given by

$$x \mapsto [x, b_1] [a_2, b_2] \ldots [a_k, b_k]$$

then $D_k(a_2, \ldots, a_k, b_1, \ldots, b_k)$ is connected. Also

$$D_{k-1} \cap D_k(a_2, \ldots, a_k, b_1, \ldots, b_k)$$

is non-empty since it contains

$$[a_2, b_2] \ldots [a_k, b_k] = [1, b_1][a_2, b_2] \ldots [a_k, b_k],$$

and

$$D_k = \bigcup_{\substack{a_2, \ldots, a_k \\ b_1, \ldots, b_k}} D_k(a_2, \ldots, a_k, b_1, \ldots, b_k).$$

Therefore D_k is connected. $\square$

Exercise 5.2. G is a *CZ*-group and H is a connected subgroup of G. Prove that $[H, {}_iG]$ is connected.

If G is any *CZ*-group and K is a normal subgroup of G then K^0 is normal in G, since for any element g of G the mapping of K into itself given by $x \mapsto g^{-1} x g$ is a homeomorphic automorphism. Suppose now that G is a subgroup of $GL(n, F)$ where F is algebraically closed and that K is a soluble normal subgroup of G. By Mal'cev's Theorem (3.6) K contains a triangularizable subgroup T of finite index. That is, there exists an element x of $GL(n, F)$ such that $T^x \subseteq \text{Tr}(n, F)$. Now $\text{Tr}(n, F)$ is a closed subgroup of $GL(n, F)$; hence $S = K \cap x \text{Tr}(n, F) x^{-1}$ is a closed subgroup of finite index in K. By 5.3, $K^0 \subseteq S$ and so K^0 is a triangularizable subgroup of finite index in K that is normal in G. This is the result that we used during the proof of 4.16 and is a special case of Exercise 3.1. We now give a proof that K^0 is triangularizable without using Mal'cev's Theorem. The result we prove is essentially Mal'cev's Theorem without the existence of the bound.

5.8. Theorem (Lie, Kolchin). *If F is algebraically closed and G is a soluble connected subgroup of $GL(n, F)$ then G is triangularizable.*

Proof. By induction we may suppose that if H is a soluble connected subgroup of $GL(m, F)$ and either $m < n$ or $m = n$ and the derived length of H is less than the derived length of G, then H is triangularizable. If G is reducible there exists representations ρ_1 and ρ_2 of G and x in $GL(n, F)$ such that

$$g^x = \begin{pmatrix} g\rho_1 & 0 \\ * & g\rho_2 \end{pmatrix} \qquad \text{for all } g \text{ in } G.$$

By 5.1, ρ_i is continuous and so $G\rho_i$ is connected. Therefore by induction $G\rho_1$ and $G\rho_2$, and hence G also, is triangularizable.

Suppose now that G is irreducible. G' is connected by 5.7 and so by induction G' is triangularizable. But by Clifford's Theorem (1.7) G' is completely reducible. Therefore G' is diagonalizable. It now follows from 1.12 and 5.4 that $C_G(G')$ is a closed subgroup of finite index in G. As G is connected $C_G(G') = G$. By 1.2, G' consists only of scalar matrices. Now every element of G' has determinant 1, so that if $g \in G'$, then $g = \alpha 1_n$

where α is an n-th root of 1. Hence G' is finite. Being connected $G' = \{1\}$ and G is abelian. 1.3 now implies that $n = 1$ and so G is triangular. ▢

5.9. Lemma. *In a CZ-group G the closure of a subgroup is a subgroup and the closure of a normal subgroup is a normal subgroup.*

Proof. Let H be a subgroup of G and $\bar{H}$ its closure in G. Now $H = H^{-1}$. Since the mapping given by $x \mapsto x^{-1}$ is a homeomorphism of G, $(\bar{H})^{-1} = \bar{H}$. Let $h \in H$. The inverse image of $\bar{H}$ in G under the mapping given by $x \mapsto h\,x$ is closed and contains H. Therefore it contains $\bar{H}$; that is, $h\bar{H} \subseteq \bar{H}$ for all h in H. If $k \in \bar{H}$, the inverse image of $\bar{H}$ under the mapping given by $x \mapsto x\,k$ contains H and so $\bar{H}$. Hence $\bar{H} \cdot \bar{H} \subseteq \bar{H}$ and $\bar{H}$ is a subgroup of G.

Suppose that H is a normal subgroup of G. Since the mapping given by $x \mapsto a^{-1}\,x\,a$ is continuous, $a\bar{H}a^{-1}$ is a closed subset of G containing H. Therefore $\bar{H} \subseteq a\bar{H}a^{-1}$ for all a in G; that is $\bar{H}$ is normal in G. ▢

5.10. Lemma. *Let A, B and C be subgroups of the CZ-group G and $\bar{A}$, $\bar{B}$ and $\bar{C}$ their closures in G. If $[A, C] \subseteq B$ then $[\bar{A}, \bar{C}] \subseteq \bar{B}$.*

Proof. Let $c \in C$. The mapping given by $x \mapsto [x, c]$ is continuous and the inverse image of $\bar{B}$ in G contains A. Therefore $[\bar{A}, c] \subseteq \bar{B}$ for all c in C. If $a \in \bar{A}$ the inverse image of $\bar{B}$ under the mapping given by $x \mapsto [a, x]$ contains C and hence $\bar{C}$. Thus $[\bar{A}, \bar{C}] \subseteq \bar{B}$. ▢

5.11. Theorem. *Let G be a CZ-group, H a subgroup of G and K the closure of H in G. Then:*

   i) *H is soluble of derived length d if and only if K is.*
   ii) *H is nilpotent of class c if and only if K is.*

Proof. Let $\{1\} = H_0 \subseteq H_1 \subseteq \cdots \subseteq H_n = H$ be a normal series of H with abelian factors, and denote the closure of H_i in G by K_i. Then

$$\{1\} = K_0 \subseteq K_1 \subseteq \cdots \subseteq K_n = K$$

is a normal series of K by 5.10. For $i = 1, 2, \ldots, n$, we have $[H_i, H_i] \subseteq H_{i-1}$; thus $[K_i, K_i] \subseteq K_{i-1}$ by 5.10. Hence if H is soluble of derived length d, so is K. If K is soluble of derived length d trivially H is soluble of derived length d_1 for some $d_1 \leq d$. But then by the above K has derived length d_1, so $d_1 = d$.

If the above series for H is a central series, i.e. if $[H_i, H] \subseteq H_{i-1}$ for each i, then $[K_i, K] \subseteq K_{i-1}$ for each i and so part ii) of the theorem follows in similar way to part i). ▢

The object of this work is to study the groups of units of a certain class of rings, namely matrix rings of finite degree over fields. We could widen our interest somewhat by considering the groups of units of subrings of these matrix rings. We have already started on this project;

in Chapter 4 we looked at the groups of units of subrings of F_n of the form R_n where R is a finitely generated subring of F. We now consider briefly division subrings of F_n.

If D is a division algebra of finite dimension n over the field F then D has a faithful representation of degree n over F. Conversely any division subring of a matrix ring (of finite degree over a field) has finite dimension over its centre ([26b] p. 226). In [53] D. A. Suprunenko studies the maximal soluble subgroups of the multiplicative subgroup of a finite-dimensional division algebra, using the techniques of Chapter 3 of this book and particularly of Chapter 1 of his book [54]. The Zariski topology gives a second approach to this problem. Since finite division rings are fields we shall confine ourselves to infinite division rings. If S is any subset of F_n put

$$S^* = \{x \in S \colon x \neq 0\}.$$

A topological space X is *irreducible* if X is not the union of two proper closed subsets. Trivially an irreducible space is connected.

5.12. Lemma. *Let D be an infinite division subring of F_n. Then D^* is irreducible.*

Proof. Suppose firstly that D is reducible. Then D contains two proper closed subsets S and T such that $D = S \cup T$. There exist elements $s \in S \smallsetminus T$ and $t \in T \smallsetminus S$. Let A be an irreducible subset of D containing at least two elements, say a and b. The mapping ϕ of D into itself given by

$$\phi \colon x \mapsto (s-t)(a-b)^{-1} x + s - (s-t)(a-b)^{-1} a$$

is continuous (5.1). Also $a\phi = s$ and $b\phi = t$. Now $A\phi$ is irreducible, $A\phi = (S \cap A\phi) \cup (T \cap A\phi)$ and $S \cap A\phi$ and $T \cap A\phi$ are proper closed subsets of $A\phi$. This contradiction proves that the only non-empty irreducible subsets of D are the one-element sets.

By the descending chain condition on closed sets D contains a minimal infinite reducible closed subset U. Since U is reducible U contains proper closed subsets V and W such that $U = V \cup W$. The minimality of U and the above paragraph imply that V and W are both finite. This contradiction proves that D is irreducible.

Suppose now that S and T are closed subsets of D such that $D^* \subseteq S \cup T$ and $D^* \nsubseteq S$. Then $D = S \cup T \cup \{0\}$ and by the irreducibility of D, $D = T$. Hence $D^* \subseteq T$. This proves that D^* is irreducible. (More generally this argument shows that every open subset of an irreducible set is irreducible.) ☐

5.13. Theorem. *Let D be an infinite division subring of F_n and H a maximal soluble subgroup of D^*.*

Then:

i) H^0 is the unique maximal abelian normal subgroup of H.

ii) $H^0 = K^*$ for some subfield K of D containing the centre of D.

iii) $(H:H^0) \le \mu(n)$ where $\mu(n)$ is the integer-valued function of Mal'cev's Theorem (3.6).

For the more detailed structure of H see [53]. Note that since a locally soluble linear group is soluble every soluble subgroup of D is contained in a maximal soluble subgroup.

Proof. If g is any unipotent element of D, $(g-1)^n = 0$. Since D is a division ring $g = 1$. Therefore every triangularizable subgroup of D^* is abelian. In particular by 5.8, H^0 is an abelian normal subgroup of H.

Let A be any maximal abelian normal subgroup of H. If K is the subfield of D generated by A then $A \subseteq K^*$, H normalizes K^* (since it normalizes A) and HK^* is soluble. By the maximality of H, $K^* \subseteq H$, so by the maximality of A, $A = K^*$. D is infinite and of finite dimension over its centre. Thus the centre of D is infinite. Trivially the centre of D^* is contained in A, so K is infinite. Therefore by 5.12, K^* is connected. This implies that $K^* \subseteq H^0$ and so $H^0 = A = K^*$.

There remains to prove that $(H:H^0) \le \mu(n)$. By Mal'cev's Theorem H contains a triangularizable normal subgroup T of finite index at most $\mu(n)$. By the above T is abelian, so $T \subseteq H^0$. Hence $(H:H^0) \le \mu(n)$. □

5.14. Lemma. *Let G be a CZ-group and N a closed normal subgroup of G. Then the induced topology makes G/N into a CZ-group in such a way that the natural map of G onto G/N is continuous.* □

Thus if G is a subgroup of GL(n, F) and N is a closed normal subgroup of G then G/N is a CZ-group. In the next chapter we shall prove that it is actually isomorphic to a linear group over F.

Question 10. *Is a locally soluble CZ-group necessarily soluble (or perhaps only hyperabelian)?* We know that this is so for linear groups. A speical case of this conjecture is proved in [63].

Question 11. *Is a locally nilpotent CZ-group necessarily hypercentral?* Again this is true for linear groups (see 8.2). Exercises 5.3 and 5.4 indicate some partial answers to this question.

Exercise 5.3. If G is a non-trivial locally nilpotent CZ-group prove that $\zeta_1(G) \ne \{1\}$.

Exercise 5.4. Prove that every locally nilpotent CZ-group is soluble.

Exercise 5.5. If G is a CZ-group prove that $\zeta_i(G)$ is closed in G for all finite i.

Exercise 5.6. Show that the derived subgroup of the subgroup

$$G = \left\langle \begin{pmatrix} 0 & 1 \\ 1 & 0 \end{pmatrix}, \begin{pmatrix} 2 & 0 \\ 0 & \frac{1}{2} \end{pmatrix} \right\rangle$$

of $GL(2, \mathbb{Q})$ is not a closed subgroup of G.

Exercise 5.7. Prove that a connected CZ-group is irreducible. Use this result to give a short proof of 5.12. (See 14.3 for a proof of the first part.)

Exercise 5.8. G is a linear group and $\mathbf{v}$ is a set of words. Prove that:

    i) If G is connected then $\mathbf{v}(G)$ is connected (cf. 5.7).

    ii) If A is a subgroup of G then $\mathbf{v}(\bar{A}) \subseteq \overline{\mathbf{v}(A)}$ (cf. 5.10), where the bar denotes the closure in G.

    iii) If $\mathbf{v}^*(G)$ is the largest $\mathbf{v}$-marginal normal subgroup of G then $\mathbf{v}^*(G)$ is closed in G (cf. 5.4).

Exercise 5.9 (Burnside). If G is a linear group with only a finite number of conjugacy classes (of elements), prove that G is a finite group. Compare this result with 5.6.

For further results on CZ-groups see Chapters 4 and 8 of [27], and [61], [62] and [63]. See also Chapter 14.

6. The Homomorphism Theorems

There exist linear groups with homomorphic images having no faithful representations of finite degree over any field whatever. For example by 2.2 every free abelian group, but not every abelian group, has faithful representations of finite degree over some field. This raises two questions. Firstly, for which classes-of-groups $\mathfrak{X}$ are homomorphic images of linear $\mathfrak{X}$-groups necessarily isomorphic to linear groups? Secondly, given an arbitrary linear group G for which normal subgroups N of G, is G/N isomorphic to a linear group? In some ways surprisingly the second question is fruitful while the first remains unproductive, except of counterexamples.

We list some answers to questions of the first type. Recall our notation of Chapter 2; $\mathscr{L}(n, p)$ is the class of all groups having a faithful representation of finite degree n over a field of characteristic p and $\mathscr{L}(n) = \bigcup_p \mathscr{L}(n, p)$. Put $\mathscr{L} = \bigcup_{n=1}^{\infty} \mathscr{L}(n)$.

If G is polycyclic-by-finite every homomorphic image of G is in $\mathscr{L}$ (2.3 and 2.5). For every characteristic $p \geq 0$ there exists a finitely generated group in $\mathscr{L}(2, p)$ having homomorphic images not in $\mathscr{L}$. For example every free group is in $\mathscr{L}(2, p)$ by 2.9 while examples of finitely generated groups not in $\mathscr{L}$ abound. There exist finitely generated infinite simple groups ([32] Vol. 2, p. 226) and finitely generated linear simple groups are finite (4.4). Any finitely generated non-Hopf group cannot belong to $\mathscr{L}$ (by 4.4), nor can any finitely generated group with insoluble word problem (4.30). There exist finitely generated soluble groups not in $\mathscr{L}$, for example a free soluble group of derived length at least three and rank at least two is not nilpotent-by-abelian-by-finite as soluble linear groups are. Further examples are the finitely generated soluble groups constructed by P. Hall in [23] (Theorems 1 and 4).

Every nilpotent image of a finitely generated group in $\mathscr{L}$ is in $\mathscr{L}$ (by 2.5). However, there even exist finitely generated soluble groups in $\mathscr{L}$ with homomorphic images not in $\mathscr{L}$. For let x be an indeterminate over $\mathbb{Z}$, $R = \mathbb{Z}[x, x^{-1}]$ and

$$G = \left\langle \begin{pmatrix} 1 & 0 \\ 1 & 1 \end{pmatrix}, \begin{pmatrix} x & 0 \\ 0 & 1 \end{pmatrix} \right\rangle \subseteq \mathrm{GL}(2, R).$$

Denote the maximal unipotent subgroup of G by U. U is isomorphic to the additive subgroup of $\mathbb{Z}[x, x^{-1}]$ (see examples before and after 4.8). Then U^p is a normal subgroup of G and U/U^p is an infinite elementary abelian p-group (for p a prime). Thus if q is a prime distinct from p, U/U^{pq} is not in $\mathscr{L}$ by 2.2. Hence G/U^{pq} is not in $\mathscr{L}$. Therefore G is a finitely generated soluble group in $\mathscr{L}$ with homomorphic images not in $\mathscr{L}$.

For every characteristic p there exist torsion-free abelian groups in $\mathscr{L}(1, p)$ with homomorphic images not in $\mathscr{L}$ (by 2.2, see above remarks).

Every homomorphic image of a periodic linear group over a field of characteristic zero is in $\mathscr{L}$. (This we shall prove in Chapter 9.) Also every homomorphic image of an abelian periodic linear group is in $\mathscr{L}$ (see 2.2). However, homomorphic images of nilpotent periodic linear groups need not be in $\mathscr{L}$.

6.1. Example. *If p is any prime there exists a p-group T in $\mathscr{L}(3, p)$ such that T is nilpotent of class 2 and has homomorphic images not in $\mathscr{L}$.*

Proof. Let F be any infinite field of characteristic p. Additively F is the direct sum of infinitely many cyclic groups of order p. Let E be an additive subgroup of F such that $F = E \oplus \langle 1_F \rangle$.

Suppose that x and z are alements of F such that for all a and c in F, $xc - az \in E$. Then $x = z = 0$; for if $xc - az = e \in E \smallsetminus \{0\}$, $x(c\,e^{-1}) - (a\,e^{-1})z = 1$. But 1 is not in E and so $xc - az = 0$ for all a and c in F. The values $a = 0$, $c = 1$ give $x = 0$ while the values $a = 1$, $c = 0$ yield $z = 0$.

In $GL(3, F)$ let

$$A = \begin{pmatrix} 1 & 0 & 0 \\ a & 1 & 0 \\ b & c & 1 \end{pmatrix}, \quad X = \begin{pmatrix} 1 & 0 & 0 \\ x & 1 & 0 \\ y & z & 1 \end{pmatrix} \quad \text{and} \quad C = \begin{pmatrix} 1 & 0 & 0 \\ 0 & 1 & 0 \\ m & 0 & 1 \end{pmatrix},$$

where a, b, c, x, y, z and m are elements of F. Put

$$T = \mathrm{Tr}_1(3, F) = \{A : a, b, c \in F\},$$
$$T = \{C : m \in F\}$$

and

$$R = \{C : m \in E\}.$$

Then T is a p-group and $T_1 = \zeta_1(T) = T'$ (Exercise 1.3). R is a normal subgroup of T and the isomorphism $T_1 \cong F^+$ yields that $(T_1 : R) = p$. A simple calculation shows that

$$[A, X] = \begin{pmatrix} 1 & 0 & 0 \\ 0 & 1 & 0 \\ xc - az & 0 & 1 \end{pmatrix}.$$

If $X \in C_T(T/R)$ then $xc - azeE$ for all a and c in F. Thus $x = z = 0$ and so $X \in T_1$. Therefore $\zeta_1(T/R) = T_1/R$.

Let K be a field and G a subgroup of $\mathrm{GL}(n, K)$ such that $G \cong T/R$. Then $G' = \zeta_1(G)$ is cyclic of order p and $G/\zeta_1(G)$ is an infinite elementary abelian p-group. If $g \in G$ the mapping of G into G' given by $x \mapsto [g, x]$ is a homomorphism with kernel $C_G(g)$. Hence $(G : C_G(g))$ is finite and so $G^0 \subseteq C_G(g)$, using that $C_G(g)$ is closed in G. This is for every g in G, so $G^0 \subseteq \zeta_1(G)$. But $(G : G^0)$ is finite and $(G : \zeta_1(G))$ is infinite. This proves that T/R does not lie in $\mathscr{L}$. □

We turn now to the second question. If V_m and V_n are row vector spaces over F of dimensions m and n respectively and T is a subset of V_m, then a rational mapping of T into V_n is a mapping ϕ of the form

$$x\phi = (r_1(x), \ldots, r_n(x))$$

where $r_1, \ldots, r_n$ are rational functions over F in m indeterminates such that $r_i(x)$ is defined for all x in T. (By this last phrase we mean that there exist polynomials p_i and q_i such that $r_i = p_i/q_i$ and $q_i(x) \neq 0$ for every x in T; or equivalently, that for each x in T there exist polynomials p_i and q_i such that $r_i = p_i/q_i$ and $q_i(x) \neq 0$, equivalently since $F[X_1, \ldots, X_m]$ is a unique factorization domain, see [72] §1.18, Theorem 13.) By 5.1 rational mappings are continuous. Let U and W be vector spaces of dimensions m and n respectively over F. Choosing bases in U and W determines isomorphisms $\lambda : U \to V_m$ and $\mu : W \to V_n$. A rational mapping of a subset T of U into W is a mapping $\phi : T \to W$ such that $\lambda^{-1}\phi\mu$ is rational. This definition is clearly independent of the choice of λ and μ.

6.2. Theorem. *Let G be a subgroup of $\mathrm{GL}(n, F) \subseteq F_n$ and H a normal subgroup of G such that $H = C_G(X)$ for some subset X of F_n. Then there exists a rational homomorphism of G into $\mathrm{GL}(n^2, F)$ with kernel H.*

Proof. Let $E = F_n$ and $V = C_E(H)$. V is a vector space over F of dimension at most n^2. If $v \in V$, $h \in H$ and $g \in G$ then

$$(g^{-1}vg)h = g^{-1}vh^{g^{-1}}g = g^{-1}h^{g^{-1}}vg = h(g^{-1}vg).$$

Thus $v^g \in V$ for all g in G. Hence V can be made into an FG-module via conjugation. This representation of G is faithful on $G/C_G(V)$ and is clearly rational. We have just to prove that $H = C_G(V)$. Since $H \subseteq G$, $H \subseteq C_G C_E(H) = C_G(V)$. But $H = C_G(X)$, so $X \subseteq C_E(H)$. Therefore

$$C_G C_E(H) \subseteq C_G(X) = H. □$$

Let V be a finite-dimensional vector space over the field F. Put $T_0 = F$, $T_i = V \otimes_F \ldots \otimes_F V$ (i times) and $T = \bigoplus_{i=0}^{\infty} T_i$. T is an F-algebra, the multiplication being given by the tensor product;

$$\text{if} \quad x \in T_i, \ y \in T_j \quad \text{then} \quad x \otimes y \in T_{i+j}$$

(we identify $F \otimes_F V$ and V, and $V \otimes_F (V \otimes_F V)$ and $(V \otimes_F V) \otimes_F V$ in the usual way). T is the tensor algebra over V.

Let t be the F-ideal of T generated by all the elements of the form $v \otimes v$ where $v \in V$, and put $E(V) = T/t$ and $E_i(V) = (T_i + t)/t$. Then $E(V) = \oplus E_i$ is the exterior algebra over V, where $E_i = E_i(V)$. Since $t \cap T_1 = \{0\}$, $E_1 \cong_F V$. We identify E_1 and V. If $u, v \in V$ we write $u \wedge v$ for their product in $E(V)$. That is $u \wedge v = (u + t) \wedge (v + t) = u \otimes v + t$.

E is in fact a functor. If ϕ is an endomorphism of V then ϕ induces an endomorphism $E(\phi)$ of $E(V)$ in the following way. Let Φ_0 be the identity mapping on F, $\Phi_i: T_i \rightarrow T_i$ be the F-linear mapping $\phi \otimes \cdots \otimes \phi$ (i times) and put $\Phi = \bigoplus\limits_{i=0}^{\infty} \Phi_i$. Φ is an F-algebra endomorphism of T. It is easy to see that $t \Phi \subseteq t$ so that Φ induces an F-algebra endomorphism $E(\phi)$ of $E(V)$. It is immediate that if ϕ_1 is a second endomorphism of V, then $E(\phi \phi_1) = E(\phi) E(\phi_1)$. In particular if ϕ is invertible so is $E(\phi)$. Note also that $(E_i) E(\phi) \subseteq E_i$.

The following is an elementary result on exterior algebras.

Lemma. *Let U be a non-zero subspace of the finite-dimensional vector space V over F and $u_1, \ldots, u_r$ a basis of U. If $\phi \in \mathrm{Aut}_F(V)$ then $U\phi = U$ if and only if $u_1 \wedge u_2 \wedge \cdots \wedge u_r$ is an eigenvector of $E(\phi)$ in $E(V)$.*

6.3. Theorem. *Let G be a subgroup of $\mathrm{GL}(n, F)$ and V the n-row vector space over F regarded as FG-module in the usual way. Suppose that H is a normal subgroup of G and U a subspace of V of dimension r such that $H = N_G(U)$ (i.e. if $g \in G$, $Ug = U$ if and only if $g \in H$). Then there exists a rational homomorphism of G into $\mathrm{GL}(({}^nC_r)^2, F)$ with kernel H (where ${}^nC_r$ denotes $n!/(r!)(n-r)!$). In particular G/H is isomorphic to a subgroup of $\mathrm{GL}(n!^2, F)$.*

Proof. Let $u_1, \ldots, u_r$ be a basis of U and put $w = u_1 \wedge u_2 \wedge \cdots \wedge u_r \in E_r \subseteq E(V)$. There exists a group homomorphism $\sigma: G \rightarrow \mathrm{Aut}_F E_r$ given by $g\sigma = E(g)|_{E_r}$ (see above). It is easy to check that σ is rational. If $g \in G$, then $g \in H$ if and only if $Ug = U$, which happens if and only if $w(g\sigma) \in Fw$ (by the above lemma). Let $W = \sum\limits_{g \in G} Fw(g\sigma)$; W is an FG-submodule of E_r. Let $\bar{\sigma}$ denote the map of G into $\mathrm{Aut}_F W$ induced by σ. If $g \in G$, $w(g\bar{\sigma})$ is a common eigenvector of H; for if $h \in H$,

$$w(g\bar{\sigma})(h\bar{\sigma}) = w(h^{g^{-1}}\sigma)(g\sigma) \in Fw(g\bar{\sigma}),$$

as H is normal in G.

Hence W has an F-basis $w = w_1, w_2, \ldots, w_t$ of common eigenvectors of H. If $g \in G$ and $w_i(g\bar{\sigma}) \in F w_i$ for each i, then $w(g\sigma) \in F w$ and so $g \in H$. Consequently $H\bar{\sigma}$ is exactly the set of diagonal elements of $G\bar{\sigma}$ with respect to the above basis of W. Moreover $\ker \bar{\sigma} \subseteq H$. Now the full diagonal group is self-centralizing. Thus $G\bar{\sigma}$ has a rational representation with kernel $H\bar{\sigma}$ on the centralizer of $H\bar{\sigma}$ in $\text{End}_F W$ (by 6.2). We have now constructed a rational homomorphism of G into $\text{GL}(t^2, F)$ with kernel H.

If $v_1, \ldots, v_n$ is a basis of V then

$$\{v_{i_1} \wedge v_{i_2} \wedge \cdots \wedge v_{i_r} : 1 \leq i_1 < i_2 < \cdots < i_r \leq n\}$$

is a basis of E_r. Hence

$$t = \dim_F W \leq \dim_F E_r = {}^nC_r. \quad \square$$

6.4. Theorem. *Let G be a subgroup of $\text{GL}(n, F)$ and H a closed normal subgroup of G. Then there exists a rational representation of G over F with kernel H. In particular G/H is isomorphic to a linear group over F.*

Proof. Let $\{X_{ij}\}$ be n^2 independent indeterminates over F and put $R = F[X_{11}, \ldots, X_{nn}]$. Write $X = (X_{ij})$, so symbolically $R = F[X]$. Denote the annihilator ideal of H in R by $\mathfrak{a}$, i.e. $\mathfrak{a} = \{f \in R : f(H) = \{0\}\}$. Since H is closed in G, $H = G \cap V(\mathfrak{a})$, see Chapter 5.

Define an action of G on R by

$$f^g(X) = f(gX) = f\left(\ldots, \sum_k g_{ik} X_{kj}, \ldots\right)$$

where $f \in R$ and $g \in G$. It is easy to check that this makes R into an FG-module. By the Hilbert Basis Theorem $\mathfrak{a}$ is finitely generated. Consider a specific finite set of generators of $\mathfrak{a}$ and denote by m the greatest total degree of any element of this generating set. Let R_m denote the F-submodule of R of all elements of total degree at most m and put $\mathfrak{a}_m = \mathfrak{a} \cap R_m$. Now for $f \in R$ and $g \in G$ the total degrees of f and f^g are the same. Hence R_m is an FG-module of finite F-dimension and $\mathfrak{a}_m$ is an F-submodule of R_m. Clearly this representation of G into $\text{End}_F R_m$ is rational. If we can prove that $H = N_G(\mathfrak{a}_m)$ then the theorem follows from 6.3.

Let $h \in H$. If $f \in \mathfrak{a}_m$ then $f^h(k) = f(hk) = 0$ for all k in H and hence $f^h \in \mathfrak{a} \cap R_m = \mathfrak{a}_m$. Thus $\mathfrak{a}_m^h \subseteq \mathfrak{a}_m$. Since h is invertible $\mathfrak{a}_m^h = \mathfrak{a}_m$. Therefore $H \subseteq N_G(\mathfrak{a}_m)$. Let $g \in N_G(\mathfrak{a}_m)$. If $f \in \mathfrak{a}_m$ then $f^g \in \mathfrak{a}_m$. Also $1 \in H$, so $f(g) = f^g(1) = 0$. This is for all f in $\mathfrak{a}_m$ and $\mathfrak{a}_m$ contains a generating set of $\mathfrak{a}$. Consequently $g \in G \cap V(\mathfrak{a}) = H$. The theorem is now proved. $\quad \square$

Suppose that G is a closed subgroup of $\text{GL}(n, \bar{F})$ and ϕ a rational homomorphism of G. If g is a unipotent (resp. diagonalizable) element of G then $g\phi$ is a unipotent (resp. diagonalizable) element of $G\phi$ (see [3] 9.3 for example, or [3a] §4.4). A fairly direct proof however can be constructed along the lines of the proof of our 7.3 and this we shall indicate in Chapter 14. Apart from trivial cases we shall only need this theorem for the rational homomorphisms of 6.2, 6.3 and 6.4, and in these cases the result is directly computable.

6.5. Lemma. *Let G be a subgroup of* $\mathrm{GL}(n, F)$, *H a normal subgroup of G and g a diagonal element of G. In the representations of* G/H *discussed in 6.2, 6.3 and 6.4 the image of g is diagonalizable over F.*

We may suppose that g is actually diagonal since the constructions in 6.2 and 6.3 are basis-free and in 6.4 for any element g of $\mathrm{GL}(n, F)$ the mapping
$$f(X) \mapsto f^g(X)$$

is a ring isomorphism. (We could have made the construction in 6.4 basis-free as well by using the symmetric algebra on F_n.)

Proof. Let $e_1, \dots, e_n$ be the standard basis of the row vector space on which $\mathrm{GL}(n, F)$ acts, let $\{e_{ij} : i, j = 1, 2, \dots, n\}$ be the basis of F_n of matrix units, and let $g = \mathrm{diag}(g_1, \dots, g_n)$.

1. F_n is an $F\langle g \rangle$-module via conjugation. Direct multiplication shows that $e_{ij}^g = (g_i^{-1} g_j) e_{ij}$. Thus F_n has a basis of eigenvectors of g. Therefore g is diagonalizable over F on any $\langle g \rangle$-invariant subspace of F_n.

2. In this case we represent G on a subspace of E_r via $\bar{\sigma}$ and then apply a representation of type 1 above. E_r is spanned by the elements of the form $e_{i_1} \wedge e_{i_2} \cdots \wedge e_{i_r}$ where $1 \le i_1 < i_2 < \cdots < i_r \le n$.
$$(e_{i_1} \wedge e_{i_2} \wedge \cdots \wedge e_{i_r})(g\,\sigma) = (e_{i_1} g \wedge e_{i_2} g \wedge \cdots \wedge e_{i_r} g)$$
$$= (g_{i_1} g_{i_2} \cdots g_{i_r})(e_{i_1} \wedge e_{i_2} \wedge \cdots \wedge e_{i_r}).$$

Thus g is diagonalizable over F on E_r (via σ) and part 2 follows.

3. Here we represent G on R_m and then apply a representation of type 2. R_m is spanned by the monomials $X_{11}^{i_{11}} X_{12}^{i_{12}} \dots X_{nn}^{i_{nn}}$ where $\sum_{jk} i_{jk} \le m$.
$$\left(\prod_{jk} X_{jk}^{i_{jk}}\right)^g = \prod_{jk} (g_j X_{jk})^{i_{jk}} = \left(\prod_{jk} g_j^{i_{jk}}\right)\left(\prod_{jk} X_{jk}^{i_{jk}}\right).$$

Therefore g is diagonalizable over F on R_m. □

6.6. Lemma. *Let G be a subgroup of* $\mathrm{GL}(n, F)$, *H a normal subgroup of G and g a unipotent element of G. In the representations of* G/H *discussed in 6.2, 6.3 and 6.4 the image of g is also unipotent.*

Proof. Let e_i and e_{ij} be as in 6.5. We can suppose that $g = (g_{ij})$ is actually unitriangular.

1. F_n is an $F\langle g\rangle$ module via conjugation. If $v\in F_n$,

$$v^{(1-g)}=v-g^{-1}vg=(1-g^{-1})v+v(1-g)-(1-g^{-1})v(1-g).$$

An inductive argument shows that

$$v^{(1-g)^r}=\sum_{\substack{i,j=1\\i+j\geq r}}^{r}n_{ij}(1-g^{-1})^i v(1-g)^j$$

for some integers n_{ij}. Since $(1-g)^n=0$, $v^{(1-g)^{2n}}=0$, so g acts unipotently on F_n.

2. If $1\leq i_1<i_2<\cdots<i_r\leq n$,

$$(e_{i_1}\wedge\cdots\wedge e_{i_r})(1-g\sigma)$$

$$=e_{i_1}\wedge\cdots\wedge e_{i_r}-(e_{i_1}+\sum_{j<i_1}g_{i_1,j}e_j)\wedge\cdots\wedge(e_{i_r}+\sum_{j<i_r}g_{i_r,j}e_j),$$

$$\in\sum_{\substack{1\leq j_1<j_2<\cdots<j_r\leq n\\j_1+\cdots+j_r<i_1+\cdots+i_r}}Fe_{j_1}\wedge e_{j_2}\wedge\cdots\wedge e_{j_r},$$

using that g is unitriangular. Hence $E_r(1-g\sigma)^{rn}=\{0\}$ and the result follows.

3. If $\sum_{j,k=1}^{n}i_{jk}\leq m$, then

$$\left(\prod_{j=1}^{n}\prod_{k=1}^{n}X_{jk}^{i_{jk}}\right)^{(1-g)}=\left(\prod_{j=1}^{n}\prod_{k=1}^{n}X_{jk}^{i_{jk}}\right)-\prod_{j=1}^{n}\prod_{k=1}^{n}\left(X_{jk}+\sum_{l=1}^{j-1}g_{jl}X_{lk}\right)^{i_{jk}},$$

$$\in\sum_{f_{jk}}\left(F\prod_{j=1}^{n}\prod_{k=1}^{n}X_{jk}^{f_{jk}}\right),$$

where the summation is taken over all sets of n^2 non-negative integers f_{jk} satisfying

$$f_{jk}+f_{j+1,k}+\cdots+f_{nk}\leq i_{jk}+i_{j+1,k}+\cdots+i_{nk}$$

for every j and k, with strict inequality for at least one pair (j,k). Therefore

$$R_m^{(1-g)^{mn}}=\{0\}.\quad\square$$

Exercise 6.1 ([69 b]). R is a finitely generated integral domain of characteristic $p\geq 0$ and $G=\mathrm{GL}(n,R)$. Prove that G contains a normal subgroup T of finite index such that

a) if $p=0$ then the closed subgroups C of T are isolated in T, that is if x is any element of T such that some positive power of x lies in C then x lies in C; and

b) if $p>0$, if C is a closed subgroup of T and if x is an element of T such that some positive power of x lies in C then

$$x^{p^{-[-\log_p n]}} \in C.$$

Hint: copy the proof of 4.24, but use 6.6 and 2.6 as well as 6.4. Notice that part b) also improves the bound in 4.24.

Exercise 6.2. R is a finitely generated integral domain of characteristic $p \geq 0$ and $G = GL(n, R)$. Prove that G contains a normal subgroup T of finite index such that

a) every subgroup of T is connected if $p = 0$; and

b) for every subgroup X of T the group X/X^0 is a p-group of finite exponent dividing $p^{-[-\log_p n]}$ if $p > 0$.

Hint: use the T of Exercise 6.1.

7. The Jordan Decomposition and Splittable Linear Groups

Firstly we recall some definitions and results. Let x be an element of $GL(n, F)$. By definition x is *unipotent* (sometimes called a u-element) if $(x - 1_n)^n = 0$. x is unipotent if and only if all the eigenvalues of x are 1, which happens if and only if there exists an element g of $GL(n, F)$ such that x^g is unitriangular. In this case x has infinite order if char $F = 0$ and is a p-element if char $F = p > 0$.

By definition x is a *d-element* if x is diagonalizable, i.e. if for some g in $GL(n, \bar{F})$, where $\bar{F}$ denotes the algebraic closure of F, x^g is diagonal. Notice that the definition applies only to invertible elements, so for example 0 is diagonalizable but is not a d-element. x is a d-element if and only if $\langle x \rangle$ is completely reducible over $\bar{F}$ by 1.3.

7.1. Lemma. *Let X be a set of commuting elements of F_n.*

i) *If each x in X is diagonalizable over F then there exists a g in $GL(n, F)$ such that x^g is diagonal for each x in X.*

ii) *If each x in X is unipotent then there exists a g in $GL(n, F)$ such that x^g is unitriangular for each x in X.*

Proof. X contains a finite spanning set and all the unipotent elements of $\mathrm{Tr}(n, F)$ are unitriangular. Hence we can suppose that $X = \{x_1, \ldots, x_r\}$ is finite. If $r = 1$ the result is trivial. The proof is by induction on r. Let V be the row vector space of dimension n over F and let F_n act on V in the usual way.

i) Suppose that $x_2, \ldots, x_r$ are simultaneously diagonalizable over F. Then V is completely reducible as $B = F\{x_2, \ldots, x_r\}$-module. If $V_1, \ldots, V_t$ are the homogeneous components of V as B-module, then $V = V_1 \oplus \cdots \oplus V_t$ and each x_i acts on each V_j as a scalar (for $i \geq 2$). The V_j are fully invariant and x_1 and B commute; hence $V_j x_1 \subseteq V_j$ for each j. V reduces completely into one-dimensional irreducible $F\{x_1\}$-submodules and so each V_j is a direct sum of one-dimensional $F\{x_1\}$-submodules (see 1.4). Therefore $V = U_1 \oplus \cdots \oplus U_s$ for some one-dimensional subspaces U_j of V such that $U_j x_i \subseteq U_j$ for each x_i. That is, the elements of X are simultaneously diagonalizable over F.

ii) Suppose that $V(Y-1)^n = 0$ where $Y = \{x_2, \cdots, x_r\}$. Let V_i be the subspace of V spanned by $V(Y-1)^i$. Then $V_i(x_j - 1) \subseteq V_{i+1}$ for all $j \geq 2$. Since x_1 and Y commute each V_i is x_1-invariant. By the Jordan-Hölder Theorem x_1 acts unipotently on V_{i-1}/V_i. Hence for some positive integer m_i $V_{i-1}(x_1 - 1)^{m_i} \subseteq V_i$. Therefore $V(X-1)^{m_1 + \cdots + m_n} = 0$ and the result follows. □

7.2. Theorem. *If F is algebraically closed and x is an element of $\mathrm{GL}(n, F)$, then $\mathrm{GL}(n, F)$ contains a unique unipotent element x_u and a unique d-element x_d satisfying $x = x_u x_d = x_d x_u$.*

This is called the Jordan (multiplicative) decomposition of x. It is only necessary to assume that F is perfect. It can also be shown that x_u and x_d lie in $F\{x\}$. (See [7] Tome II, §1.8 or [3a] §4.) Note that x_u and x_d depend only on x; if we consider x as an element of $\mathrm{GL}(n, K)$ where K is some extension field of F, then the Jordan decomposition of x in $\mathrm{GL}(n, K)$ is the same as the Jordan decomposition of x in $\mathrm{GL}(n, F)$.

Proof. There exists g in $\mathrm{GL}(n, F)$ such that x^g is in Jordan normal form, say

$$x^g = \mathrm{diag}\left(J_{n_1}(\lambda_1), \ldots, J_{n_r}(\lambda_r)\right)$$

where $J_m(\lambda)$ denotes the $m \times m$ Jordan matrix with eigenvalue λ. Put

$$x_d = g\left(\mathrm{diag}(\lambda_1 1_{n_1}, \ldots, \lambda_r 1_{n_r})\right) g^{-1} \quad \text{and} \quad x_u = x x_d^{-1}.$$

Trivially $x = x_u x_d = x_d x_u$. If $x = y_u y_d = y_d y_u$ where y_u and y_d are unipotent and diagonalizable respectively then $[x, y_u] = 1_n = [x, y_d]$. If we can show that $[x_d, y_d] = 1_n = [x_u, y_u]$ then by 7.1 $y_u^{-1} x_u = y_d x_d^{-1}$ is both unipotent and diagonalizable, and hence is 1.

The n-row vector space V over F on which $\mathrm{GL}(n, F)$ acts is a finitely generated torsion $F[X]$-module via x (here X is a single indeterminate). V is a direct sum of its primary components as $F[X]$-module. Since F is algebraically closed x has a unique eigenvalue on each primary component and thus x_d acts as a scalar on each of these primary components. Also since $[x, y_d] = 1$, y_d is an $F[X]$-automorphism of V, and so leaves the primary components invariant. Hence $[x_d, y_d] = 1$. In the same way $[x_d, y_u] = 1$. Since also $[x, y_u] = 1$, so $[x_u, y_u] = 1$ and the proof of 7.2 is complete. □

Remark. If x is an element of $\mathrm{GL}(n, F)$ then the following two statements are equivalent.

    a) *x is a d-element.*

    b) *There exists an element $f(X)$ of $F[X]$ such that*

$$f(x) = 0 \quad \text{and} \quad (f, f') = 1.$$

(The final condition here just says that f has no repeated roots; again X is a single indeterminate.)

For suppose that x is a d-element and let $\lambda_1, \lambda_2, \ldots, \lambda_r$ be the distinct eigenvalues of x in some algebraic extension of F. Put

$$f(X)=(X-\lambda_1)(X-\lambda_2)\ldots(X-\lambda_r).$$

Clearly $(f, f')=1$ and it is easy to see that $f(x)=0$. In fact f is the minimal polynomial of x, so the coefficients of f all lie in F. Thus b) holds. If b) is satisfied then the minimal polynomial of x has no repeated roots and consequently x is diagonalizable by the Jordan normal form theorem.   □

7.3. Theorem. *If F is algebraically closed and x is an element of* $\mathrm{GL}(n, F)$, *then* $\{x_u, x_d\} \subseteq \mathscr{A}_F(x)$.

Proof. Suppose first that x has finite order. If char $F=0$, then x is diagonalizable (1.6 and 1.3). Thus $x_u=1$, $x_d=x$ and $\{x_u, x_d\} \subseteq \langle x \rangle$. If

$$\text{char } F = p > 0,$$

there exist integers r, s, h and k satisfying

$$(p, s)=1, \quad |x|=p^r s \quad \text{and} \quad 1=hp^r+ks.$$

Then x^{ks} is a p-element and so is unipotent, and x^{hp^r} has order prime to p and so is diagonalizable. Also $x=x^{hp^r} \cdot x^{ks}$ and thus $x_u=x^{ks}$ and $x_d=x^{hp^r}$. Consequently $\{x_u, x_d\} \subseteq \langle x \rangle$.

We consider now the general case. There exists g in $\mathrm{GL}(n, F)$ such that $x_d{}^g$ is diagonal. Let $f_1, \ldots, f_r$ be polynomials over F in n^2 indeterminates generating the annihilator ideal of $\mathscr{A}_F(x)$. Denote by R the subring of F generated by all the coefficients of the f_i and all the entries in the matrices $x_u, x_u^{-1}, x_d, x_d^{-1}, g$ and g^{-1}. Then x_u, x_d, g and x all lie in $\mathrm{GL}(n, R)$. Let $\{m_i: i \in I\}$ be the set of all maximal ideals of R. Each R/m_i is a finite field and $\bigcap_I m_i = \{0\}$ by 4.1. Denote by ϕ_i the ring homomorphism of R_n onto $(R/m_i)_n$ induced by the natural projection of R onto R/m_i. Now $x_u \phi_i$ is unipotent (4.2) and trivially $(x_d \phi_i)^{g\phi_i} = (x_d{}^g) \phi_i$ is diagonal. Also

$$x \phi_i = (x_u \phi_i)(x_d \phi_i) = (x_d \phi_i)(x_u \phi_i),$$

so $x_u \phi_i = (x \phi_i)_u$ and $x_d \phi_i = (x \phi_i)_d$. By the paragraph above

$$\{x_u \phi_i, x_d \phi_i\} \subseteq \langle x \phi_i \rangle.$$

Therefore

$$\{x_u, x_d\} \subseteq \bigcap_I (\ker \phi_i + \langle x \rangle).$$

Let $y \in \bigcap_I (\ker \phi_i + \langle x \rangle)$. Then there exists y_i in $\ker \phi_i$ and an integer k_i such that $y = y_i + x^{k_i}$. Clearly

$$f_j(y_i + x^{k_i}) \equiv f_j(x^{k_i}) \equiv 0, \quad \text{modulo } \mathfrak{m}_i.$$

Hence $f_j(y) \in \bigcap_I \mathfrak{m}_i = \{0\}$. This is for $j = 1, 2, \ldots, r$ and so

$$\bigcap_I (\ker \phi_i + \langle x \rangle) \subseteq V(f_1, \ldots, f_r).$$

Therefore
$$\{x_u, x_d\} \subseteq V(f_1, \ldots, f_r) \cap \mathrm{GL}(n, F) = \mathscr{A}_F(x). \quad \square$$

Let G be a subgroup of $\mathrm{GL}(n, F)$. G is said to be *locally completely reducible* (as a subgroup of $\mathrm{GL}(n, F)$) if every finitely generated subgroup of G is completely reducible. Since every subgroup of G contains a finite spanning subset this is equivalent to saying that every subgroup of G is completely reducible. G is called a *splittable* linear group if for each element g of G both g_u and g_d lie in G.

There are two limiting cases of splittable groups. G is called a *unipotent group* if every element of G is unipotent and a *d-group* if every element of G is diagonalizable. G is a unipotent subgroup of $\mathrm{GL}(n, F)$ if and only if $\mathrm{GL}(n, F)$ contains an element x such that G^x is unitriangular (1.21). We now consider *d-groups*. Note that if F is algebraically closed G is a *d-group* if and only if every cyclic subgroup of G is completely reducible. Thus in this case every locally completely reducible subgroup of $\mathrm{GL}(n, F)$ is a *d-group*. (In fact this holds if F is merely perfect, see 1.24. This same remark applies to 7.4, 7.5, 7.6 and 7.7 below.)

7.4. Lemma. *If is algebraically closed and G is an abelian subgroup of $\mathrm{GL}(n, F)$ then the following are equivalent.*

a) *G is a d-group.*
b) *G is locally completely reducible.*
c) *G is completely reducible.*
d) *G is diagonalizable.*

Proof. Now a) implies d) by 7.1 and trivially d) implies c). Since G is abelian c) implies b) by Clifford's Theorem, and b) implies a) as above. $\quad \square$

7.5. Lemma. *If F is algebraically closed and G is a periodic subgroup of $\mathrm{GL}(n, F)$ then the following are equivalent.*

a) *G is a d-group.*
b) *G is locally completely reducible.*
c) *If char $F = p > 0$, G contains no elements of order p.*

Proof. That a) implies c) and b) implies a) is trivial. If G satisfies c) so does every subgroup of G. Thus c) implies b) by 1.6 and 4.9. $\quad \square$

A subgroup G of $GL(n, F)$ is called a *modular subgroup* of $GL(n, F)$ if char $F = p > 0$ and p divides $(G : G^0)$. Otherwise it is called *non-modular*.

7.6. Theorem [60]. *If F is algebraically closed and G is a soluble subgroup of $GL(n, F)$ then the following are equivalent.*

a) *G is a d-group.*
b) *G is locally completely reducible.*
c) *G is completely reducible and non-modular.*

Proof. Suppose that a) holds. G^0 is triangularizable by 5.8. Thus $(G^0)'$ is a unipotent d-group and so is $\{1\}$; that is G^0 is abelian. By 7.4, G^0 is completely reducible. Now G/G^0 is isomorphic to a finite linear d-group over F (6.4 and 6.6). Hence if char $F = p > 0$, 7.5 implies that G/G^0 is a p'-group and so G is non-modular. By 1.5, G is completely reducible. Therefore a) implies c). Since every subgroup of a d-group is a d-group this also proves that a) implies b). We know that b) implies a). It remains to show that c) implies a).

Suppose that G is non-modular and completely reducible. By Clifford's Theorem G^0 is completely reducible and by 5.8, G^0 is triangularizable. Hence G^0 is diagonalizable (and so is a d-group). Let $r = (G : G^0)$ and $x \in G$. There exists a polynomial $f(X) \in F[X]$ such that $(f, f') = 1$ and $f(x^r) = 0$. Let $g(X) = f(X^r)$. F contains r distinct r-th roots of unity and so $(g, g') = 1$. Hence x is a d-element (see remark on p. 91) and therefore G is a d-group. $\square$

7.7. Theorem ([66]). *If F is algebraically closed and G is a locally nilpotent subgroup of $GL(n, F)$ then the following are equivalent.*

a) *G is a d-group.*
b) *G is locally completely reducible.*
c) *G is completely reducible.*

A consequence of this theorem and 7.6 is that if F is algebraically closed and G is a locally nilpotent completely reducible subgroup of $GL(n, F)$ then G is non-modular.

Proof. Suppose that G is completely reducible. Let H be any subgroup of G, $h_1, \ldots, h_r$ elements of H spanning H linearly and $g_1, \ldots, g_s$ elements of G spanning G linearly. Put $H_1 = \langle h_1, \ldots, h_r \rangle$ and $G_1 = \langle g_1, \ldots, g_s, H_1 \rangle$. G_1 is completely reducible since G is completely reducible. Also G_1 is nilpotent, so H_1 is subnormal in G_1. Therefore by 1.8, H_1 is completely reducible and thus H is completely reducible. This proves that c) implies b). Trivially b) implies a) and a) implies c) by 7.6. $\square$

The non-modularity condition in 7.6c) is necessary since $GL(2, 2) \subseteq GL(2, 2^\infty)$ is soluble, modular and irreducible as a subgroup of $GL(2, 2^\infty)$, but clearly is not a d-group.

For the moment suppose that F is algebraically closed and that G is a subgroup of $GL(n, F)$. If G is locally completely reducible and either soluble or periodic then G is abelian-by-finite. However, the unitary group

$$U(n, \mathbb{C}) = \{x \in GL(n, \mathbb{C}): x \bar{x}^T = 1_n\}$$

is locally completely reducible and, if $n \geq 2$, is not soluble-by-finite. (For if V is the n-row vector space over $\mathbb{C}$, H a subgroup of $U(n, \mathbb{C})$ and W an FH-submodule of V, then $W^\perp$ is H-invariant and $V = W \oplus W^\perp$. Thus H is completely reducible.) Another example is furnished by 7.8 below.

If G is locally completely reducible then G is a d-group. Conversely if $n \leq 2$ and G is a d-group then G is locally completely reducible. For if H is a subgroup of G either H is irreducible or H is triangularizable. In the latter case H is soluble and so is completely reducible by 7.6. In [46a] V. P. Platonov and A. E. Zalesskii give an example (on which 7.8 is based) which shows that this is false for $n \geq 3$.

7.8. Example. *Let p be zero or a prime and F any field of characteristic p with transcendence degree at least two over its prime field. Then $GL(2, F)$ contains a locally completely reducible subgroup that is not soluble-by-finite and $GL(3, F)$ contains a d-subgroup that is not completely reducible.*

Proof. Let P be a prime field of characteristic p; ξ, η independent indeterminates over P and put

$$G_p = \left\langle c = \begin{pmatrix} \xi & 0 \\ \eta & \xi^{-1} \end{pmatrix}, \ d = \begin{pmatrix} \xi & \eta \\ 0 & \xi^{-1} \end{pmatrix} \right\rangle.$$

G_p is a free group of rank 2, see Exercise 2.2.

Suppose that $p > 0$ and let $a \in G_p$. Since $\det a = 1$, over a suitable extension field a is conjugate to an element of the form $\begin{pmatrix} \alpha & 0 \\ \beta & \alpha^{-1} \end{pmatrix}$. If $\alpha = \alpha^{-1}$, then $\alpha = \pm 1$ and a has finite order ($1, 2, p$ or $2p$). But G is a free group, so $a = 1$. If $\alpha = 1$ then $\alpha = \alpha^{-1}$ and $a = 1$. Therefore every non-trivial element of G_p has two distinct eigenvalues both distinct from 1.

Now $G_0 \subseteq GL(2, \mathbb{Z}[\xi, \xi^{-1}, \eta])$ and the natural ring homomorphism of $(\mathbb{Z}[\xi, \xi^{-1}, \eta])_2$ into $(\mathbb{F}_q(\xi, \eta))_2$ maps G_0 onto G_q, q some prime. Hence every non-trivial element of G_0 has two distinct eigenvalues both distinct from 1. In particular we have shown now that for all $p \geq 0$, G_p is a d-group. It follows from the above remarks that G_p is locally completely reducible.

Let F be any field containing $P(\xi, \eta)$ and put

$$H = \left\langle h = \begin{pmatrix} \xi & 0 & 0 \\ \eta & \xi^{-1} & 0 \\ \xi & 0 & 1 \end{pmatrix}, \ k = \begin{pmatrix} \xi & \eta & 0 \\ 0 & \xi^{-1} & 0 \\ 0 & 0 & 1 \end{pmatrix} \right\rangle \subseteq GL(3, F).$$

The map ψ of H onto G_p induced by $h \mapsto c$ and $k \mapsto d$, is a well-defined homomorphism and G_p is free on c and d. Hence ψ is an isomorphism. Consequently every non-trivial element of H has three distinct eigenvalues (1 and the two of its image under ψ) and so is diagonalizable. Therefore H is a d-group.

Let V be the 3-row vector space over F regarded as H-module in the usual way. Denote the standard basis of V by (e_1, e_2, e_3). $U = F e_1 \oplus F e_2$ is an FH-submodule of V. If H is completely reducible over F there exists an FH-submodule Fv of V such that $V = U \oplus Fv$. Then v is an eigenvector of k, and the eigenspaces of k all have dimension 1. It follows that $Fv = F e_3$, and yet $e_3 h \notin F e_3$. This contradiction shows that H is not completely reducible. □

If F is algebraically closed every algebraic (i.e. closed) d-subgroup of $GL(n, F)$ is abelian-by-finite ([3] 19.5 or [3a] 11.5) and thus is locally completely reducible. Every d-group is non-modular (see the proof of 7.6) but since every connected subgroup of $GL(n, F)$ is non-modular there can be no question of non-modular groups necessarily being d-groups.

Question 12. *If G is a completely reducible d-subgroup of $GL(n, F)$, F algebraically closed, is G necessarily locally completely reducible?* See [46a] Theorem in this connection.

We add one further result on d-groups.

7.9. *Every d-group is locally torsion-free-by-finite.*

For if G is a finitely generated d-subgroup of $GL(n, F)$, G contains a normal subgroup T of finite index such that every element of T of finite order is unipotent (4.8). Clearly T is torsion-free. □

Suppose now that F is an arbitrary field and let G be a subgroup of $GL(n, F)$. Write

$$G_u = \{x \in G: x \text{ is unipotent}\}$$

and

$$G_d = \{x \in G: x \text{ is diagonalizable}\}.$$

7.10. Lemma. *If G is a splittable nilpotent subgroup of $GL(n, F)$, then $[G_u, G_d] = \{1\}$.*

Proof. Clearly we can suppose that F is algebraically closed. The proof is by induction on the class of G; if G is abelian the result is trivial. $Z = \zeta_1(G)$ is a closed subgroup of G, so by 6.4, 6.5 and 6.6 there exists a rational homomorphism ϕ of G into $GL(m, F)$, for some m, such that $\ker \phi = Z$ and $\operatorname{Im} \phi$ is splittable. By induction

$$[(G\phi)_u, (G\phi)_d] = \{1\}.$$

Let $x \in G_u$ and $y \in G_d$. Then $x\phi \in (G\phi)_u$ and $y\phi \in (G\phi)_d$, so $[x,y] = z \in Z$. Since Z is closed in G and G is splittable both z_u and z_d lie in Z (7.3). Therefore

$$x^y = (x\,z_u)\,z_d = z_d(x\,z_u) \quad \text{and} \quad (y^{-1})^x = z_u(z_d\,y^{-1}) = (z_d\,y^{-1})\,z_u.$$

Now x and z_u commute and are unipotent, so by 7.1, $x\,z_u$ is unipotent. Thus $x^y = (x\,z_u)\,z_d$ is the Jordan decomposition of x^y. But x^y is unipotent and so $z_d = (x^y)_d = 1$. In a similar fashion $(y^{-1})^x = z_u(z_d\,y^{-1})$ is the Jordan decomposition of $(y^{-1})^x$ and $z_u = ((y^{-1})^x)_u = 1$. Therefore $[x,y] = 1$. $\square$

The following theorem is due to D. A. Suprunenko and R. I. Tyškevič [56]. Proofs of the connected case had already been given by A. Borel [3] and Tôgô Shigeaki [58].

7.11. Theorem (Suprunenko and Tyškevič). *If G is a locally nilpotent subgroup of $\mathrm{GL}(n, F)$ then G_u and G_d are subgroups of G and $\langle G_u, G_d \rangle = G_u \times G_d$.*

Proof ([66]). We can clearly suppose that F is algebraically closed. Also it suffices to prove 7.11 for finitely generated groups. Thus we may suppose that G is nilpotent. But then the closure of G in $\mathrm{GL}(n, F)$ is also nilpotent (5.11) and so by 7.3 we can assume that G is splittable. Note also that if G is abelian the result follows from 7.1.

There exists an element h of $\mathrm{GL}(n, F)$ and irreducible representations $\rho_1, \ldots, \rho_t$ of G such that

$$g^h = \begin{pmatrix} g\,\rho_1 & & 0 \\ & \ddots & \\ * & & g\,\rho_t \end{pmatrix} \quad \text{for all } g \text{ in } G.$$

By 7.7, $G\rho_i$ is a d-group. Also by 1.14, $G\rho_i$ is monomial and so contains a closed diagonalizable normal subgroup of finite index. Put

$$T = \bigcap_{i=1}^{t} (G\rho_i)^0\, \rho_i^{-1}.$$

Then T is a closed triangularizable normal subgroup of finite index in G.

Let U be the maximal unipotent subgroup of T. Trivially $U \subseteq G_u$. But $G\rho_i$ is a d-group and $(G_u)\rho_i \subseteq (G\rho_i)_u$. Hence $G_u = U$ is a normal subgroup of G. T/U is abelian. If $g \in T$ and $x \in T_d$ then $g^{-1} x g = x u$ for some u in U. Since by 7.10, $[x,u] = 1$, $x^g = xu$ is the Jordan decomposition of x^g. But x^g is a d-element, so $x^g = (x^g)_d = x$. Therefore T_d is central in T and consequently (by 7.1) $T_d = (\zeta_1(T))_d$ is a central subgroup of T. Since G is splittable and T is closed in G it follows that $T = T_u \times T_d$.

We complete the proof by induction on $(G:T)$. Suppose that H is a normal subgroup of G containing T such that H_u and H_d are subgroups of H satisfying $H = H_u \times H_d$, and that G/H is cyclic. Since $G_u = U \subseteq H$

there exists a g in G_d such that $G=\langle H, g \rangle$. Let $r=(G:H)$. Then $g^r \in H$, so $g^r \in H_d$. Also g normalizes H_d and so $(\langle g, H_d \rangle : H_d)=r$. By 7.6 and 7.7, $G\rho_i$ is non-modular. Thus if char $F=p>0$, then $(p, (G:T))=1=(p, r)$. It follows (cf. the proof of 7.6) that $\langle g, H_d \rangle$ is a d-group. By 7.10, $\langle g, H_d \rangle \subseteq C_G(G_u)$ and consequently $G=G_u \times \langle g, H_d \rangle$. It is now a triviality that $G_d=\langle g, H_d \rangle$ and so G_d is a subgroup of G. ☐

7.12. Corollary. *Let F be an algebraically closed field and G a locally nilpotent subgroup of $GL(n, F)$. If G is generated by completely reducible subgroups then G is completely reducible. Conversely if G is completely reducible then every subgroup of G is completely reducible.*

Proof. If G is generated by completely reducible subgroups G is generated by d-elements (7.7). Hence $G=G_d$ by 7.11 and so is completely reducible by 7.7. The converse is just 7.7, c) implies b). ☐

7.13. Corollary. *If G is a locally nilpotent, splittable linear group, then G_u and G_d are subgroups of G and $G=G_u \times G_d$.* ☐

If G is any subgroup of $GL(n, F)$ put $\mu(G)=\langle g_u, g_d : g\in G \rangle$. $\mu(G)$ is a subgroup of $GL(n, \bar{F})$, where $\bar{F}$ is the algebraic closure of F.

7.14.

  a) $G \subseteq \mu(G)$;

  b) *G is soluble of derived length d if and only if $\mu(G)$ is soluble of derived length d;*

  c) *G is nilpotent of class c if and only if $\mu(G)$ is nilpotent of class c;*

  d) *G is triangularizable if and only if $\mu(G)$ is triangularizable;*

  e) *G is diagonalizable if and only if $\mu(G)$ is diagonalizable;*

  f) *G is locally nilpotent if and only if $\mu(G)$ is locally nilpotent;*

  g) *if G is locally nilpotent $\mu(G)$ is splittable.*

Proof. $G \subseteq \mu(G) \subseteq \mathscr{A}_F(G)$ by 7.3. Thus a) is trivial and b), c), d) and e) follow from the corresponding properties of $\mathscr{A}_F(G)$.

  f) Let G be locally nilpotent and H any finitely generated subgroup of $\mu(G)$. There exists a finitely generated subgroup K of G such that $H \subseteq \mu(K)$. By c) above $\mu(K)$ is nilpotent. Therefore $\mu(G)$ is locally nilpotent. The converse is trivial.

  g) By f) $\mu(G)$ is locally nilpotent and trivially $\mu(G)=\langle \mu(G)_u, \mu(G)_d \rangle$. Therefore $\mu(G)$ is splittable (7.11). ☐

7.15. Theorem (Suprunenko and Tyškevič [54] Theorem 45). *Let F be an algebraically closed field, V a finite-dimensional F-module and G a locally nilpotent group of automorphisms of V. Then V has a direct decomposition*

$$V=V_1 \oplus V_2 \oplus \cdots \oplus V_r$$

*as FG-module such that the composition factors of each V_i (as FG-module)
are all isomorphic.*

Proof. Suppose firstly that G is splittable, i.e. that $G=G_u\times G_d$. G_d is
completely reducible, let $V_1, \ldots, V_r$ be the homogeneous components of
V as FG_d-module. Since G_u and G_d commute, G_u is a group of FG_d-auto-
morphisms of V. Hence each V_i is actually a G-submodule of V. Let

$$\{0\}=M_0\subset M_1\subset\cdots\subset M_t=V_i.$$

Be an FG-composition series of V_i. Since G_u is normal in G, M_j/M_{j-1} is
a trivial G_u-module by Clifford's Theorem and so is irreducible as FG_d-
module. Therefore all the composition factors of V_i are FG-isomorphic.

We now consider the general case. We claim that a subspace U of V
is a G-submodule if and only if it is a $\mu(G)$-submodule. One way round
this is trivial. Suppose that U is a G-submodule. Then $Ug=U$ for all g
in G. The set $\{x\in\mathrm{Aut}_F V: Ux=U\}$ is a closed subset of $\mathrm{Aut}_F V$ and so
contains $\mathscr{A}_F(G)$. Since $G\subseteq\mu(G)\subseteq\mathscr{A}_F(G)$ this proves the point. By the
above V has a decomposition of the required type with respect to $\mu(G)$
and this decomposition will be a decomposition of the required type with
respect to G. □

We shall require the following two lemmas later. (The first is not
strictly necessary, but I think it clarifies the situation somewhat.)

7.16. Lemma ([66]). *Let G be a subgroup of $\mathrm{GL}(n, F)$, $K=\mu\eta(G)$ and
$\bar{G}=\langle K, G\rangle$. Then $K=\eta(\bar{G})$, $K\cap G=\eta(G)$ and $\eta_1(G)\subseteq\eta_1(\bar{G})\cap G$. If $\eta_1(\bar{G})$
is nilpotent then $\eta_1(G)=\eta_1(\bar{G})\cap G$.*

One use of this lemma will be to prove that $\eta_1(G)$ is always nilpotent.

Proof. Clearly G normalizes K, so $K\subseteq\eta(\bar{G})$ and $\bar{G}=KG$. Hence $\eta(\bar{G})=$
$K(G\cap\eta(\bar{G}))=K$. Trivially $\eta(G)\subseteq K\cap G\subseteq\eta(G)$.

Let N be a nilpotent normal subgroup of G. Then $\mathscr{A}_F(N)$ is a
nilpotent normal subgroup of $\mathscr{A}_F(G)$ by 5.10. But $\bar{G}\subseteq\mathscr{A}_F(G)$ and so
$N\subseteq\mathscr{A}_F(N)\cap\bar{G}\subseteq\eta_1(\bar{G})$. Therefore $\eta_1(G)\subseteq\eta_1(\bar{G})\cap G$. If $\eta_1(\bar{G})$ is nilpotent
trivially $\eta_1(\bar{G})\cap G\subseteq\eta_1(G)$. □

7.17. Lemma ([66]). *Let G be a subgroup of $\mathrm{GL}(n, F)$, $K=\mu\zeta(G)$ and
$\bar{G}=\langle K, G\rangle$. Then $K=\zeta(\bar{G})$, $K\cap G=\zeta(G)$ and $\zeta_i(G)=\zeta_i(\bar{G})\cap G$ for
$i=0, 1, 2, \ldots, \omega$.*

Proof. G normalizes K, so $\bar{G}=KG$. K is splittable (7.14) and thus $K=$
$K_u\times K_d$. Clearly K_u is normal in $\bar{G}$ and $K=\zeta(G)\cdot K_u$. Thus K/K_u and
$\zeta(G)/\zeta(G)\cap K_u$ are G-isomorphic. Therefore K/K_u is a hypercentral
factor of $\bar{G}$. In the same way K/K_d is hypercentral in $\bar{G}$. Since $K_u\cap K_d=\{1\}$,
we have $K\subseteq\zeta(\bar{G})$. Trivially $\zeta(\bar{G})\subseteq K$ and $\zeta(\bar{G})\cap G\subseteq\zeta(G)$. Thus $\zeta(\bar{G})=$
$K(\zeta(\bar{G})\cap G)=K$ and $K\cap G=\zeta(G)$.

Trivially $\zeta_i(\bar{G}) \cap G \subseteq \zeta_i(G)$ for every i. It follows from 5.10 (by induction on i) that for each $i < \omega$, $\mathscr{A}_{\bar{F}}\,\zeta_i(G) \subseteq \zeta_i \mathscr{A}_{\bar{F}}(G)$. But $\bar{G} \subseteq \mathscr{A}_{\bar{F}}(G)$ and so $\zeta_i(G) = \zeta_i(\bar{G}) \cap G$ for $i = 0, 1, 2, \dots$. Since $\zeta_\omega(G) = \bigcup_{i < \omega} \zeta_i(G)$ and ditto for $\bar{G}$, so $\zeta_\omega(G) = \zeta_\omega(\bar{G}) \cap G$. $\square$

Exercise 7.1. G is a finitely generated nilpotent subgroup of $GL(n, F)$. Prove that $\mu(G)$ is finitely generated.

Exercise 7.2. F is algebraically closed and G is a subgroup of $GL(n, F)$. Prove that

i) $G \subseteq \mu(G) \subseteq \mathscr{A}_F(G)$,
   $G^0 \subseteq \mu(G^0) \subseteq \mu(G)^0 \subseteq \mathscr{A}_F(G^0) = \mathscr{A}_F(G)^0$;

ii) $G \cdot \mu(G)^0 = \mu(G)$, $G \cdot \mathscr{A}_F(G)^0 = \mathscr{A}_F(G)$;

iii) $G \cap \mathscr{A}_F(G)^0 = G^0$, $\mu(G) \cap \mathscr{A}_F(G)^0 = \mu(G)^0$.

The statements about $\mathscr{A}_F$ above not involving μ do not depend on the algebraic closedness of F.

For further results on Jordan decomposition in linear groups see Chapter 14, particularly 14.20 and 14.22. A detailed study of locally nilpotent linear groups is contained in Chapter 8.

8. The Upper Central Series in Linear Groups

In [15] M.S. Garaščuk proves that every locally nilpotent linear group is hypercentral. K.W. Gruenberg ([19]) gave another proof of this and, amongst other things, proved that the Fitting subgroup of a linear group is nilpotent and that the central height of a linear group is less than 2ω. In [20] he sharpened the latter result; there exists an integer-valued function $\psi(n)$ such that if G is any subgroup of $\mathrm{GL}(n, F)$, then G has central height at most $\omega + \psi(n)$. This bound is reduced in [66], the correct bound being given for the locally nilpotent case. Chapter 8 consists of an exposition of the results of these papers together with some related theorems (on the size of the upper central factors of a linear group) and examples.

8.1. Lemma (Gruenberg [18] Prop. 2.2 ii). *If A is an abelian normal subgroup of the group G such that $A \mathbin{\mathrm{e}} G$ and if $(G : C_G(A))$ is finite, then $A \subseteq \zeta_\omega(G)$.*

Proof. Let $a \in A$ and put $B = \langle a^G \rangle$. B is a finitely generated abelian group. Hence B contains a characteristic series

$$\{1\} = B_0 \subseteq B_1 \subseteq \cdots \subseteq B_s \subseteq B_{s+1} = B$$

such that for $i < s$, B_{i+1}/B_i is an elementary abelian group of finite rank, n_i say, and B/B_s is free abelian of finite rank, n_s say. Since $A \mathbin{\mathrm{e}} G$ and n_i is finite, each element of G acts unipotently on the factor B_{i+1}/B_i. Now every unipotent linear group is unitriangularizable over the ground field (1.21). Therefore $B_{i+1}^{(G-1)^{n_i}} \subseteq B_i$ for $i = 0, 1, 2, \ldots, s$ and hence $B \subseteq \zeta_n(G)$ where $n = n_0 + n_1 + \cdots + n_s$. Consequently $a \in \zeta_\omega(G)$ and the lemma is proved. $\square$

Denote by $c(n, 0)$ the maximal class of a nilpotent subgroup of S_n, and if p is a prime, denote by $c(n, p)$ the maximal class of a nilpotent p'-subgroup of S_n.

If $n \geq 2$,

$$c(n, 0) = \left(\max_{\text{primes } q \leq n} \{q^{\{[\log_q n] - 1\}}\} \right), \qquad \leq [\tfrac{1}{2} n],$$

for $n+p>4$,

$$c(n, p)=\left(\max_{\text{primes } q \leq n, q \neq p} \{q^{\{[\log_q n]-1\}}\}\right), \qquad \leq c(n, 0) \leq [\tfrac{1}{2} n],$$

and $c(2, 2)=0$.

To prove this it suffices to show that a Sylow q-subgroup of $\mathbf{S}_n$ has nilpotency class exactly $q^{\{[\log_q n]-1\}}$. The class of a Sylow q-subgroup of $\mathbf{S}_n$ is equal to the class of a Sylow q-subgroup of $\mathbf{S}_{q^m}$ where $m=[\log_q n]$, see [21] p. 82. This Sylow q-subgroup is isomorphic to $((C_q \wr C_q) \wr C_q \cdots) \wr C_q$ (m-times) and its class is exactly q^{m-1}, see [70] or [26a] p. 379. $\quad\square$

Let p be zero or a prime and n any positive integer. Define the ordinal number $\gamma(n, p)$ by

$$\gamma(n, p)=\begin{cases} 1 & \text{if } n=1 \text{ or if } n=p=2, \\ \omega+c(n, p), & \text{otherwise}; \end{cases}$$

and the integer-valued function $f(n, p)$ by

$$f(n, p)=\begin{cases} n!, & \text{if } p=0, \\ (n!)_{p'}, & \text{the largest } p'\text{-factor of } n!, \text{ if } p>0. \end{cases}$$

8.2. Theorem ([66]). *Let G be a subgroup of $\mathrm{GL}(n, F)$ where char $F=p\geq 0$. Then*

i) $\eta(G)$ *is hypercentral (Garaščuk-Gruenberg) of central height at most $\gamma(n, p)$;*

ii) $\eta_1(G)$ *is nilpotent (Gruenberg) and $(\eta(G); \eta_1(G))$ divides $f(n, p)$;*

iii) $\gamma^s \eta(G)$ *is a diagonalizable $n!$-group where*

$$s=\begin{cases} 2, & \text{if } n=1, \\ n+c(n, p), \ \leq[3n/2], & \text{otherwise}. \end{cases}$$

Proof. We can clearly suppose that F is algebraically closed and $n\geq 2$. Since $\eta_1(G)\subseteq\eta_1\eta(G)$, with equality if $\eta_1\eta(G)$ is nilpotent, we may also suppose that G is locally nilpotent. Let $K=\mu(G)$. K_d is completely reducible (7.14, 7.12 and 7.7) and so is monomial (1.14). Hence K_d contains a diagonalizable closed normal subgroup D such that K_d/D is isomorphic to a subgroup of $\mathbf{S}_n$. $(K_d)^0\subseteq D$. Hence if $p>0$, K_d/D is a p'-group (7.6). Therefore $K/K_u\times D\cong K_d/D$ is nilpotent of class at most $c(n, p)$.

If $n=p=2$, $K_d=D$ and both K_u and K_d are abelian. Consequently G is abelian. In general $K_u\subseteq\zeta_{\frac{1}{2}n(n-1)}(K)\subseteq\zeta_\omega(K)$ (4.13; although $n-1$ will suffice by Exercise 1.3). Also $(K: C_K(D))$ is finite (1.12), so by 8.1, $D\subseteq\zeta_\omega(K)$. Hence K (and therefore G also) is hypercentral with central height at most $\omega+c(n, p)$.

$K_u\times D$ is nilpotent, so $K_u\times D\subseteq\eta_1(K)$. Therefore $(K: \eta_1(K))$ divides $f(n, p)$. Also since $(\eta_1(K): K_u\times D)$ is finite, $\eta_1(K)$ is generated by $K_u\times D$

and a finite number of nilpotent normal subgroups of K. Thus Fitting's Theorem implies that $\eta_1(K)$ is nilpotent. By 7.16, $\eta_1(G)=G\cap\eta_1(K)$ and ii) follows.

Clearly $[G,_{c(n,\,p)}G]\subseteq K_u\times D$. Since

$$K_u\subseteq\zeta_{(n-1)}(K),\qquad\gamma^s G\subseteq[D,_{(n-1)}G]\subseteq D.$$

Therefore $\gamma^s G$ is diagonalizable. For $n\geq2$, $n-1\geq1$, so $\gamma^s G\subseteq[D,G]$ and by 4.14, $[D,G]$ is an $n!$-group. Therefore $\gamma^s G$ is an $n!$-group. $\quad\square$

Exercise 8.1. If G is a linear group prove that $\eta_1(G)$ and $\eta(G)$ are closed subgroups of G. (Use 5.9, 5.11 and 8.2; see [69 a] p. 49.)

8.3. Theorem. *Let F be an algebraically closed field of characteristic $p\geq0$ and n a positive integer. Then $\mathrm{GL}(n,F)$ contains a locally nilpotent subgroup with central height exactly $\gamma(n,p)$. If $n\geq2$ and if $n=2$ assume $p\neq2$, then $\mathrm{GL}(n,F)$ contains a nilpotent subgroup of every class. There exist no locally nilpotent linear groups with central height ω.*

Proof. If $n=1$ or if $n=p=2$ everything is trivial, so suppose otherwise. There exists a prime $q\neq p$ and a q-subgroup Q of S_n such that Q is nilpotent of class $c(n,p)$. As $n\geq2$ and $p\neq2$ whenever $n=2$, $c(n,p)\geq1$. Also since F is algebraically closed F^* contains a C_{q^∞}-subgroup. Therefore $\mathrm{GL}(n,F)$ contains an isomorphic copy of $C_{q^\infty}\wr Q$ where Q is regarded as a permutation group on n-symbols in the obvious way. It is easy to see that $\zeta_\omega(C_{q^\infty}\wr Q)$ is just the base group and that $C_{q^\infty}\wr Q$ has central height $\gamma(n,p)$.

Let r be any positive integer and suppose that $p\neq2$. Then F contains a primitive 2^r-th root of 1, say ξ_r. Put

$$G_r=\left\langle\begin{pmatrix}0&1\\1&0\end{pmatrix},\begin{pmatrix}\xi_r&0\\0&\xi_r^{-1}\end{pmatrix}\right\rangle.$$

Then G_r is isomorphic to the dihedral group of order 2^{r+1} and so is nilpotent of class r. Trivially for $n\geq2$, $G_r\subseteq\mathrm{GL}(2,F)\hookrightarrow\mathrm{GL}(n,F)$.

Now let $p=2$, so that $n\geq3$. Denote by η_r a primitive 3^r-th root of 1 and put

$$H_r=\left\langle\begin{pmatrix}0&1&0\\0&0&1\\1&0&0\end{pmatrix},\begin{pmatrix}\eta_r&0&0\\0&1&0\\0&0&1\end{pmatrix}\right\rangle.$$

Then H_r is isomorphic to $C_{3^r}\wr C_3$ and so H_r is nilpotent of class $2r+1$. It is easy to see that $H_r\cap\mathrm{SL}(3,F)\cong H_r/\zeta_1(H_r)$, thus $H_r\cap\mathrm{SL}(3,F)$ is nilpotent of class $2r$. Trivially $\mathrm{GL}(3,F)$ contains nilpotent subgroups of class 1. The proof of the second part of the theorem is now complete.

Let G be a linear group and suppose that $G = \zeta_\omega(G)$. For each $i < \omega$, $\zeta_i(G)$ is nilpotent and normal in G. Thus $G = \bigcup_{i<\omega} \zeta_i(G) \subseteq \eta_1(G)$, which is nilpotent by 8.2 ii). Therefore G is nilpotent and so there exist no locally nilpotent linear groups with central height ω. $\square$

8.4. Lemma ([66]). *Let F be algebraically closed, G a subgroup of $GL(n, F)$ and N a normal d-subgroup of G lying in $\zeta(G)$. Then N contains a closed diagonalizable subgroup A such that A is normal in G and N/A is isomorphic to a subgroup of S_n.*

Proof. Let V be the space of n-row vectors over F regarded as FG-module in the normal way. Suppose that $V = V_1 \oplus \cdots \oplus V_r$ where $\{V_1, \ldots, V_r\}$ is a minimal system of imprimitivity for G. If $H = N_G(V_i)$, H acts primitively on V_i; let $\rho: H \to \mathrm{Aut}_F(V_i)$ be the induced homomorphism. By 1.12 and 7.4 the centre of $H\rho$ contains every abelian normal d-subgroup of $H\rho$.

Trivially $(N \cap H)\rho \subseteq \zeta(H\rho)$. If $(N \cap H)\rho$ is not central in $H\rho$ then it contains an abelian subgroup which is normal but not central in $H\rho$. But $(N \cap H)\rho$ is a d-group, for if $h \in (N \cap H)$, h is a d-element and V_i is an $F\langle h \rangle$-submodule. Thus V, and so V_i, is completely reducible as $F\langle h \rangle$-module and consequently $h\rho$ is a d-element. This is a contradiction and therefore $(N \cap H)\rho$ is central in $H\rho$. In particular $(N \cap H)\rho$ is abelian and so is diagonalizable (7.4). Put $A = N \cap \left(\bigcap_{i=1}^{r} N_G(V_i) \right)$. Then A has all the required properties. $\square$

8.5. Corollary ([66]). *Let G be a subgroup of $GL(n, F)$. Then $\zeta(G)$ contains a triangularizable subgroup T such that T is normal in G and $(\zeta(G):T)$ divides $n!$.*

Proof. Clearly we can suppose that F is algebraically closed. There exist integers $r, n_1, \ldots, n_r$, irreducible representations $\rho_1, \ldots, \rho_r$ of G where ρ_i has degree n_i, and an element x of $GL(n, F)$ such that

$$ g^x = \begin{pmatrix} g\rho_1 & & 0 \\ & \ddots & \\ * & & g\rho_r \end{pmatrix} \qquad \text{for all } g \text{ in } G. $$

By Clifford's Theorem $\zeta(G)\rho_i$ is completely reducible and so is a d-group (7.7). Trivially $\zeta(G)\rho_i \subseteq \zeta(G\rho_i)$, so $G\rho_i$ contains a diagonalizable normal subgroup D_i such that $D_i \subseteq \zeta(G)\rho_i$ and $(\zeta(G)\rho_i: D_i)$ divides $n_i!$ Put $T = \bigcap_{i=1}^{r} D_i \rho_i^{-1}$. Then T is a subgroup of the required type. $\square$

If q is any prime $(n!)_q$ is the order of a Sylow q-subgroup of S_n. The length of a composition series of such a Sylow subgroup is $\log_q(n!)_q$. Note that if r is a prime with $r \leq q$ then $\log_q(n!)_q \leq \log_r(n!)_r$. If p is zero or a prime and n is an integer with $n \geq 2$ define

$$d(n, p) = \begin{cases} \log_2(n!)_2, & \text{if } p \neq 2, \\ \log_3(n!)_3, & \text{if } p = 2. \end{cases}$$

Clearly $d(n, p) \leq d(n, 0) \leq \frac{1}{2}(n!)$. Define the ordinal number $\delta(n, p)$ by

$$\delta(n, p) = \begin{cases} 1, & \text{if } n = 1 \text{ or } n = p = 2, \\ \omega + d(n, p), & \text{otherwise}. \end{cases}$$

8.6. Theorem ([66]). *Let G be a subgroup of $\mathrm{GL}(n, F)$ where char $F = p \geq 0$.*
Then

   i) *G has central height at most $\delta(n, p)$;*
   ii) *$(\zeta(G): \zeta_\omega(G))$ is finite and divides $f(n, p)$;*
   iii) *$[\zeta(G), {}_s G]$ is a diagonalizable $n!$-group where*

$$s = \begin{cases} 1, & \text{if } n = 1, \\ d(n, p) + \frac{1}{2}n(n-1), & \leq \frac{1}{2}(n! + n^2 - n) \quad \text{otherwise}. \end{cases}$$

Recall that

$$f(n, p) = \begin{cases} n!, & \text{if } p = 0, \\ (n!)_{p'}, & \text{if } p > 0. \end{cases}$$

Proof. If $n = 1$ everything is trivial so suppose that $n \geq 2$. Let $K = \mu\zeta(G)$
and $\bar{G} = KG$. Then $\zeta(G) \subseteq K = \zeta(\bar{G})$ by 7.17. Also $K = K_u \times K_d$ and K_d
contains a closed diagonalizable subgroup D such that D is normal in $\bar{G}$
and K_d/D is isomorphic to a subgroup of $\mathbf{S}_n$ by 8.4. By 4.13 and 8.1
$K_u \times D \subseteq \zeta_\omega(\bar{G})$. If $p > 0$, 7.6 implies that K_d/D is a p'-group. The central
height of K_d/D in $\bar{G}/D$ is equal to the maximum of the central heights in
$\bar{G}/D$ of the Sylow subgroups of K_d/D, and this cannot exceed $d(n, p)$.
Hence K has central height in $\bar{G}$ at most $\omega + d(n, p)$. Since $\zeta(G) \subseteq K$ this
proves i) except for the $n = p = 2$ case which we postpone for the moment.

    Now $(K: K_u \times D)$ is finite and divides $f(n, p)$. But by 7.17, $K \cap G = \zeta(G)$
and $(K_u \times D) \cap G \subseteq \zeta_\omega(\bar{G}) \cap G = \zeta_\omega(G)$. Therefore $(\zeta(G): \zeta_\omega(G))$ divides
$f(n, p)$. By 4.13, $K_u \subseteq \zeta_{\frac{1}{2}n(n-1)}(\bar{G})$. Consequently

$$[\zeta(G), {}_s G] \subseteq [D, {}_{\frac{1}{2}n(n-1)}\bar{G}] \subseteq [D, \bar{G}] \subseteq D \qquad \text{since } n \geq 2.$$

By 4.14, $[D, \bar{G}]$ is an $n!$-group and so $[\zeta(G), {}_s G]$ is a diagonalizable
$n!$-group.

    Suppose now that $n = p = 2$. Then $s = 1$. Since F has characteristic 2,
a diagonalizable $2!$-group is trivial. Therefore $[\zeta(G), G] = \{1\}$; that is G
has central height at most $\delta(2, 2)$. $\square$

    It seems unlikely that $\delta(n, p)$ is the true bound.

Question 13. *Does there exist a linear group of degree n over a field of characteristic p whose central height exceeds* $\gamma(n, p)$? *There definitely are differences between the locally nilpotent case and the general case as the following example of P. Hall shows.*

8.7. Example (P. Hall). *GL*(4, $\mathbb{C}$) *contains a subgroup with central height* ω.

Proof. Let

$$
a = \begin{pmatrix} 0 & 1 & 0 & 0 \\ 1 & 0 & 0 & 0 \\ 0 & 0 & 1 & 0 \\ 0 & 0 & 2 & 1 \end{pmatrix}, \quad b = \begin{pmatrix} 0 & 1 & 0 & 0 \\ 1 & 0 & 0 & 0 \\ 0 & 0 & 1 & 2 \\ 0 & 0 & 0 & 1 \end{pmatrix} \quad \text{and} \quad c_i = \begin{pmatrix} \varepsilon_i & 0 & 0 & 0 \\ 0 & 1 & 0 & 0 \\ 0 & 0 & 1 & 0 \\ 0 & 0 & 0 & 1 \end{pmatrix}
$$

where ε_i is a primitive 2^i-root of unity and put $G = \langle a, b, c_i : i = 1, 2, \ldots \rangle$. $\langle a, b \rangle$ is a free subgroup of G of rank 2 (Exercise 2.3) and G contains a diagonal normal subgroup D that is isomorphic to the direct product of two $\mathbf{C}_{2^\infty}$-groups. $G = D] \langle a, b \rangle$ where a and b act on D by permuting the two given factors. $\langle a, b \rangle$ contains a normal subgroup N of G such that $G/N \cong \mathbf{C}_{2^\infty} \wr \mathbf{C}_2 \cdot \mathbf{C}_{2^\infty} \wr \mathbf{C}_2$ has central height $\omega + 1$ and $\zeta_\omega(G/N) = DN/N$. Hence $D \subseteq \zeta_i(G)$ for $i = \omega$ but not for $i < \omega$. Since G/D is free of rank 2, $D = \zeta_\omega(G) = \zeta(G)$. Therefore G has central height exactly ω. ☐

Exercise 8.2. F is the field of rational functions in a single indeterminate (two if $p = 2$) over $GF(p^\infty)$. Prove that i) if $p \neq 2$, GL(4, F) contains a subgroup with central height ω and ii) if $p = 2$, GL(5, F) contains a subgroup with central height ω. (Use 2.8 and Exercise 2.2. In fact one indeterminate will suffice even if $p = 2$, but then Exercise 2.2 is not applicable.)

8.8. Corollary ([66]). *Let G be a linear group containing no $\mathbf{C}_{q^\infty}$-subgroups for any prime q. Then G has finite central height and nilpotent Hirsch-Plotkin radical.*

Proof. By 8.2 iii) and 8.3 iii) there exists a positive integer r such that both $\gamma^r \eta(G)$ and $[\zeta(G), {}_r G]$ are diagonalizable $n!$-groups where n is the degree of G. Such a group is an abelian group satisfying the minimal condition on subgroups. Since G contains no $\mathbf{C}_{q^\infty}$-subgroups $\gamma^r \eta(G)$ and $[\zeta(G), {}_r G]$ are both finite. Therefore $\eta(G)$ is nilpotent and G has finite central height. ☐

8.9. Corollary (Gruenberg [20]). *If G is a finitely generated linear group (or more generally if G is a subgroup of some GL(n, R) where R is a finitely generated integral domain), then G has finite central height and nilpotent Hirsch-Plotkin radical.*

For G is residually finite (4.2) and so can contain no $\mathbf{C}_{q^\infty}$-subgroups. See 4.23 for a different proof of this result. ☐

8.10. Corollary ([66]). *A locally nilpotent linear group satisfying the maximal condition on abelian normal subgroups is nilpotent.*

This is false for arbitrary locally nilpotent groups since there exist locally finite p-groups whose only normal (even ascendent) abelian subgroup is the trivial group ([47] p. 108). If $G \subseteq GL(n, F)$ is the group in question then by 8.2 iii) $A = \gamma^{[3n/2]}(G)$ is an abelian group with minimal condition. Since A is normal in G, A must satisfy the maximal condition on characteristic subgroups. Thus A is finite and so G is nilpotent. ☐

8.11. Corollary (Gruenberg [20]). *A torsion-free linear group has finite central height and nilpotent Hirsch-Plotkin radical.*

This follows trivially from 8.8. ☐

The following theorem shows that we can be more precise in this case.

8.12. Theorem. *Let G be a subgroup of $GL(n, F)$ containing no non-trivial diagonalizable $n!$-elements. Then $\eta(G)$ is nilpotent of class at most $[3n/2] - 1$ and G has central height at most $\frac{1}{2}(n! + n^2 - n)$. If in fact G contains no non-unipotent element g satisfying $g_d^{n!} = 1$, then $\eta(G)$ is tri-angularizable and nilpotent of class at most $e = \max\{1, n-1\}$ and G has central height at most $f = \max\{1, \frac{1}{2}n(n-1)\}$.*

Notice that the second restriction on G is equivalent to the first for char $F \neq 0$. A torsion-free, locally nilpotent linear group over a field of characteristic zero need not be triangularizable, an example being the subgroup

$$
\left\langle
\begin{pmatrix}
0 & 1 & 0 & 0 & 0 \\
1 & 0 & 0 & 0 & 0 \\
0 & 0 & 1 & 0 & 0 \\
0 & 0 & 1 & 1 & 0 \\
0 & 0 & 0 & 0 & 1
\end{pmatrix},
\begin{pmatrix}
i & 0 & 0 & 0 & 0 \\
0 & -i & 0 & 0 & 0 \\
0 & 0 & 1 & 0 & 0 \\
0 & 0 & 0 & 1 & 0 \\
0 & 0 & 0 & 1 & 1
\end{pmatrix}
\right\rangle
$$

of $GL(n, \mathbb{C})$.

Proof. The first part of the theorem is a trivial consequence of 8.2 and 8.6. Assume now that $g_d^{n!} \neq 1$ for every g in G that is not unipotent.

There exist absolutely irreducible representations ρ_i of $\eta(G)$ of degree n_i and an element g of $GL(n, \bar{F})$ such that

$$
x^g =
\begin{pmatrix}
x\rho_1 & & 0 \\
 & \ddots & \\
* & & x\rho_r
\end{pmatrix}
$$

for every x in $\eta(G)$. Denote the maximal unipotent subgroup of $\eta(G)$ by U, and suppose that x lies in the second centre of $\eta(G)$ modulo U

and that $y \in \eta(G)$. Then for each i, Schur's Lemma implies that

$$[x, y] \rho_i = \alpha_i 1_{n_i}$$

for some $\alpha_i \in \bar{F}$. Clearly $\det([x, y] \rho_i) = 1$, so $\alpha_i^{n_i} = 1$. Hence $[x, y]^{n'} \in U$. By hypothesis this implies that $[x, y] \in U$. Since this is for every y in $\eta(G)$, it follows that $\eta(G)/U$ is abelian. Hence $\eta(G)$ is triangularizable by 1.3.

By 7.14, $K = \mu \eta(G) = K_u \times K_d$ is also triangilarizable and thus K_d is diagonalizable. K_u is nilpotent of class at most $n-1$ by Exercise 1.3 and K_d is abelian. Therefore $\eta(G)$ is nilpotent of class at most e.

Further $L = \mu \zeta(G)$ lies in K. Hence L_d is also diagonalizable. By 1.12, $(LG: C_{LG}(L_d))$ divides $n!$ and by 4.13 we have $L_u \subseteq \zeta_f(LG)$. Therefore $[L, {}_f LG]$ is diagonalizable $n!$-group by 4.14 and so $[\zeta(G), {}_f G] = \{1\}$. That is, G has central height at most f. $\quad \square$

Example ([69 b]). *There exists a linear group G each of whose subgroups has finite central height and such that there is no bound on the central heights of the subgroups of G.*

In the terminology of [32 b] the latter part of the statement of this example says that G is not centrally stunted. An immediate corollary of 8.6 iii) is that a linear group of degree n, whose diagonalizable $n!$-subgroups have order at most k, is centrally stunted of height at most

$$\tfrac{1}{2}(n! + n^2 - n) + \log_2 k.$$

Examples of linear groups satisfying this hypothesis are, finitely generated linear groups (by 4.8), more generally any $GL(n, R)$ for R a finitely generated integral domain, $GL(n, \mathbb{Q})$, $GL(n, \mathbb{Q}_p)$ and torsion-free linear groups; see [69 b].

Proof. Let $p \geq 0$ be any characteristic and n any integer not less that two (three if $p = 2$). For $i = 1, 2, \ldots$ there exists a linear group G_i of degree n over a field of characteristic p that is nilpotent of class i (by 8.3). By 2.13 there exists a linear group G of degree n over a field of characteristic p such that $G/Z \cong \overset{\infty}{\underset{i=1}{\bigstar}} G_i/Z_i$, where Z is a central subgroup of G and Z_i is a central subgroup of G_i. It follows easily from the Kuroš Subgroup Theorem ([21] §17.3 or [32] §34) that every subgroup of G has finite central height. $\quad \square$

We have already seen (8.6) that in a linear group G the order of the factor $\zeta(G)/\zeta_\omega(G)$ is finite and bounded by $n!$. We prove now that the "majority" of the factors of the upper central series of G are finite and bounded in terms of n. We deal first with the irreducible case where the bounds are smaller. The proof of 8.13 below is essentially the same as that of 3.13.

8.13. Theorem. *Let G be an irreducible subgroup of* $\mathrm{GL}(n, F)$. *Then:*

i) $\zeta_2(G)/\zeta_1(G)$ *has order at most* n^2 *and exponent dividing* n.

ii) $\zeta_l(G)/\zeta_1(G)$ *has order dividing* $n!\,n^{(l-1)(n-1)}$ *and exponent dividing* $n^{(l-1)}$, *for* $l = 1, 2, \ldots$.

Proof. By 1.19 we can assume that G is absolutely irreducible. Part i) follows at once from 3.1, taking $H = \zeta_2(G)$, $L = G$ and $K = C_H(L) = \zeta_1(G)$.

$\zeta(G)$ is monomial and so contains a diagonalizable normal subgroup A of finite index dividing $n!$. Since the exponent of $\zeta_2(G)/\zeta_1(G)$ divides n, $\zeta_{i+1}(G)^n \subseteq \zeta_i(G)$ for $i = 1, 2, \ldots$. In particular the exponent of $\zeta_l(G)/\zeta_1(G)$ divides $n^{(l-1)}$. By Schur's Lemma $\zeta_1(G) = G \cap F^* 1_n$, so $\zeta_l(G) \cap A/\zeta_1(G) \cap A$ is isomorphic to a periodic subgroup of $D(n, F)/F^* 1_n$. Hence

$$\zeta_l(G) \cap A/\zeta_1(G) \cap A$$

has rank at most $n-1$ and therefore its order divides $n^{(l-1)(n-1)}$. Consequently $(\zeta_l(G) : \zeta_1(G))$ divides $n!\,n^{(l-1)(n-1)}$. $\quad\square$

8.14. Theorem. *Let G be a subgroup of* $\mathrm{GL}(n, F)$ *and* $e = \max\{1, \frac{1}{2}n(n-1)\}$. *Then for* $e \leq l < \omega$, $\zeta_l(G)/\zeta_e(G)$ *has order dividing* $(n!)^{1+(l-e)(n-1)}$ *and exponent dividing* $(n!)^{(l-e)}$.

Proof. Clearly we can suppose that F is algebraically closed. Let $K = \mu\zeta(G)$ and $\bar{G} = KG$. Then $K = \zeta(\bar{G})$, $K = K_u \times K_d$ and $\zeta_i(G) = \zeta_i(\bar{G}) \cap G$ for $i \leq \omega$ by 7.17. By 4.13, $K_u \subseteq \zeta_e(\bar{G})$. Let $Z_i = \zeta_i(\bar{G}) \cap K_d$. Since $K = K_u \cdot \zeta(G)$,

$$\frac{\zeta(G)}{\zeta_e(G)} \cong_G \frac{\zeta(\bar{G})}{\zeta_e(\bar{G})} \cong_G \frac{K_d}{Z_e},$$

(the isomorphisms are all G-maps). Therefore $\zeta_l(G)/\zeta_e(G) \cong Z_l/Z_e$. Suppose that Z_2/Z_1 has exponent dividing $n!$. Then Z_l/Z_e has exponent dividing $(n!)^{l-e}$. (If $Z_i^{n!} \subseteq Z_{i-1}$ and $x \in Z_{i+1}$, $g \in \bar{G}$, then

$$[x^{n!}, g] \equiv [x, g]^{n!} \equiv 1 \bmod Z_{i-1}.$$

Thus $x^{n!} \in Z_{i+1} \cap \zeta_i(\bar{G}) = Z_i$ and $Z_{i+1}^{n!} \subseteq Z_i$.) Also K_d is monomial and so Z_l/Z_e has order dividing $(n!)^{1+(l-e)(n-1)}$, just as in 8.13. It remains to prove that $Z_2^{n!} \subseteq Z_1$.

There exist irreducible representations $\rho_1, \ldots, \rho_r$ of $\bar{G}$ of degrees $n_1, \ldots, n_r$ respectively and an x in $\mathrm{GL}(n, F)$ such that

$$g^x = \begin{pmatrix} g\,\rho_1 & & 0 \\ & \ddots & \\ * & & g\,\rho_r \end{pmatrix} \qquad \text{for all } g \text{ in } \bar{G}.$$

By 8.13, $(Z_2^{n!})\,\rho_i \subseteq \zeta_1(\bar{G}\,\rho_i)$. Hence $[Z_2^{n!}, \bar{G}] \subseteq \bigcap_i \ker \rho_i$ and this is a unipotent group. Since K_d is normal in $\bar{G}$, $[Z_2^{n!}, \bar{G}] \subseteq K_d$, which is a d-group. Therefore $Z_2^{n!} \subseteq \zeta_1(\bar{G}) \cap Z_2 = Z_1$. The theorem is now proved. $\quad\square$

Let A and B be subsets of the arbitrary group G. We write $A \mathbf{e} B$ if for each a in A and b in B there exists a positive integer $r = r(a, b)$ such that $[a, {}_r b] = 1$. If for each a in A we can choose the r independent of the b in B we write $A \mathbf{e} | B$ and if for each b in B we can choose the r independent of the a in A we write $A | \mathbf{e} B$.

Let

$$L(G) = \{a \in G : G \mathbf{e} a\},$$

$$\bar{L}(G) = \{a \in G : G | \mathbf{e} a\},$$

$$R(G) = \{a \in G : a \mathbf{e} G\},$$

and

$$\bar{R}(G) = \{a \in G : a \mathbf{e} | G\},$$

$$(a \mathbf{e} B \text{ is short for } \{a\} \mathbf{e} B, \text{ etc.}).$$

These are, respectively, the sets of left, bounded left, right and bounded right Engel elements of G. Put

$\sigma(G) = \{a \in G : \langle a \rangle \text{ is an ascendant subgroup of } G\}$,
$\bar{\sigma}(G) = \{a \in G : \langle a \rangle \text{ is a subnormal subgroup of } G\}$,
$\rho(G) = \{a \in G : \forall x \in G, \langle x \rangle \text{ is an ascendant subgroup of } \langle x, a^G \rangle\}$,
$\bar{\rho}(G) = \{a \in G : (\exists k \in \mathbb{N})(\forall x \in G, \langle x \rangle \text{ is subnormal in } \langle x, a^G \rangle \text{ in } k \text{ steps}\}$.

In any group G, $\sigma(G), \bar{\sigma}(G), \rho(G)$ and $\bar{\rho}(G)$ are subgroups of G, see Gruenberg [17].

Linear groups have a very well behaved Engel structure. The following result is due to K. W. Gruenberg [19], though some partial results had been obtained previously by M. S. Garaščuk and D. A. Suprunenko ([16] and [16a]).

8.15. Theorem (Gruenberg [19]). *Let G be a linear group. Then:*

i) $L(G) = \sigma(G) = \eta(G)$,
ii) $\bar{L}(G) = \bar{\sigma}(G) = \eta_1(G)$,
iii) $R(G) = \rho(G) = \zeta(G)$,
iv) $\bar{R}(G) = \bar{\rho}(G) = \zeta_\omega(G)$.

Proof. The proof uses more abstract group theory than it would be appropriate to develop here. We confine ourselves to indicating those parts of the proof that use the linearity and quote whatever abstract group-theoretic results we need.

Let $\mathfrak{X}$ be the class of groups G satisfying $G = \langle L(G), R(G) \rangle$. Trivially $Q\mathfrak{X} = \mathfrak{X}$ and $\mathfrak{X} \subseteq L(\mathfrak{G} \cap \mathfrak{X})$. Also $\mathfrak{X} \cap \mathfrak{F} \subseteq \mathfrak{N}$ since for a finite group G, $L(G) = \eta_1(G)$ and $R(G) = \zeta(G)$ (this is due to R. Baer). By 4.3 it follows that linear $\mathfrak{X}$-groups are soluble. (From 4.16 we can even deduce that linear $\mathfrak{X}$-groups are locally nilpotent, but this will follow anyway.) Hence

110

by Theorem 1.5 of [18]

$$L(G)=\sigma(G)=\eta(G), \quad \bar{L}(G)=\bar{\sigma}(G), \quad R(G)=\rho(G) \quad \text{and} \quad \bar{R}(G)=\bar{\rho}(G)$$

for any linear group G.

Suppose that we can prove the following statement.

($*$) *If G is a subgroup of* $GL(n, F)$ *and N is a normal subgroup of G then*

    i) *$N \mathbf{e} G$ implies that $N \subseteq \zeta(G)$ and*

    ii) *$N | \mathbf{e} G$ implies that $N \subseteq \zeta_h(G)$ for some finite h.*

Let G be any linear group. Since $\rho(G)\mathbf{e} G$, so by ($*$) we have $\rho(G)\subseteq\zeta(G)$ and thus $\rho(G)=\zeta(G)$. If $a\in\bar{\rho}(G)$ then $\langle a^G\rangle|\mathbf{e} G$ and hence $\langle a^G\rangle\subseteq\zeta_h(G)$ for some finite h. Therefore $\bar{\rho}(G)\subseteq\zeta_\omega(G)\subseteq\bar{\rho}(G)$. Since $G|\mathbf{e}\bar{\sigma}(G)$, so $\bar{\sigma}(G)|\mathbf{e}\bar{\sigma}(G)$ and consequently $\bar{\sigma}(G)\subseteq\zeta_h\bar{\sigma}(G)$ for some finite h. Therefore $\bar{\sigma}(G)$ is nilpotent. But $\eta_1(G)\subseteq\bar{\sigma}(G)$ and so $\eta_1(G)=\bar{\sigma}(G)$.

To complete the proof of 8.15 it remains only to prove ($*$). The proof of ($*$) is by induction on n. Trivially we can suppose that F is algebraically closed.

i) Since N is an $\mathfrak{X}$-group N is soluble. Suppose firstly that G is irreducible. By Clifford's Theorem and 5.8, N^0 is diagonalizable. Hence $(G: C_G(N^0))$ is finite (1.12). By 8.1, $N^0\subseteq\zeta_\omega(G)$. Since N/N^0 is finite and soluble repeated application of 8.1 yields that $N\subseteq\zeta(G)$.

Suppose now that G is reducible. Let V be the n-row vector space over F regarded as FG-module and let U be a proper FG-submodule of V. By the inductive hypothesis $NC_G(U)/C_G(U)$ is hypercentral in $G/C_G(U)$ and $NC_G(V/U)/C_G(V/U)$ is hypercentral in $G/C_G(V/U)$. Put $A=C_G(U)\cap C_G(V/U)$. Then $N/N\cap A$ is a hypercentral factor of G. By 4.12 we have $N\cap A\subseteq\zeta_{n^2}(G)$. Therefore $N\subseteq\zeta(G)$.

ii) The proof of this is similar to that of i) above using the following result ([18] Proposition 2.2i) in place of 8.1. If A is a normal subgroup of the arbitrary group G such that $A|\mathbf{e} G$ and $(G: C_G(A))$ is finite then $A\subseteq\zeta_h(G)$ for some finite h. $\quad\square$

9. Periodic Linear Groups

In this chapter we study the periodic (that is, torsion) subgroups of $GL(n, F)$. For most of the chapter we are interested in the conjugacy of the maximal π-subgroups in various situations. For example we prove extensions of the Sylow and Schur-Zassenhaus Theorems to the class of periodic linear groups. Then we study serial subgroups of periodic linear groups and (briefly) simple periodic linear groups. The chapter closes with a proof of the finiteness of rational and p-adic periodic linear groups.

Certain aspects of the theory of periodic linear groups are closely tied to the theory of finite linear groups, and we have to assume that the reader is familiar with a little of the latter. Full references are given.

We have already proved the following.

9.1. *Let G be a periodic subgroup of* $GL(n, F)$ *and q some prime.*

i) (Schur [49].) *G is locally finite* (4.9).

ii) (Maschke-Schur.) *If* char $F = 0$ *or if* char $F = p > 0$ *and G contains no elements of order p, then G is completely reducible* (1.6).

iii) (Burnside.) *If G has finite exponent e and is completely reducible then G is finite of order at most* e^{n^3} *(see 1.23).*

iv) *If* $q \neq$ char F *and G is a q-group then G satisfies the minimal condition on subgroups and is monomial over the algebraic closure of F (see 2.6 and 1.14).*

v) *If* $q =$ char F *and G is a q-group then G is nilpotent of class at most* $n - 1$, *of finite exponent dividing* q^n *and is unitriangularizable over F (see 2.6 and 1.21).*

9.2. Theorem (C. Jordan). *There exists an integer-valued function* $\beta(n)$ *of n only such that every finite subgroup of* $GL(n, F)$, *where* char $F = 0$, *contains an abelian normal subgroup of finite index at most* $\beta(n)$.

It suffices to consider the case $F = \mathbb{C}$ since $\mathbb{C}$ contains an isomorphic copy of every countable field of characteristic zero, which is in fact the case that Jordan proved. See [10] 36.13 for a proof. The function $\beta(n)$ given there is $(\sqrt{8n}+1)^{2n^2}-(\sqrt{8n}-1)^{2n^2}$. In Speiser's book [52] a slightly different proof gives the bound $n!\,12^{n(\pi_n-1)}$ where π_n is the number of primes less than $n+2$. A less economical proof is given in [13b].

9.3. Theorem (L. E. Dickson). *Let G be a finite subgroup of* $GL(n, F)$ *where* char $F>0$ *and* $(|G|,\ \text{char}\ F)=1$. *Then G is isomorphic to a subgroup of* $GL(n, \mathbb{C})$.

A proof of this is easily derived from §83 of [10] (see particularly 83.9). Proofs are given in [52] (Satz 208) and [13b] (Corollary 3.8).

9.4. Corollary (I. Schur [49]). *Let G be a periodic subgroup of* $GL(n, F)$, *where if* char $F=p>0$, *G contains no elements of order p. Then G contains an abelian normal subgroup of finite index at most $\beta(n)$.*

Proof. Let I be the set of all finite subgroups of G. If $H\in I$ denote by S_H the set of all abelian normal subgroups of H with index at most $\beta(n)$. By 9.2 and 9.3, S_H is not empty. If H and K are elements of I with $H\subseteq K$ define the mapping $\lambda_H^K\colon S_K\to S_H$ by $A_K\lambda_H^K=H\cap A_K$ for $A_K\in S_K$. It is easy to check that $\Lambda=(S_H,\ \lambda_H^K\colon H,\ K\in I)$ is an inverse system of finite non-empty sets. Then $\varprojlim \Lambda$ is non-empty (cf. [31a] §1.K): let $(B_H)\in\varprojlim\Lambda$ where $B_H\in S_H$ and put $B=\bigcup_{H\in I}B_H$. A simple check shows that $H\cap B=B_H$ for every H in I. It is now easy to see that B is an abelian normal subgroup of G of finite index at most $\beta(n)$. ☐

The assumption that G contains no elements of order p cannot be dropped since, for example, $PSL(n, F)$ is an infinite simple periodic linear group for $n\geq 2$ and for any infinite locally finite field F. However, we can say a little about the general case.

9.5. Theorem (D. J. Winter [71]). *A periodic linear group is a countable extension of a unipotent (and so nilpotent) group.*

Proof. 9.5 is clearly equivalent to the statement "an absolutely irreducible periodic linear group is countable". Let G be an irreducible periodic subgroup of $GL(n, F)$ where F is algebraically closed, and let P be the (maximal) absolutely algebraic subfield of F. P is countable. We prove that G is conjugate to a subgroup of $GL(n, P)$.

G contains a finite subgroup H spanning G linearly. Let $g\in G$. By 29.21 of [10] there exist elements x and y of $GL(n, F)$ such that

$$H^x\cup\langle H, g\rangle^y\subseteq GL(n, P).$$

Since H contains n^2 linearly independent elements it follows that $g^y\in P\{H^y\}$. Hence $g\in P\{H\}$. Therefore $G\subseteq P\{H\}$ and so $G^x\subseteq GL(n, P)$. ☐

9.6. Theorem (R. Brauer and W. Feit [4]). *Let F be a field of characteristic p>0 and G a finite subgroup of GL(n, F). If the Sylow p-subgroups of G have order at most p^m then G contains an abelian normal subgroup of finite index at most $f(m, n, p)$, where $f(m, n, p)$ is an integer-valued function of m, n and p only.*

This theorem can be thought of as a modular form of Jordan's Theorem. The proof depends on the theory of modular representations and uses Jordan's Theorem.

9.7. Corollary. *Let F be a field of characteristic p>0 and G a periodic subgroup of GL(n, F) containing a finite Sylow p-subgroup P of order at most p^m. Then G contains an abelian normal subgroup of finite index at most $f(m, n, p)$.*

By a *Sylow p-subgroup* of an arbitrary group G we mean a maximal p-subgroup. This is not the most useful definition for groups in general, see [31a] Chapter 3, but it is the simplest, and 9.14 will imply that the two definitions are equivalent for the groups that we are interested in here.

Proof. If Q is any finite p-subgroup of G then $\langle P, Q \rangle$ is a finite group and P is a Sylow p-subgroup of $\langle P, Q \rangle$. Therefore some conjugate of Q lies in P and so $|Q| \leq p^m$. The usual inverse limit argument completes the proof (cf. the proof of 9.4). ⬜

9.8. Corollary. *A linear group satisfying the minimal condition on subgroups is abelian-by-finite.*

9.8 is a special case, in view of 9.1, of the very much deeper result that every locally finite group satisfying the minimal condition on subgroups is abelian-by-finite. This result too can be generalized so as to embrace 9.4 and 9.7 as well. See Chapter 5 of [31a].

Proof. Let G be a subgroup of GL(n, F) satisfying the minimal condition. Then G is periodic. If char $F=0$, G is abelian-by-finite by 9.4. Suppose that char $F=p>0$ and let P be a Sylow p-subgroup of G. P is nilpotent and so by Cernikov's Theorem ([32] Vol. 2, p. 191) P is a finite extension of a divisible abelian group. But P has finite exponent, so P is finite. The result now follows from 9.7.   ⬜

9.9. If π is any set of primes and if the group G contains a unique maximal π-subgroup we denote this subgroup by G_π. We never use this notation unless the maximal π-subgroup of G is unique.

Let G be a subgroup of GL(n, F) and S any subset of G. Put

$$\mathcal{L}_G(S) = \left\langle \sum_{i=1}^{n^2} \alpha_i s_i : \alpha_i \in F, s_i \in S, \sum \alpha_i s_i \in G \right\rangle.$$

This definition is easily seen to be independent of F, that is, extending the ground-field does not enlarge $\mathscr{L}_G(S)$. If S is diagonalizable then so is $\mathscr{L}_G(S)$ and $\mathscr{L}_G(S)_\pi$ is defined for every set of primes π. If S is triangularizable then so is $\mathscr{L}_G(S)$ and $\mathscr{L}_G(S)_\pi$ is defined for all sets of primes π containing char F. For the remainder of this chapter $p(\geq 0)$ denotes the characteristic of F.

9.10. Theorem. *If G is a periodic linear group, then for every prime q the Sylow q-subgroups of G are all conjugate.*

Proofs of 9.10 are given in [44], [64] and [68]. In [61] and [62] a proof of the Sylow theorem is given for an abstract class of groups that (properly) contains the class of periodic linear groups.

Proof ([68]). Let G be a periodic subgroup of $\mathrm{GL}(n, F)$ and P and Q Sylow q-subgroups of G. There are two cases.

i) $q = p$. Let $x_1, x_2, \ldots, x_{n^2}$ (resp. $y_1, y_2, \ldots, y_{n^2}$) be elements of P (resp. Q) spanning P (resp. Q) linearly. Put $\Gamma = \langle x_i, y_i : i = 1, 2, \ldots, n^2 \rangle$ and let Π and K be Sylow q-subgroups of Γ containg $P \cap \Gamma$ and $Q \cap \Gamma$ respectively. Since Γ is a finite group there exists a g in $\Gamma \subseteq G$ such that $\Pi^g = \mathrm{K}$. Now each $x_i \in P \cap \Gamma \subseteq \Pi$, and Π is triangularizable. Hence $P \subseteq \mathscr{L}_G(\Pi)_q$ which is a q-subgroup of G. Therefore $P = \mathscr{L}_G(\Pi)_q$. In the same way $Q = \mathscr{L}_G(\mathrm{K})_q$ and so $P^g = \mathscr{L}_G(\Pi^g)_q = Q$.

ii) $q \neq p$. P and Q satisfy the minimal condition and so contain divisible abelian normal subgroups P_1 and Q_1 of finite index. P_1 contains elements $x_1, \ldots, x_n$ such that $x_1^{n!}, \ldots, x_n^{n!}$ span P_1 linearly. Let $h_1, \ldots, h_r$ be a transversal of P_1 to P. Let $y_1, \ldots, y_n$ and $k_1, \ldots, k_s$ be similar elements of Q_1 and Q. Put
$$\Gamma = \langle x_i, y_i, h_j, k_l : i = 1, 2, \ldots, n; j = 1, 2, \ldots, r; l = 1, 2, \ldots, s \rangle$$
and denote by Π and K Sylow q-subgroups of Γ containing $P \cap \Gamma$ and $Q \cap \Gamma$ respectively. Since Γ is a finite group there exists an element g of $\Gamma \subseteq G$ such that $\Pi^g = \mathrm{K}$.

Π is monomial (over $\bar{F}$) and so contains a diagonalizable normal subgroup Π_1 such that $(\Pi : \Pi_1)$ divides $n!$. $\mathrm{K}_1 = \Pi_1^g$ is a similar such subgroup of K. Since each x_i lies in $P \cap \Gamma \subseteq \Pi$, each $x_i^{n!} \in \Pi_1$. Thus $P_1 \subseteq \mathscr{L}_G(\Pi_1)_q$ and Π contains a transversal of P_1 to P. Therefore $P \subseteq \Pi \mathscr{L}_G(\Pi_1)_q$. The latter is a q-subgroup of G and consequently $P = \Pi \mathscr{L}_G(\Pi_1)_q$. In the same way $Q = \mathrm{K} \mathscr{L}_G(\mathrm{K}_1)_q$ and thus $P^g = \Pi^g \mathscr{L}_G(\Pi_1^g) = Q$. $\square$

The discerning reader will see that there is a quite general localizing process at work here. Chapter 12 is devoted to an exposition of this technique.

9.11. Lemma. *Let G be a soluble subgroup of $\mathrm{GL}(n, F)$ and N a unipotent normal subgroup of G. Then G contains a triangularizable normal subgroup*

T containing *N* such that the index of *T* in *G* is finite and divides $\lambda(n) = \mu(n)!$, where $\mu(n)$ is the function of Mal'cev's Theorem (3.5 and 3.6).

Proof. Let ρ be an irreducible representation of *G* determined by one of the composition factors of the given representation. $N\rho$ is unipotent and completely reducible (by Clifford's Theorem). Therefore $N \subseteq \ker \rho$ by 1.21. The lemma now follows from the proof of 3.6. □

9.12. Lemma. *Let T be a triangularizable periodic subgroup of* GL(*n, F*). *Then for any maximal d-subgroup D of T, $T_u D = T$.*

A periodic subgroup of GL(*n, F*) is a *d*-group if and only if it is a *p'*-group (7.5). T_u denotes the set of unipotent elements of *T*.

Proof. Note that since *T* is triangularizable T_u is a subgroup of *T*,

$$T_u \cap D = \{1\}$$

and *D* is abelian. Let $x_1, \ldots, x_n$ be elements of *D* spanning *D* linearly, let *y* be any element of *T* and put $S = \langle x_1, \ldots, x_n, y \rangle$. If *B* is a maximal *d*-subgroup of *S* containing $D \cap S$ then clearly $D = \mathscr{L}_T(B)$ and $B = D \cap S$. *S* is finite and $(|S_u|, |S/S_u|) = 1$. Hence by the Schur-Zassenhaus Theorem $S = S_u B \subseteq T_u D$. Since $y \in S$, $T = T_u D$. □

Following P. Hall [22] we say that a group *G* satisfies $\mathbf{D}_\pi^s$ if every π-subgroup of *G* is soluble and the maximal π-subgroups of *G* are all conjugate.

9.13. Lemma. *Let π be a set of primes and G a locally finite group such that every subgroup of G satisfies $\mathbf{D}_\pi^s$. If ϕ is any homomorphism of G the maximal π-subgroups of $G\phi$ are exactly the images under ϕ of the maximal π-subgroups of G.*

Proof. It suffices to suppose that $G\phi$ is a π-group and to prove that $P\phi = G\phi$ for any maximal π-subgroup *P* of *G*. *P* is soluble, of derived length *d* say. Since the maximal π-subgroups of *G* are all conjugate every π-subgroup of *G* is soluble with derived length at most *d*. Let *Y* be any finite subgroup of $G\phi$. There exists a finite subgroup *X* of *G* such that $X\phi = Y$. Since *Y* is a π-group *X* contains a π-subgroup *Q* such that $Q\phi = Y$ ([22] *E*1*, for example). Therefore *Y* is soluble with derived length at most *d*. This applies to any finite subgroup of $G\phi$ and consequently $G\phi$ is soluble. The proof uses induction on the derived length of $G\phi$.

If *g* is any element of *G* there exists a π-element *x* of *G* such that $x\phi = g\phi$. Some conjugate of *x* lies in *P*, say $x^y \in P$ where $y \in G$. Let $N = \ker \phi$. Then

$$gNG' = xNG' = x^y NG' \subseteq PNG'.$$

This is for all g in G and so $PNG' = G$. $(NG')\phi = G'\phi$ is a soluble π-group whose derived length is less than that of $G\phi$. Also $P \cap NG'$ is a maximal π-subgroup of NG'. By induction $NG' \subseteq PN$ and so $G = PN$, that is $P\phi = G\phi$. $\square$

9.14. Corollary. *If G is any homomorphic image of a periodic linear group and q is any prime then the Sylow q-subgroups of G are all conjugate.* $\square$

Not every homomorphic image of a periodic linear group is linear (6.1) so 9.14 is a genuine extension of 9.10.

9.15. Theorem ([64] 2.4). *If G is a periodic linear group each of whose finite subgroups satisfies $\mathbf{D}_\pi^s$, then G satisfies $\mathbf{D}_\pi^s$. If ϕ is any homomorphism of G then the maximal π-subgroups of $G\phi$ are exactly the images under ϕ of the maximal π-subgroups of G. $G\phi$ satisfies $\mathbf{D}_\pi^s$.*

Proof. Let T be any maximal π-subgroup of G and Q a maximal unipotent subgroup of T. Every finite subgroup of T satisfies $\mathbf{D}_\pi^s$ so T is locally soluble and thus soluble (3.8). Since $Q \cap T^0$ is a closed subgroup of finite index in Q, $Q^0 \subseteq T^0$. T^0 is triangularizable (5.8), so $Q^0 \subseteq ((T^0)_u)^0$. Now $((T^0)_u)^0$ is normal in T and consequently is contained in Q. Therefore $Q^0 = ((T^0)_u)^0$ and in particular Q^0 is normal in T. Let $\bar{Q}^0$ be the closure of Q^0 in G. Then $T \subseteq N_G(\bar{Q}^0)$. $\bar{Q}^0$ is unipotent and thus $T\bar{Q}^0$ is a π-group (for if $p \notin \pi, Q = \{1\}$). T is a maximal π-subgroup of G and consequently $\bar{Q}^0 \subseteq T$. But then $Q\bar{Q}^0$ is unipotent and so by the maximality of Q, $\bar{Q}^0 = Q^0$. That is, Q^0 is a closed subgroup of G.

If $p = \mathrm{char}\, F \neq 0$ and $p \in \pi$ let P be a maximal unipotent subgroup of G. Otherwise let $P = \{1\}$. Let S be a maximal π-subgroup of G containing P. We prove that S and T are conjugate. Just as above $P^0 = ((S^0)_u)^0$ and P^0 is normal in S. By 9.10 we may suppose without loss of generality that $Q^0 \subseteq P^0$. We prove that $Q^0 = P^0$.

Suppose that $Q^0 \neq P^0$. P^0 is nilpotent, so there exists a finite integer r such that
$$Q^0 \lhd N_1 \lhd N_2 \lhd \cdots \lhd N_r = P^0$$

where $N_1 = N_{P^0}(Q^0)$ and $N_{i+1} = N_{P^0}(N_i)$. Then
$$Q^0 \lhd N_1^0 \lhd N_2^0 \lhd \cdots \lhd N_r^0 = P^0.$$

Let i be the least integer such that $N_i^0 \neq Q^0$. Q^0 has infinite index in N_i^0 since Q^0 is a closed connected subgroup of the connected group N_i^0. Also Q^0 is normal in N_i^0. Hence Q^0 has infinite index in $N_1 = N_{P^0}(Q^0)$.

Let $n_1, n_2, \ldots, n_l$ be distinct coset representatives of Q^0 in N_1 where $l = 1 + (Q : Q^0)$ and let $a_1, \ldots, a_{n^2}$ be elements of Q^0 spanning Q^0 linearly. By 9.12, $T^0 = (T^0)_u D$ where D is a maximal d-subgroup of T^0. D is the

117

direct product of its primary components D_q. D_q satisfies the minimal condition (since $D_p = \{1\}$) and so contains a divisible subgroup C_q of finite index. Let

$$C = \langle\langle C_q : q \mid \lambda(n) \rangle, \langle D_q : q \nmid \lambda(n) \rangle\rangle$$

where $\lambda(n)$ is the function of 9.11. We shall write $\lambda = \lambda(n)$ since we shall not vary the n. C is λ-divisible so there exist elements $h_1, \ldots, h_n$ of C such that $h_1^\lambda, \ldots, h_n^\lambda$ span C linearly. Also $Q^0 C$ has finite index in T; let $x_1, \ldots, x_t$ be a left transversal of $Q^0 C$ to T. Put

$$G_1 = \langle n_1, \ldots, n_l, a_1, \ldots, a_{n^2}, h_1, \ldots, h_n, x_1, \ldots, x_t \rangle.$$

G_1 is a finite group and so satisfies $\mathbf{D}_\pi^s$. Also $Q^0 \cap G_1$ is normal in G_1. Let H be a maximal π-subgroup of G_1 containing $T \cap G_1$. By 9.11, H contains a triangularizable normal subgroup K of finite index dividing λ and containing $Q^0 \cap G_1$.

Now $h_1, \ldots, h_n \in H$ so $h_1^\lambda, \ldots, h_n^\lambda \in K$. Also $a_1, \ldots, a_{n^2} \in K$. Therefore $Q^0 C \subseteq \mathscr{L}_G(K)_\pi$. (Note that $\mathscr{L}_G(K)_\pi$ always exists since K is triangularizable and if $p \notin \pi$, K is even diagonalizable.) Hence $T \subseteq H \mathscr{L}_G(K)_\pi$, which is a π-subgroup of G. Consequently $T = H \mathscr{L}_G(K)_\pi$.

The maximal π-subgroups of G_1 are conjugate and thus there exists g in $G_1 \subseteq N_G(Q^0)$ such that $(N_1 \cap G_1)^g \subseteq H$. Hence

$$M = \langle n_1, \ldots, n_l, Q^0 \rangle \subseteq g\, T g^{-1}.$$

But $M \subseteq P^0$ and so is unipotent, Q is a maximal unipotent subgroup of T and $P = \{1\}$ if $p \notin \pi$. Therefore by 9.10 there exists h in $g T g^{-1} \subseteq N_G(Q^0)$ such that $M \subseteq Q^{g^{-1}h}$. By the choice of l, $(M : Q^0) > (Q : Q^0)$. This contradiction implies that $Q^0 = P^0$.

By 9.12, $S^0 = (S^0)_u\, B$ where B is a maximal d-subgroup of S^0. Let A_q be the minimal subgroup of finite index in the q-primary component B_q of B and put

$$A = \langle\langle A_q : q \mid \lambda \rangle, \langle B_q : q \nmid \lambda \rangle\rangle.$$

There exist $k_1, \ldots, k_n$ in A such that $k_1^\lambda, \ldots, k_n^\lambda$ span A linearly. Also $Q^0 A$ has finite index in S; let $y_1, \ldots, y_s$ be a left transversal of $Q^0 A$ to S. Put $G_2 = \langle G_1, k_1, \ldots, k_n, y_1, \ldots, y_s \rangle \subseteq N_G(Q^0)$, and let S_2 (respectively T_2) be a maximal π-subgroup of G_2 containing $S \cap G_2$ resp. $T \cap G_2$). There exists a triangularizable normal subgroup S_3 of S_2 of finite index dividing λ and containing $Q^0 \cap G_2$ by 9.11. The maximal π-subgroups of G_2 are conjugate since G_2 is finite, so there exists z in G_2 such that $S_2^z = T_2$. Then $T_3 = (S_3)^z$ is a triangularizable normal subgroup of T_2 of finite index dividing λ and containing $Q^0 \cap G_2$. In the same way as above we obtain

$$S = S_2 \mathscr{L}_G(S_3)_\pi \quad \text{and} \quad T = T_2 \mathscr{L}_G(T_3)_\pi.$$

Hence $S^z = S_2^z \mathscr{L}_G(S_3^z)_\pi = T$.

We have now proved that G satisfies $\mathbf{D}_\pi^s$. Thus every subgroup of G satisfies $\mathbf{D}_\pi^s$. The remainder of the theorem now follows from 9.13. □

Let π be any set of primes and G any group. A π-subgroup H of G is a *quasi-Hall π-subgroup* of G if for every prime q in π each Sylow q-subgroup of H is a Sylow q-subgroup of G. If H is also a maximal π-subgroup of G, H is called a *Hall π-subgroup* of G.

If G is a soluble periodic linear group it follows from 9.15 that the maximal π-subgroups of G are all conjugate and are Hall π-subgroups of G. It can also be shown ([65] 2.8 and comments) that a quasi-Hall π-subgroup of such a group is a Hall π-subgroup. More generally these same remarks apply to a locally π-separable periodic linear group and in particular to a locally π-soluble periodic linear group. (A finite group is π-separable if each of its composition factors involves at most one prime from π. See [32] Vol. 2, § 60.)

The following lemma is a special case of more general results.

9.16. Lemma. *Let G be a locally finite group and K a normal Hall π'-subgroup of G. If either*

i) $(G:K)$ *is finite, or*

ii) $\big(G: C_G(K)\big)$ *is finite,*

then K has complements in G and all the complements of K in G are conjugate.

Note that the second condition is satisfied if K is finite or if G is linear and K is diagonalizable (1.19).

Proof. i) Since G is locally finite there exists a finite subgroup G_1 of G such that $G = G_1 K$. By the Schur-Zassenhaus Theorem G_1 contains a complement H of $G_1 \cap K$. Trivially H is a complement of K in G. If L is a second complement of K in G then $G_2 = \langle H, L \rangle$ is a finite group and H and L complement $G_2 \cap K$ in G_2. Therefore H and L are conjugate.

ii) Let $C = C_G(K)$. If X is any finite subgroup of C then $K \cap X$ has a complement Y in X. But $[K, X] = \{1\}$, so Y is normal in X and thus consists of precisely the π-elements of X. Therefore the set L of all π-elements of C is a subgroup and $C = (C \cap K) \times L$. $(G:KL)$ is finite so G/L satisfies condition i). If H/L is a complement of KL/L in G/L then H is a complement of K in G (being a π-group). Also if M is a second complement of K in G then $L \subseteq M$ since L is a normal π-subgroup. Therefore H and M are conjugate from i) above. □

9.17. Lemma. *If G is a periodic d-subgroup of $\mathrm{GL}(n, F)$ then G contains a characteristic abelian subgroup of finite index.*

Proof. By 9.4, G contains an abelian normal subgroup A of finite index r say. Let $A_1 = \{a \in A: q \in \pi(a) \Rightarrow q > r\}$. Then A_1 is a normal Hall subgroup of G and A/A_1 satisfies the minimal condition on subgroups. Denote its minimal subgroup of finite index by B/A_1. $B \subseteq A$ so B is abelian. Also A_1 is characteristic in G and B/A_1 is characteristic in G/A_1, so B is characteristic in G. ☐

9.18. Theorem ([44] and [64]). *If G is a periodic subgroup of $\mathrm{GL}(n, F)$ and K is a normal Hall π'-subgroup of G, then K has complements in G, all these complements are conjugate and are exactly the maximal π-subgroups of G.*

Proof. We have two cases to consider.

Case 1, $p \in \pi$. In this case K is a d-group and so contains an abelian characteristic subgroup A of finite index by 9.17. There exists a subgroup H_1 of G such that $H_1 K = G$ and $H_1 \cap K = A$, (9.16 ii) applied to G/A).

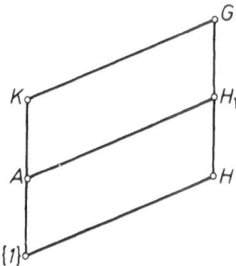

H_1/A is a π-group. Hence by 9.16 ii) and 1.12 there exists a π-subgroup H of G such that $HA = H_1$. Clearly H is a complement of K in G.

If M is a second complement of K in G, MA/A is a complement of K/A in G/A. Thus $M^g A = H_1$ for some g in G. But then M^g is a complement of A in H_1, so M and H are conjugate.

Case 2, $p \notin \pi$. If X/K is any finite subgroup of G/K, X/K is isomorphic to a finite d-subgroup of G by 9.16 i) and so by 9.4, X/K contains an abelian normal subgroup of finite index at most $\beta(n)$. The usual inverse limit argument shows that G/K contains an abelian normal subgroup A/K of finite index. Now A is locally π-separable and so there exists a

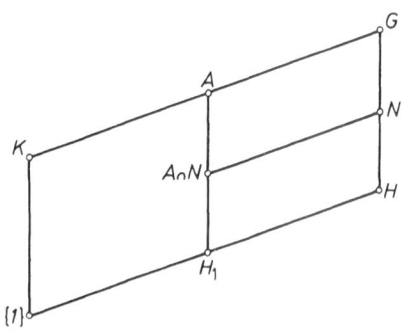

maximal π-subgroup H_1 of A such that $H_1 K = A$ by 9.15. The maximal π-subgroups of A are all conjugate. Hence $G = AN$ where $N = N_G(H_1)$. $(A \cap N)/H_1$ is a normal Hall π'-subgroup of N/H_1 and $(N : A \cap N)$ is finite. Thus by 9.16 i) there exists a subgroup H of G such that $A \cap N \cap H = H_1$ and $(A \cap N) H_1 = N$. Clearly H is a complement of K in G.

If M is a second complement of K in G then $A = (A \cap M) K$. Therefore there exists g in G such that $(A \cap M)^g = H_1$. Since A is normal in G, $M^g \subseteq N$. Also $N = (A \cap N) M^g$ so M^g/H_1 is a complement of $(A \cap N)/H_1$ in N/H_1. Therefore M and H are conjugate.

Clearly every complement of K in G is a maximal π-subgroup of G. It only remains to show that a maximal π-subgroup P of G is also a complement of K. Since $HK = G$, $PK = (PK \cap H) K$. Both P and $PK \cap H$ are complements of K in PK. The above applied to PK yields that P and $PK \cap H$ are conjugate. Since P is maximal, P and H are conjugate. The proof is now complete. $\square$

9.19. Lemma. *Let G be a periodic subgroup of $\mathrm{GL}(n, F)$ such that every element of $G \smallsetminus \{1\}$ has order greater than n. Then G is triangularizable, and if G is also a d-group G is diagonalizable.*

Proof. G contains a finite subgroup spanning G linearly. Therefore we may suppose that G is a finite group. Also it clearly suffices to suppose that G is absolutely irreducible and to prove that $n = 1$. If Q is a Sylow q-subgroup of G where $q \neq p$, $\left(N_G(Q) : C_G(Q)\right)$ divides $n!$ by 1.12. Hence $N_G(Q) = C_G(Q)$. By Burnside's Transfer Theorem ([21] Theorem 14.3.1 or [50] 6.2.9), G contains a normal q-complement $G_{q'}$. Put $P = \bigcap_{q \neq p} G_{q'}$. P is a unipotent normal subgroup of G and G/P is abelian. Since G is irreducible $P = \{1\}$ by Clifford's Theorem. But then G is abelian so by Schur's Lemma $n = 1$. $\square$

It is not in fact necessary to reduce to the finite case since Burnside's Transfer Theorem can be extended to periodic linear groups ([68] PC 1, but see also Exercise 12.3). 9.19 becomes false if G is just assumed to contain no non-trivial d-elements of order less than or equal to n, and in this case G need not even be soluble. The simplest counter example is perhaps $\mathrm{GL}(2, 4)$.

9.20. Theorem ([64] 3.4). *Let G be a periodic subgroup of $\mathrm{GL}(n, F)$ and K a normal subgroup of G such that G/K is a π-group (π some set of primes). Then there exists a π-subgroup L of G such that $KL = G$ and $K \cap L$ is locally nilpotent.*

Proof. Let $\pi_1(K) = \{q \in \pi(K) : q \leq n \text{ or } q = p\}$, this may be empty. The proof uses induction on the cardinality or $\pi_1(K)$. If $|\pi_1(K)| = 0$, K is abelian by 9.19 and so $K=K_\pi \times K_{\pi'}$. Since $K_{\pi'}$ is the unique maximal π'-subgroup of G there exists by 9.18 a π-subgroup L of G complementing $K_{\pi'}$. Then $KL=G$ and $K\cap L$ is abelian. From now on suppose that $\pi_1(K)>0$.

As an inductive hypothesis suppose the following. For every periodic subgroup A of $GL(n,F)$ and every normal subgroup B of A with $|\pi_1(B)|<|\pi_1(K)|$ there exists a subgroup C of A such that $A=BC$, $B\cap C$ is locally nilpotent and $\pi(C)=\pi(A/B)$.

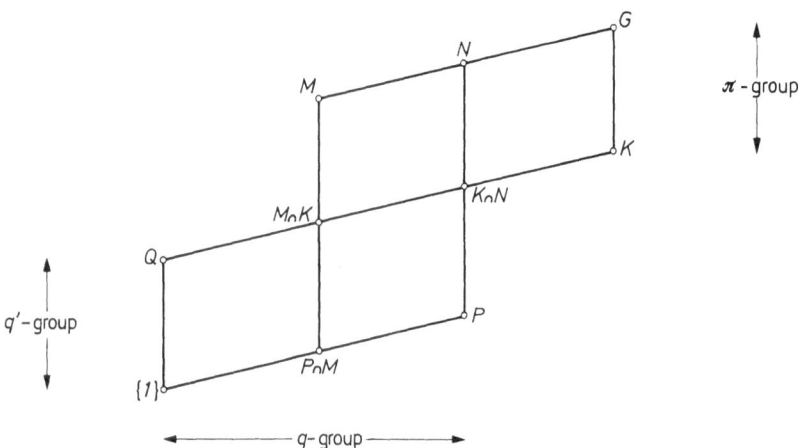

Let $q\in\pi_1(K)$, P be a Sylow q-subgroup of K and $N=N_G(P)$. By the Frattini argument $G=NK$. P is the unique maximal q-subgroup of $K\cap N$, and so by 9.18, P has a complement Q in $K\cap N$ and all such complements are conjugate. Let $M=N_N(Q)$. Since $K\cap N$ is normal in N, $N=M(K\cap N)=MP$. We have therefore shown that $G=MK$ and $M/K\cap M\cong G/K$. Also $Q\subseteq K\cap N$ so $K\cap M=(P\cap M)Q$. $P\cap M$ and Q are normal Hall subgroups of $K\cap M$ and in particular centralize each other.

Now Q is normal in M and M/Q is a $\{\pi,q\}$-group. Also

$$\pi_1(Q)\subseteq\pi_1(K)\smallsetminus\{q\}$$

so there exists by induction a $\{\pi,q\}$-subgroup J of M such that $JQ=M$ and $J\cap Q$ is locally nilpotent. $J\cap K=J\cap(K\cap M)\subseteq(P\cap M)(J\cap Q)$ since $P\cap M$ and Q are normal Hall subgroups of $K\cap M$. Now $P\cap M$ and $J\cap Q$ are both locally nilpotent and centralize each other. Hence $J\cap K$ is locally nilpotent.

There are now two cases to consider.

Case 1, $q\in\pi$. Put $L=J$. Then L is a π-group, $L\cap K$ is locally nilpotent and $LK=JQK=MK=G$.

122

Case 2, $q \notin \pi$. Since $N/K \cap N$ is a q'-group, $P \cap J$ is a normal Sylow q-subgroup of J. By 9.18 there exists a complement L of $P \cap J$ in J. L is a π-group since J is a $\{\pi, q\}$-group. $(L \cap K) \subseteq (J \cap K)$ and so is locally nilpotent. $LK = L(P \cap J) QK = JQK = MK = G$. □

9.21. Corollary ([64] H 5). *A locally soluble homomorphic image of a periodic linear group is a homomorphic image of a soluble periodic linear group, and in particular is soluble.*

Proof. Let G be a periodic linear group and K a normal subgroup of G such that G/K is locally soluble. By 9.20 there exists a subgroup L of G such that $KL = G$ and $K \cap L$ is locally nilpotent. $L/K \cap L \cong G/K$ and so is locally soluble. Since G is locally finite L is locally soluble and hence soluble (3.8). □

9.22. Corollary ([64]). *Let G be a soluble homomorphic image of a periodic linear group. For every set of primes π the maximal π-subgroups of G are all conjugate.*

This follows at once from 9.21 and 9.15. □

From this result one can develop a theory for the class $\mathfrak{C}$ of all soluble homomorphic images of periodic linear groups similar to the theory of finite soluble groups with Sylow bases, basis normalizers, Carter subgroups etc. having all the expected properties, see [65]. We should perhaps point out that 6.1 shows that $\mathfrak{C}$ is strictly larger than the class of all soluble periodic linear groups. It follows from the existence and conjugacy of the Carter subgroups of the members of $\mathfrak{C}$ that the members of $\mathfrak{C}$ have $\mathfrak{X}$-injectors and that all the $\mathfrak{X}$-injectors of a given member of $\mathfrak{C}$ are conjugate whenever $\mathfrak{X}$ is closed under taking subnormal subgroups, products of finitely many normal $\mathfrak{X}$-groups and unions of ascending chains of $\mathfrak{X}$-groups. Recent work by A. Gardiner, B. Hartley and M. J. Tomkinson shows that formation theory can be extended to the class $\mathfrak{C}$ for locally defined formations, see [16 b].

9.23. Corollary. *A homomorphic image of a periodic linear group satisfying the minimal condition on subgroups is abelian-by-finite.*

Proof. Let G be a periodic linear group and K a normal subgroup of G such that G/K satisfies the minimal condition. There exists by 9.20 a subgroup L of G such that $KL = G$ and $K \cap L$ is locally nilpotent. The closure H of $K \cap L$ in L is soluble (5.11) and L/H is isomorphic to a linear group (6.4). But $L/K \cap L \cong G/K$ so L/H satisfies the minimal condition and so is abelian-by-finite (9.8). Hence G/K is soluble-by-finite and thus is abelian-by-finite by Černikov's Theorem ([32] § 59 or [31a] $1 \cdot E \cdot 6$). □

A *poly-$\pi\,\pi'$ group* is a group with a series of finite length whose factors are either π-groups or π'-groups. (Unfortunately such a group these days is increasingly being called a π-separable group, which clashes with the earlier accepted use of the term π-separable.) The series can always be taken to be a normal series.

9.24. Theorem. *If G is a poly-π π' homomorphic image of a periodic linear group then the maximal π-subgroups of G are all conjugate. Also if ϕ is any homomorphism of G then the maximal π-subgroups of $G\phi$ are exactly the images under ϕ of the maximal π-subgroups of G.*

This is the generalization to periodic linear groups of the strong form of the Schur-Zassenhaus Theorem. The conditions of the theorem are symmetric in π and π' so the same statements are valid for the maximal π'-subgroups of G.

Proof. If Y is a normal subgroup of G and X/Y is a maximal π-subgroup of G/Y, then by 9.20 there exists a π-subgroup Z of X such that $X = YZ$ and clearly we may choose Z to be a maximal π-subgroup of G. Thus the second part of the theorem follows from the first.

G has a series of normal subgroups

$$\{1\} = J_0 \subseteq G_1 \subseteq J_1 \subseteq \cdots \subseteq J_r \subseteq G_{r+1} = G$$

where each J_i/G_i is a π-group and each G_i/J_{i-1} is a π'-group. If $r = 0$ the result is trivial. If P and Q are maximal π-subgroups of G then by induction on r we may assume that there exists an element g of G such that $\langle P, Q^g \rangle J_1/J_1$ is a π-group. Since J_1/G_1 is also a π-group, it follows that $\langle P, Q^g \rangle G_1/G_1$ is a π-group.

There exists by 9.20 a periodic linear group H, of characteristic p say, and a homomorphism ϕ of H onto $\langle P, Q^g \rangle G_1$ such that $K = \ker \phi$ is locally nilpotent. Let L denote the inverse image of G_1 in H. Then we have a series

$$\{1\} \subseteq K_\pi \subseteq L \subseteq H$$

of normal subgroups of H, where K_π and H/L are π-groups and L/K_π is a π'-group.

Again by 9.20 there exist maximal π-subgroups P_1 and Q_1 of H such that $P_1\phi = P$ and $Q_1\phi = Q^g$. Hence it suffices to prove that the maximal π-subgroups of H are conjugate. (Since these maximal π-subgroups of H all contain K_π we are really reduced to extending the Schur-Zassenhaus Theorem to cover the group H/K_π.) As with 9.18 the proof divides into two cases.

Case 1, $p \in \pi$. There exists a complement of K_π in L by 9.18, so by 9.17 there exists a normal subgroup A of H between K_π and L such that A/K_π is abelian and L/A is finite. By 9.18 the maximal π'-subgroups B of A are all conjugate in A and complement K. Hence $H = N_H(B)A$ by the Frattini argument. Now 1.12 implies that the index of $C_H(B)$ in $N_H(B)$ is finite. Thus $C_H(A/K_\pi)$ has finite index in H and the result follows as in Case 1 of the proof of 9.18.

Case 2, $p \notin \pi$. By 9.20 there exists a π-subgroup R of H such that $H = RL$. Hence in this case there exists a normal subgroup A of H containing L such that A/L is abelian and H/A is finite. Here 9.15 is applicable to A and the result now follows as in Case 2 of the proof of 9.18. □

9.25. Corollary. *Let G be a homomorphic image of a periodic linear group and K a normal subgroup of G such that G/K is a π-group. Then there exists a π-subgroup L of G such that $KL = G$ and $K \cap L$ is locally nilpotent.*

This corollary may be proved exactly as 9.20, using 9.24 in place of 9.18; in fact the proof may be shortened since now we can apply the induction hypothesis to any factor of G and not just to the subgroups of G. □

The following two results extend theorems of H. Wielandt and P. Hall, see [50] 9.3.14 and [23].

9.26. Theorem. *Let G be a homomorphic image of a periodic linear group and H a locally nilpotent quasi-Hall π-subgroup of G. Then H is a Hall π-subgroup of G and the maximal π-subgroups of G are all conjugate (to H).*

Proof. Let K be any π-subgroup of G. It suffices to prove that some conjugate of K lies in H. We prove this first for the case where K is finite by induction on the order of K. Thus we assume that $K \neq \{1\}$ and that every proper subgroup of K has a conjugate in H. Therefore every proper subgroup of K is nilpotent, and so K contains a normal Sylow q_1-subgroup K_0 for some prime q_1 in $\pi(K)$ and a complement K_1 of K_0 in K, see [50] 6.5.7. By 9.14 we may assume that $K_0 \subseteq H$.

Since H is locally nilpotent H is the direct product over i of Sylow q_i-subgroups P_i of G, where q_i runs through the elements of π. Set $\pi_1 = \{q_i : i \neq 1\}$ and $H_1 = \langle P_i : i \neq 1 \rangle$. Then H_1 is a locally nilpotent quasi-Hall π_1-subgroup of $N = N_G(K_0)$ and $K_1 \subseteq N$. Therefore by the induction hypothesis $K_1^g \subseteq H_1$ for some g in N, from which it clearly follows that $K^g \subseteq H$.

Suppose now that K is any π-subgroup of G. By the above some conjugate of every finite subgroup of K lies in H and therefore K is locally nilpotent. Hence $K = \underset{i}{\times} Q_i$ where Q_i is a Sylow q_i-subgroup of K.

We now treat the case where G is actually linear. In this case G satisfies the minimal condition on centralizers by 5.4. Hence we may assume that the statement of the theorem is valid in the centralizers of the non-central subgroups of G. Now if $K \subseteq \zeta_1(G)$ then $K \subseteq H$. Assume otherwise. Then one of the Q_i, say Q_1 is not central in G. By 9.14 again we may suppose that $Q_1 \subseteq P_1$. Then both $K_1 = \langle Q_i : i \neq 1 \rangle$ and $H_1 = \langle P_i : i \neq 1 \rangle$ lie in $C = C_G(Q_1)$ and H_1 is a locally nilpotent quasi-Hall π_1-subgroup of

C, where again $\pi_1 = \{q_i \colon i \neq 1\}$. By our inductive assumption $K_1^g \subseteq H_1$ for some g in C and consequently $K^g \subseteq H$.

In general there exists by 9.20 a periodic linear group J, of degree n and characteristic p say, and a homomorphism ϕ of J onto G such that $\ker \phi$ is locally nilpotent. Put $\mu = \{q \in \pi \colon q > n + p\}$. The inverse image $H \phi^{-1}$ of H in J is (locally) soluble. Hence $H \phi^{-1}$ has a Hall μ-subgroup A by 9.15 and

$$A\phi = H_\mu = \langle P_i \colon q_i \in \mu \rangle.$$

In the same way $K \phi^{-1}$ contains a μ-subgroup B such that $B\phi = K_\mu$.

By 9.19, A is abelian, and also A is a quasi-Hall μ-subgroup of J. For if R is a Sylow q-subgroup of J containing a Sylow q-subgroup of A for some q in μ, then by 9.13, $R\phi$ is a Sylow q-subgroup of G containing a Sylow q-subgroup of H, and therefore $R \subseteq H \phi^{-1}$. But A is a Hall μ-subgroup of $H \phi^{-1}$, so $R \subseteq A$. Hence by the part of the theorem that we have already proved the maximal μ-subgroups of J are all conjugate. Hence some conjugate of B lies in A; that is some conjugate of K_μ lies in H_μ.

Let λ be a subset of π containing μ that is maximal subject to some conjugate of K_λ lying in H_λ; λ exists since $\pi \smallsetminus \mu$ is finite. We may assume that $K_\lambda \subseteq H_\lambda$. If $q_1 \in \pi \smallsetminus \lambda$ then $Q_1 \cup P_1 \subseteq C_G(K_\lambda)$ and so $Q_1^g \subseteq P_1$ for some $g \in C_G(K_\lambda)$ by 9.14. Thus $K_{\{\lambda, q_1\}}^g \subseteq H_{\{\lambda, q_1\}}$, a contradiction which shows that $\lambda = \pi$ and completes the proof of the theorem. $\quad\square$

9.27. Corollary. *Let G be a homomorphic image of a periodic linear group, and H and K locally supersoluble quasi-Hall π-subgroups of G. Then H and K are conjugate in G.*

Proof. There exists by 9.20 a periodic linear group J, of degree n and characteristic p say, and a homomorphism ϕ of J onto G such that $\ker \phi$ is locally nilpotent. Set $\mu = \{q \in \pi \colon q > n + p\}$. It is easy to see (use [21] Theorem 10.5.3) that any periodic locally supersoluble group has a descending Sylow tower. Thus H and K have unique maximal μ-subgroups H_μ and K_μ, and these will be quasi-Hall μ-subgroups of G. As in the proof of 9.26 there exists quasi-Hall μ-subgroups A and B of J lying respectively in $H \phi^{-1}$ and $K \phi^{-1}$ such that $A\phi = H_\mu$ and $B\phi = K_\mu$. Now A and B are abelian and so are conjugate in J by 9.26. Hence H_μ and K_μ are conjugate in G.

Let λ be a right segment of the naturally ordered set π containing μ that is maximal subject to H_μ and K_μ being conjugate in G; λ exists since $\pi \smallsetminus \mu$ is finite. We may assume that $H_\lambda = K_\lambda$. If $H \neq K$ let q be the largest prime of $\pi \smallsetminus \lambda$. Then $H_{\{\lambda, q\}} = H_\lambda P$ and $K_{\{\lambda, q\}} = K_\lambda Q$ for some Sylow q-subgroups P and Q of G and hence of $N_G(H_\lambda)$. By 9.14 there exists g in G normalizing H_λ such that $P^g = Q$. Then $H_{\{\lambda, q\}}^g = K_{\{\lambda, q\}}$, contradicting the choice of λ, and therefore H and K are conjugate in G. $\quad\square$

126

Special cases of 9.24, 9.26 and 9.27 are proved in [64] and were some-what extended in various preliminary versions of this chapter. For these three results in their full generality I am indebted to Brian Hartley.

In case the above results have given the impression that all the properties of maximal π-subgroups of finite groups extend to periodic linear groups we conclude this topic with the following example, which shows that periodic linear groups can have some very unpleasant properties.

9.28. Example. *There exists an infinite simple periodic linear group that has a metabelian quasi-Hall π-subgroup, where π is the set of all primes.*

Proof. Let F be an algebraic closure of the field of two elements and put $G = \mathrm{SL}(2, F) = \mathrm{PSL}(2, F)$, $U = \mathrm{Tr}_1(2, F)$ and $D = G \cap D(2, F)$. G is a simple periodic linear group, D normalizes U and $T = UD$ is metabelian. We claim that T is a quasi-Hall π-subgroup of G.

Any Sylow 2-subgroup P of G is unitriangularizable, that is there exists some g in $\mathrm{GL}(2, F)$ such that $P^g \subseteq U$. Clearly we can choose g to lie in G, so U is a Sylow 2-subgroup of G. If Q is a Sylow q-subgroup of G and $q > 2$, then Q is diagonalizable by 9.19, so Q is conjugate in G to D_q. In particular D_q is a Sylow q-subgroup of G. Hence by 9.10 it follows that T is a quasi-Hall π-subgroup of G. □

In the above example T is clearly not a Hall π-subgroup of G. Notice that D is a Hall 2'-subgroup of G and that every 2'-subgroup of G is conjugate to some subgroup of D; this follows from either 9.19 or 9.26.

9.29. Lemma (Kargapolov [30]). *There exists an integer-valued function $k(n)$ such that any finite linear group of degree n has at most $k(n)$ non-abelian chief factors and at most $n \cdot k(n)$ non-abelian composition factors.*

Proof. If p denotes a prime, let

$$\{1\} \subseteq N_1 \subseteq M_1 \subseteq N_2 \subseteq \cdots \subseteq N_r \subset M_r \subseteq G$$

be a normal series of the finite group G such that for $i = 1, 2, \ldots, r$, M_i/N_i is a non-abelian chief factor of G such that for every prime $q \neq p$ its Sylow q-subgroups are all abelian. We prove by induction on r that G contains a soluble subgroup S containing a normal series

(1) $\begin{cases} \{1\} \subseteq K_1 \subset L_1 \subseteq \cdots \subseteq K_r \subset L_r \subseteq S \\ \text{such that for each } i = 1, 2, \ldots, r \text{ there exists a prime } q_i \neq p \\ \text{such that } L_i/K_i \text{ contains a normal but non-central Sylow } q_i\text{-subgroup.} \end{cases}$

If $r = 0$ there is nothing to prove. By repeated use of the Frattini argument (or by 9.20) there exists a subgroup G_1 of G such that $G_1 N_1 = G$ and $K_1 = G_1 \cap N_1$ is nilpotent. Replacing G by G_1/K_1 we may assume that $N_1 = \{1\}$. M_1 is then the direct product of copies of a perfect simple

127

group. Thus there exists a prime $q_1 \neq p$ for which a Sylow q_1-subgroup P_1 of M_1 is non-trivial. If $H_1 = N_G(P_1)$ then $H_1 M_1 = G$ by the Frattini argument again.

By assumption P_1 is abelian, so if $[P_1, H_1 \cap M_1] = \{1\}$ then M_1 contains a normal p_1-complement by Burnside's p-complement theorem ([21] Theorem 14.3.1 or [50] 6.2.9), which is manifestly false. Hence $H_1 \cap M_1$ contains a non-trivial Sylow subgroup Q_1 that does not centralize P_1. Once more by the Frattini argument, if $J_1 = N_{H_1}(P_1 Q_1)$, then $J_1 M_1 = G$. Choose $L_1 = P_1 Q_1$ and apply induction to the series

$$\{1\} \subseteq (J_1 \cap N_2)/L_1 \subset (J_1 \cap M_2)/L_1 \subseteq \cdots \subseteq (J_1 \cap M_r)/L_1 \subseteq J_1/L_1$$

to obtain (1).

Now let G be a finite subgroup of $GL(n, F)$ such that a chief series of G contains at least $n \cdot n! + \mu(n)$ non-abelian factors where μ is the function of Mal'cev's Theorem (3.6), and put $p = \operatorname{char} F$. If q is a prime other than p then a q-subgroup of G contains an abelian normal subgroup of index dividing $n!$, and in particular is abelian if q exceeds n. Hence more than $\mu(n)$ chief factors of G are non-abelian and yet have their Sylow q-subgroups abelian for all $q \neq p$. By the above G contains a soluble subgroup S containing a normal series satisfying (1) in which $r > \mu(n)$.

By Mal'cev's Theorem S contains a triangularizable normal subgroup T of index in S at most $\mu(n)$. Hence there exists i such that $L_i \subseteq K_i T$. Thus $L_i/K_i \cong (L_i \cap T)/(K_i \cap T)$. But if $q \neq p$ the group T contains a normal q-complement and an abelian Sylow q-subgroup, so in every factor of T the normalizer and the centralizer of a given Sylow q-subgroup coincide. This contradicts the structure of L_i/K_i. Therefore we may take $k(n) = n \cdot n! + \mu(n)$; in fact, considerably less than this will do. Finally, if $q \neq p$ any elementary abelian q-factor of G has rank at most n, so each non-abelian chief factor of G is the direct product of at most n (isomorphic) simple groups. $\square$

Before we can state the next theorem we have to introduce some notation. By a *series* we mean a generalized series, that is a series in the sense of [47] or [31 a] §1.A, e.g. normal system in the sense of [32] §56. If H is a subgroup of G such that G has a series of which both H and G are members, then H is called a *serial subgroup* of G. If G is a group and X and Y are subgroups of G write

$$\delta X = [X, X], \quad \gamma X, Y = [X, Y],$$

and for any integer r, write $_r Y$ as an abbreviation of $Y, Y, \ldots, Y$, where Y appears r times. For example $\delta^2 X = [[X, X], [X, X]]$,

$$[X, Y, Y] = \gamma^2 X, Y, Y = \gamma^2 X, {}_2 Y, \qquad [X, Y, Y, Y] = \gamma^3 X, {}_3 Y, \qquad etc.$$

9.30. Theorem. *Let G be a periodic linear group of degree n, put $k = n \cdot k(n)$ where $k(n)$ is the function of 9.29, and let $d = d(n)$ denote the maximal derived length of any soluble linear group of degree n (d exists by 3.7). Then:*

   i) *Any series of G contains at most k non-soluble factors.*

   ii) *If H is any serial subgroup of G then*

$$\delta^d (\gamma\, \delta^d)^k\, G,\, {}_k H \subseteq H\,.$$

   iii) *If G is simple then $\{1\} \subset G$ is the only composition series of G. That is G is absolutely simple.*

Proof. If S is any finite subgroup of G there exists a soluble subgroup T_i of S such that $S = T_i \delta^i S$. Since by assumption $\delta^d T_i = \{1\}$ it follows that $\delta^d S \subseteq \delta^i S$ for all i.

   i) Let $\{(V_\alpha, \Lambda_\alpha)\}$ be a series of G. If for $i = 1, 2, \ldots, r$ the factor $\Lambda_{\alpha_i}/V_{\alpha_i}$ is not soluble, there exist 2^d elements x_{ij} in Λ_{α_i} such that

$$\delta^d < x_{ij} : j = 1, 2, \ldots, 2^d > \nsubseteq V_{\alpha_i}\,.$$

Put

$$H = \langle x_{ij} : i = 1, 2, \ldots, r;\ j = 1, 2, \ldots, 2^d \rangle\,.$$

By 9.29 at most k of the factors in the series $\{(H \cap V_\alpha, H \cap \Lambda_\alpha)\}$ are *not* soluble. By the above remark the soluble factors all have derived length at most d, so it follows that each factor $(H \cap \Lambda_{\alpha_i})/(H \cap V_{\alpha_i})$ is not soluble. Therefore $r \le k$ and i) follows.

   ii) Let $x \in \delta^d (\gamma\, \delta^d)^k\, G,\, {}_k H$. There exists a finite subgroup J of G such that if $K = H \cap J$ then $x \in \delta^d (\gamma\, \delta^d)^k\, J,\, {}_k K$. Since J is finite, K is subnormal in J. Thus there exists a series

$$K = K_0 \lhd K_1 \lhd \cdots \lhd K_s = K$$

in which at most k of the factors are not soluble. If K_{i+1}/K_i is soluble then its derived length is at most d by our initial remark, and $\delta^d K_{i+1} \subseteq K_i$. If K_{i+1}/K_i is not soluble then $\gamma K_{i+1}, K_i \subseteq K_i$. Therefore $\delta^d (\gamma\, \delta^d)^k\, J,\, {}_k K \subseteq K$ and so $x \in K \subseteq H$.

   iii) Since G is simple $\delta^d G = G$. Further if H is any non-trivial subgroup of G then $[G, H]$ is normal in G. Thus $\delta^d (\gamma\, \delta^d)^k\, G,\, {}_k H = G$ for any non-trivial subgroup H of G. Therefore G has no proper serial subgroups by part ii). □

   For linear groups in general parts i) and ii) of 9.30 become false, clearly since free groups have faithful representations of finite degree. About part iii) I do not know.

Question 14. *Is every simple linear group absolutely (strictly?) simple?*

One can prove the following without too much difficulty.

Theorem. *Let k denote the maximal number of non-soluble factors in a series of any finite linear group of degree n, let c denote the maximal number of non-soluble factors in a series of normal subgroups of any finite linear group of degree n and let d denote the maximal derived length of any soluble linear group of degree n.*

Let G be a subgroup of GL (n, F) *and H any ascendant subgroup of G.*

i) *Any generalized series of closed (resp. closed normal) subgroups of G has at most k (resp. c) non-soluble factors.*

ii) $\delta^d (\gamma \, \delta^d)^k \, G, \, _k H \subseteq G \cap \mathscr{A}_F(H).$

iii) *If G is simple then the closure in G of any non-trivial ascendant subgroup of G is G itself.*

iv) $G \cap \mathscr{A}_F(H^0)$ *is subnormal in G.*

9.31. Corollary (Kargapolov [30]). *If G is a periodic linear group having a series with finite factors that contains both {1} and G, then G is soluble-by-finite.*

Proof. Let $\{(V_\alpha, \Lambda_\alpha)\}$ be the given series. By 9.30 this series has at most $k = n \cdot k(n)$ non-soluble factors. Let $\Lambda_{\alpha_i}/V_{\alpha_i}$ be the non-soluble factors of this series for $i = 1, 2, \ldots, r \leq k$, where $\alpha_1 < \alpha_2 < \cdots < \alpha_r$. Put $m_i = (\Lambda_{\alpha_i} : V_{\alpha_i})$. We prove by induction on r that G has a soluble normal subgroup of finite index at most

$$t_r = ((\ldots ((m_1 ! \cdot m_2)! \cdot m_3)! \cdot \ldots)! \cdot m_r)!.$$

If $r = 0$ then since G is locally finite, G is locally soluble and hence soluble. By induction we may assime that V_{α_r} contains a soluble normal subgroup of finite index at most t_{r-1}.

Let H be any finite subgroup of G. By the usual inverse limit argument it suffices to prove that the unique maximal soluble subgroup S of H has finite index at most t_r in H. Now intersecting the members of the given series above α_r with H and using the induction hypothesis on V_{α_r} we obtain a series

$$\{1\} \lhd H_1 \lhd H_2 \lhd \cdots \lhd H_l = H$$

where H_1 is soluble, $(H_2 : H_1) \leq t_{r-1}$, $(H_3 : H_2) \leq m_r$ and H_i/H_{i-1} is soluble for $i = 4, 5, \ldots, l$. By the structure of finite semi-simple groups (e.g. [32] § 61), H/S contains a characteristic subgroup U/S such that U/S is a direct product of non-abelian simple groups and H/S is isomorphic to a subgroup of the automorphism group of U/S. Therefore

$$(H : S) \leq (t_{r-1} \cdot m_r)! = t_r.$$

The corollary is now proved. $\square$

The theory of simple periodic linear groups has yet really to be started. The only known ones are those of Lie type over locally finite fields, and the finite simple groups of course. By groups of Lie type I refer, of course, to the simple groups associated with the simple complex Lie algebras, including the twisted types. That is $PSL(n, F)$, $PSU(n, F)$, orthogonal groups, sympletic groups, Suzuki groups, Ree groups etc., where the field F is suitably restricted. By choosing F to be locally finite and infinite obtain a locally finite infinite simple linear group.

Question 15. *Is every infinite simple periodic linear group isomorphic to one of Lie type?*

Question 16. *If G is an infinite simple periodic linear group, is the centralizer of every element of G infinite?*

We close this topic by using the above machinery to prove a result due to O. H. Kegel; see [31 a] 4.6 for a very different proof. Perhaps one should remark that no locally finite simple group is known (to me) that is not locally finite-simple.

9.32. Corollary (O. H. Kegel). *A simple periodic linear group G is locally finite-simple.*

Proof. By 9.5 the group G is countable, and clearly we may assume that G is an absolutely irreducible subgroup of $GL(n, F)$. Some finite subgroup of G is absolutely irreducible. Thus, by 4.5 of [31 a] there exist finite subgroups G_i, M_i of G satisfying

$$G = \bigcup_{i=1}^{\infty} G_i, \qquad M_i \lhd G_i, \qquad G_i/M_i \text{ simple}$$

$$G_i \cap M_{i+1} = \{1\}, \quad G_i \subseteq G_{i+1}$$

and G_i absolutely irreducible for $i = 1, 2, \dots$.

Put $M = \langle M_i : i = 1, 2, \dots \rangle$ and $N_i = \langle M_j : j > i \rangle$. Then $\bigcap_i N_i = \{1\}$, $N_0 = M$ and N_{i+1} is normal in N_i for each i. By 9.30 only finitely many of the factors N_i/N_{i+1} are not soluble and hence N_j is soluble for some j. Then N_j contains a triangularizable normal subgroup T of finite index (3.6).

It follows that for some $k \geq j$ we have $N_k \subseteq T$ and hence that $M_i \subseteq T$ for all $i > k$. Now G_i is absolutely irreducible and T is triangularizable. Therefore by Clifford's Theorem M_i is diagonalizable for $i > k$ and thus $(G_i : (C_i(M_i)) \leq n!$ for $i > k$ by 1.12. But G_i/M_i is simple and its order tends to infinity with i. Since M_i is abelian for $i > k$ it follows that for some $l > k$ we have $[G_i, M_i] = \{1\}$ for all $i \geq l$. By Schur's Lemma $M_i \subseteq F 1_n$ for $i \geq l$

and hence M_i is central in G. But G is a simple group; therefore $M_i = \{1\}$ for all $i \geq l$ and consequently the sequence $\{G_i\}_{i \geq l}$ is a local system of G consisting of finite simple groups. $\square$

Exercise 9.1. G is a periodic linear group with no non-trivial soluble normal subgroups. Prove that:

i) If the ground-field characteristic is not 2 then G contains a simple subnormal subgroup.

ii) If G contains a simple subnormal subgroup S then S has subnormal depth at most 2 and $(G: N_G(S))$ is finite.

In the theory of abstract infinite groups one frequently encounters linear groups over the field of rational numbers and the field of p-adic numbers. We conclude this chapter with the very useful result that periodic linear groups over these fields are finite. We shall say that a field F satisfies $(*)$ if there exists an integer-valued function $s(n)$ such that for $n = 1, 2, \ldots$ the number of roots of unity lying in an extension field of F of degree $\leq n$ divides $s(n)$.

9.33. Theorem. i) *If F satisfies $(*)$ so does every finite extension field of F.*

ii) *If F satisfies $(*)$ and char $F = 0$ then the periodic subgroups of $\mathrm{GL}(n, F)$ are finite with order at most $\beta(n) \cdot s(n)^n$ where $\beta(n)$ is the function of 9.2.*

iii) *$\mathbb{Q}$ satisfies $(*)$.*

iv) *The field $\mathbb{Q}_p$ of p-adic numbers satisfies $(*)$.*

Proof. i) is trivial.

ii) If G is a periodic subgroup of $\mathrm{GL}(n, F)$ then by 9.4 there exists an abelian normal subgroup A of G of finite index at most $\beta(n)$. By 1.6, A is diagonalizable. If $a \in A$ then the eigenvalues of a all lie in extension fields of F of degree $\leq n$ and so are $s(n)$-th roots of unity. Hence $a^{s(n)} = 1$. But A has rank at most n by 2.2 and therefore $|G| = (G: A)|A| \leq \beta(n) \cdot s(n)^n$.

iii) A primitive m-th root of unity has degree $\phi(m)$ over $\mathbb{Q}$ where $\phi(m)$ is the Euler function (e.g. see [10] 21.13), and $\phi(m) \to \infty$ as $m \to \infty$. Define $s(n)$ to be the product of all $\phi(m)$ for which $\phi(m) \leq n$, $m = 1, 2, 3, \ldots$.

iv) We shall have to use some number theory to prove this part. Any standard text on algebraic number theory should suffice, but we shall refer the reader to [6b].

Let K be any extension of $\mathbb{Q}_p$ of degree n, J the ring of integers of K; k the residue class field of J, f the residue class degree (so $k \cong GF(p^f)$), e the ramification index of the extension, and π an irreducible element of J such that $\pi^e = p$ (see [6b] § 1.5). For $i = 1, 2, \ldots$ put $U_i = 1 + \pi^i J$. U_i is a subgroup of the group of units of J since J is a local ring ([6b] p. 14, Prop. 2). By [6b] p. 19, Prop. 3, $ef = n$.

We shall prove that $s(n)$ can be taken to be $p^{[n/(p-1)]}\prod_{i=1}^{n}(p^i-1)$. Put $t(n)=p^{[n/(p-1)]}(p^n-1)$. Let α be a root of unity in K of order r. It suffices to prove that r divides $t(n)$. Since α is a root of X^r-1 in K, so $\alpha\in J$. Thus $\alpha+\pi J\in k$ and consequently $\alpha^{|k^*|}-1\in\pi J$. Thus $\alpha^{p^f-1}\in U_1$. For each $i\geq 1$ the group U_i/U_{i+1} is isomorphic to the additive group of k via the map

$$1+\pi^i j+U_{i+1}\mapsto j+\pi J$$

and hence has exponent p. Let $t=[e/(p-1)]+1$. Then $\alpha^{p^{t-1}(p^f-1)}\in U_t$. We shall show that U_t is torsion-free. Once we have done this, we shall have shown that r divides $p^{[e/(p-1)]}(p^f-1)$, which divides $t(n)$ since $ef=n$, and the proof will be complete.

$\bigcap_i U_i=\{1\}$, so U_t is residually a p-group. Hence any element of U_t of finite order has order a power of p. Let β be a non-trivial element of U_t. There exists $i\geq t$ such that $\beta\in U_i\smallsetminus U_{i+1}$. Thus $\beta=1+\pi^i j$ where π does not divide j. By the binomial theorem

$$\beta^p=1+p\,\pi^i j+p\,\pi^{2i}h+\pi^{pi} \quad \text{for some } h\in J,$$
$$=1+\pi^{i+e}j+\pi^{2i+e}h+\pi^{pi},$$
$$=1+\pi^{i+e}j+\pi^{i+e+1}l \quad \text{for some } l\in J,$$

since $i\geq t>e/(p-1)$ implies that $p\,i>i+e$. But $j\neq 0$, so $\beta^p\neq 1$. Thus no element of U_t has order p and therefore U_t is torsion-free. $\square$

Exercise 9.2. G is a homomorphic image of a periodic linear group. Prove that the hypercentre of G is equal to the intersection of all the normalizers of the Sylow subgroups of G. (See [64].)

Exercise 9.3. For each prime p construct a soluble periodic linear group of characteristic p with an infinite chief factor and a maximal subgroup of infinite index. What happens when the ground-field characteristic is zero?

Hint: consider $\mathrm{Tr}(2,F)$ for F an infinite locally finite field.

Exercise 9.4. G is a soluble homomorphic image of a periodic linear group and M is a maximal subgroup of G containing no proper normal subgroups of G. Prove that:

a) G has a unique minimal normal subgroup K.

b) K lies in every proper normal subgroup of G.

c) If L/K is a non-trivial normal q-subgroup of G/K, then K is not a q-group. (See [65].)

For further properties of periodic linear groups, particularly properties associated with normal p-complements and maximal subgroups, see the second half of Chapter 12. For soluble periodic linear groups see [65].

10. Rank Restrictions, Varietal Properties and Wreath Products

This chapter is a bit of a hotch-potch. In the first part we give an account of Platonov's proof [43] of the nilpotence of the Frattini subgroup of a finitely generated linear group (4.17). Its main ingredient is a generalization (10.4) of Mal'cev's Theorem (4.2) on the residual finiteness of a finitely generated linear group. This result is, I think, of independent interest and may well have many applications yet to be discovered. It will enable us to give, for example, an elementary proof of another result of Platonov on linear groups of finite rank ([46 b]). We also include in this section some simple structure theorems, taken from [69 c], for linear groups satisfying certain 2-generator solubility conditions.

The latter part of the chapter starts with a brief discussion of two very important results 10.15 and 10.16. The first due to Platonov [43] states that any linear group either is soluble-by-finite or generates the variety of all groups. The second result is very closely associated with the first and is due to Tits [57 a]. It states that any finitely generated linear group either is soluble-by-finite or contains a non-cyclic free subgroup. We then give Merzljakov's solution [36] for linear groups of P. Hall's three fundamental problems for verbal and marginal subgroups. The chapter concludes with the conditions for a wreath product of linear groups to be isomorphic to a linear group. These final results are taken from Vapne's papers [58 a], [58 b] and from [69 d] and [69 e].

Our first objective is to prove the generalization of Mal'cev's Theorem (4.2) due to V. P. Platonov [43]. To do this we need some more commutative ring theory.

Let S be a commutative ring and R a subring of S. Suppose that S is integral over R, that is each element of S is a root of some monic polynomial with coefficients in R. Then if $\mathfrak{p}$ is any prime ideal of R and $\mathfrak{a}$ any ideal of S such that $R \cap \mathfrak{a} \subseteq \mathfrak{p}$, there exists a prime ideal $\mathfrak{q}$ of S containing $\mathfrak{a}$ such that $R \cap \mathfrak{q} = \mathfrak{p}$. Moreover $\mathfrak{q}$ is the only prime ideal of S with this property that contains $\mathfrak{q}$. (See [29] Theorem 44, or [72] Vol. 1, pp. 257–259.)

Thus, for example, if S is also an integral domain and $\mathfrak{a}$ is an ideal of S intersecting R in $\{0\}$, then $\mathfrak{a} = \{0\}$. For $\{0\}$ is prime in R, so there exists a prime ideal q of S containing a such that $R \cap q = \{0\}$. But now $\{0\}$ and q are both prime ideals of S intersecting R in $\{0\}$ and consequently $q = a = \{0\}$.

10.1. Lemma. *Let S be a finitely generated integral domain of characteristic $q > 0$ and $\{m_i: i = 1, 2, ...\}$ a strictly ascending sequence of positive integers such that m_i divides m_{i+1} for each i. Then there exists an integer l depending only on S such that for every finite subset X of S there exists an i and a ring homomorphism ϕ of S into $GF(q^{l m_i})$ such that $|X| = |X\phi|$.*

Proof. Let P denote the prime subfield of S. By Noether's Normalization Lemma ([36e] p. 4, or [72] Chap. VII, Th. 25) there exist elements $x_1, x_2, ..., x_r$ of S algebraically independent over P, such that S is integral over $R = P[x_1, x_2, ..., x_r]$. (Possibly $r = 0$; here set $R = P$.) Since S is finitely generated as a ring it is finitely generated as R-module ([72] Chap. V, Th. 1, or [29] Theorem 17). Hence there exist elements $y_1, y_2, ..., y_d$ of S such that $S = \sum_1^d y_i R$.

Let m be a maximal ideal of R where $R/m \cong GF(q^m)$. By the remarks above there exists a maximal ideal n of S satisfying $m = R \cap n$. Also as $(R + n)/n$-module, S/n can be generated by d elements and so is isomorphic to an extension field of $GF(q^m)$ of degree at most d. Thus if $l = d!$, S/n is isomorphic to a subfield of $GF(q^{l m})$.

Let $\{m_j: j \in J\}$ be the set of maximal ideals of R such that each R/m_j is isomorphic to a subfield of $GF(q^{m_i})$ for at least one i. Since

$$L = \varinjlim GF(q^{m_i})$$

is an infinite field every non-zero polynomial over P in $x_1, ..., x_r$ takes a non-zero value in L. Therefore $\bigcap_J m_j = \{0\}$. For each j in J there exists a maximal ideal n_j of S such that $m_j = R \cap n_j$. By the second remark above $\bigcap_J n_j = \{0\}$. Also for each j in J there exists an i such that S/n_j is isomorphic to a subfield of $GF(q^{l m_i})$.

Finally let X be any finite non-empty subset of S and put

$$z = \prod_{\substack{x, y \in X \\ x \neq y}} (x - y), \quad \text{or} \quad 1 \text{ if } |X| = 1.$$

Since S is an integral domain z is non-zero. There exists j in J such that $z \notin n_j$. Then $x n_j \neq y n_j$ for all x, y in X with $x \neq y$ and so

$$|X| = |\{x n_j: x \in X\}|. \quad \square$$

10.2. Lemma. *Let S be a finitely generated integral domain of characteristic zero and π an infinite set of positive primes. Then there exists an integer l*

depending only on S such that for every finite subset X of S there exists q in π and a ring homomorphism φ of S into GF(q^l) such that $|X|=|X\phi|$.

Proof. By Noether's Normalization Lemma there exists elements $x_1, x_2, \ldots, x_r$ of $S^{\mathbb{Q}}=S\otimes_{\mathbb{Z}}\mathbb{Q}$ algebraically independent over $\mathbb{Q}$ such that $S^{\mathbb{Q}}$ is integral over $\mathbb{Q}[x_1, \ldots, x_r]$; again possibly $r=0$. Clearly we may choose the x_i all to lie in S. Put $R=\mathbb{Z}[x_1, \ldots, x_r]$. There exist elements $z_1, \ldots, z_s$ of S such that $S=R[z_1, \ldots, z_s]$. Each z_i is integral over $R^{\mathbb{Q}}$, so there exist elements $\alpha_{1i}, \ldots, \alpha_{n_i i}$ of R and a non-zero integer t_i such that

$$t_i z_i^{n_i} + \alpha_{1i} z_i^{n_i-1} + \cdots + \alpha_{n_i i} = 0.$$

Let $I=\mathbb{Z}[t_1^{-1}, \ldots, t_s^{-1}]\subseteq\mathbb{Q}$. Then $S^I=S\otimes_{\mathbb{Z}}I$ is a finitely generated integral domain that is integral over R^I.

Let M denote the set of maximal ideals $\mathfrak{m}$ of R^I such that $R^I/\mathfrak{m}$ is isomorphic to some GF(q) for $q\in\pi$. If $f(x_1, \ldots, x_r)$ is a non-zero element of R^I (identified with $I[x_1, \ldots, x_r]$) then there exist elements $a_1, \ldots, a_r$ of I such that $f(a_1, \ldots, a_r)\neq 0$. Clearly $\bigcap_{q\in\pi} qI=\{0\}$, so there exists some q in π with $f(a_1, \ldots, a_r)\notin qI$. It follows that the kernel of the homomorphism of R^I defined by $x_i\mapsto a_i \pmod q$ for $i=1, 2, \ldots, r$, lies in M and does not contain f. Therefore $\bigcap_{\mathfrak{m}\in M}\mathfrak{m}=\{0\}$.

S^I is finitely generated as R^I-module, say $S^I=\sum_{i=1}^{d}y_i R^I$. Let $l=d!$. If $\mathfrak{m}\in M$ there exists a prime ideal $\mathfrak{n}$ of S^I satisfying $\mathfrak{m}=R^I\cap\mathfrak{n}$. Thus $S^I/\mathfrak{n}$ is a finitely generated integral extension domain of $(R^I+\mathfrak{n})/\mathfrak{n}$ that can be generated as $(R^I+\mathfrak{n})/\mathfrak{n}$-module by d elements, and this latter ring is isomorphic to GF(q) for some $q\in\pi$. Thus $S^I/\mathfrak{n}$ is isomorphic to a subfield of GF(q^l). Finally,

$$R\cap\bigcap_{\mathfrak{m}\in M}\mathfrak{n}=\{0\}, \quad\text{so}\quad \bigcap_{\mathfrak{m}\in M}\mathfrak{n}=\{0\}$$

and the proof of the lemma is completed exactly as in the final paragraph of the proof of 10.1. □

We need several results from number theory. In the results below numbered 10.3.x, if n is an integer and p is a prime, then $(n)_p$ denotes the largest power of p to divide n.

10.3.1. *If p is a (positive) prime and r and m are positive integers such that p is prime to m then*

$$1+X^{r(p-1)}+\cdots+X^{(m-1)r(p-1)}$$

has no root in GF(p).

For let $r=r_1\,p^t$ where p does not divide r_1. Then

$$\left(\sum_{i=0}^{m-1} X^{ir(p-1)}\right)(1-X^{r_1(p-1)})^{p^t}\equiv(1-X^{mr_1(p-1)})^{p^t}\ (\mathrm{mod}\ p).$$

Clearly $1-X^{r_1(p-1)}$ and $1-X^{mr_1(p-1)}$ are separable over $\mathrm{GF}(p)$ and consequently if a is any non-zero element of $\mathrm{GF}(p)$, a is a root of multiplicity p^t of both $(1-X^{r_1(p-1)})^{p^t}$ and $(1-X^{mr_1(p-1)})^{p^t}$. Therefore a is not a root of $\sum_{i=0}^{m-1} X^{ir(p-1)}$. Trivially neither is 0. ☐

10.3.2. *Let* $p_1,p_2,\ldots,p_r,q$ *be primes,* l *and* n *positive integers and put*

$$c=\prod_{j=1}^{r}(p_j-1).$$

Then there exist non-negative integers $n_1,n_2,\ldots,n_r$ *such that for every positive integer m prime to $p_1\,p_2\,\ldots\,p_r$ and for each $j=1,2,\ldots,r$, the largest power of p_j to divide $\delta_m=\prod_{t=1}^{n}(q^{lmct}-1)$ is $p_j^{n_j}$.*

For $(q^{lmct}-1)=(q^{lct}-1)(1+q^{lct}+\cdots+q^{(m-1)lct})$. By 10.3.1, p_j does not divide $\sum_{i=0}^{m-1} q^{ilct}$. Hence

$$(\delta_m)_{p_j}=\prod_{t=1}^{n}(q^{lct}-1)_{p_j}$$

which is independent of m. ☐

10.3.3. *Let* p *be a prime and* m,a *and* b *positive integers with* a *prime to* p. *Put* $q=1+a\,p^m+b\,p^{m+1}$.[7] *If t is any positive integer with $t\le m$, then p^{2m} does not divide q^t-1.*

For

$$q^t-1=\sum_{i=1}^{t}{}^tC_i\cdot p^{mi}(a+b\,p)^i$$

$$\equiv t\,p^m(a+b\,p)\quad\text{modulo }p^{2m}.$$

Now $t\le m\le p^m$ and a is prime to p, so the result follows. ☐

10.3.4. *For any set of primes* $p_1,p_2,\ldots,p_r$ *and positive integers l and n there exists an infinite set π of primes and a positive integer h such that p_j^h does not divide*

$$\delta(q)=\prod_{t=1}^{n}(q^{lt}-1)$$

for each q in π and each $j=1,2,\ldots,r$.

---

[7] This definition of q, due to J.D. Dixon, simplifies Platonov's original argument. Platonov takes $q=1+a\,p+b\,p^2$, but with this choice of q I run into difficulties if $p=2$. With $q=1+a\,p^2+b\,p^3$ Platonov's argument goes through without trouble.

Put $m = l\,n$, $h = 2\,l\,n^2$ and let π denote the set of all primes q of the form

$$q = 1 + (p_1 p_2 \cdots p_r)^m + s(p_1 p_2 \cdots p_r)^{m+1}$$

where s is a positive integer. By Dirichlet's Theorem (on primes in an arithmetic progression, see [51a] p. 122, Corollaire, for example) π is infinite. By 10.3.3 we have that p_j^{2m} does not divide $q^{l^t} - 1$ for any $t \le n$ and hence for $j = 1, 2, \ldots, r$ it follows that p_j^{2nm} does not divide $\delta(q)$.

10.4. Theorem (Platonov [43]). *Let R be a finitely generated integral domain, $G = \mathrm{GL}(n, R)$ and $p_1, p_2, \ldots, p_r$ a finite collection of positive primes.*

i) *If char $R = q > 0$ and $q \neq p_j$ for $j = 1, 2, \ldots, r$ then there exist integers $n_1, n_2, \ldots, n_r$ and a set $\{N_i : i \in I\}$ of normal subgroups of G such that*

a) *G/N_i is isomorphic to a finite linear group of degree n over a field of characteristic q,*

b) *for $j = 1, 2, \ldots, r$ the Sylow p_j-subgroups of G/N_i have order at most $p_j^{n_j}$ and*

c) *for every finite subset X of G there exists an i in I such that*

$$|X| = |\{x N_i : x \in X\}|.$$

ii) *If char $R = 0$ there exist integers $n_1, n_2, \ldots, n_r$ and a set $\{N_i : i \in I\}$ of normal subgroups of G such that*

a) *G/N_i is isomorphic to a linear group of degree n over a finite field of characteristic not equal to any of the p_j,*

b) *for $j = 1, 2, \ldots, r$ the Sylow p_j-subgroups of G/N_i have order at most $p_j^{n_j}$ and*

c) *for every finite subset X of G there exists an i in I such that*

$$|X| = |\{x N_i : x \in X\}|.$$

Proof. i) For $k = 1, 2, \ldots$ let $m_k = q^k \prod_{j=1}^{r} (p_j - 1)$. By 10.1 there exists an integer l and a set $\{N_i : i \in I\}$ of normal subgroups of G such that for each i in I there exists a positive integer k such that G/N_i is isomorphic to a subgroup of $\mathrm{GL}(n, q^{l m_k})$ and for every finite subset X of G there exists an i in I such that $|X| = |\{x N_i : x \in X\}|$. Now

$$|\mathrm{GL}(n, q^{l m_k})| = q^{\frac{1}{2} l m_k n(n-1)} \prod_{t=1}^{n} (q^{l m_k t} - 1)$$

(e.g. [26a] II.6.2). By 10.3.2 there exist positive integers n_j such that $|\mathrm{GL}(n, q^{l m_k})|_{p_j} = p_j^{n_j}$ for $j = 1, 2, \ldots, r$ and $k = 1, 2, \ldots$.

ii) Now let l denote the l of 10.2 with R playing the role of S. By 10.3.4 there exists an infinite set π of positive primes and positive integers $n_1, n_2, \ldots, n_r$ such that $p_j \notin \pi$ for each j and for every q in π and $j = 1, 2, \ldots, r$

the largest power of p_j to divide $|\mathrm{GL}(n, q^l)|$ is at most $p_j^{n_j}$. The theorem now follows from 10.2. ☐

10.5. Corollary (Platonov [43]). *Let R be a finitely generated integral domain, G a subgroup of $\mathrm{GL}(n, R)$ and $p_1, p_2, \ldots, p_r$ those prime divisors of $n!$ not equal to the characteristic of R. If for each of the normal subgroups N_i of $\mathrm{GL}(n, R)$ constructed in 10.4 the group $G/(G \cap N_i)$ is nilpotent, then G is nilpotent.*

This is essentially Platonov's proof of the nilpotence of the Frattini subgroup of a finitely generated linear group. Specifically 4.16 is a special case of 10.5 and 4.17 follows as before.

Proof. $G_i = G/(G \cap N_i)$ is isomorphic to a finite nilpotent linear group of degree n over a field of positive characteristic q_i say. The Sylow q_i-subgroup of G_i is nilpotent of class at most $n - 1$ by 2.6, and by 3.12 (and 1.6) the Hall q_i'-subgroup of G_i is an $n!$-group modulo its centre. Thus G_i is nilpotent of class at most $c = n + \prod_{j=1}^{n} p_j^{n_j}$, where the n_j are the n_j of 10.4. But $\bigcap_i N_i = \{1\}$, and so G is also nilpotent of class at most c. ☐

10.6. Corollary *Let R be a finitely generated integral domain of characteristic zero and G a subgroup of $\mathrm{GL}(n, R)$. If there exists an integer t such that $(H : H'H^n) \leq t$ for every subgroup H of finite index in G then G is soluble-by-finite.*

Proof. If $\pi = \{p_1, p_2, \ldots, p_r\}$ is the set of primes not exceeding n, there exist integers $n_1, n_2, \ldots, n_r$ and a set $\{N_i : i \in I\}$ of normal subgroups of $\mathrm{GL}(n, R)$ satisfying 10.4. Denote by $\{M_k : k \in K\}$ the set of the intersections of the members of the finite subsets of $\{G \cap N_i : i \in I\}$. For $j = 1, 2, \ldots, r$ the Sylow p_j-subgroups of G/M_k have exponent at most $p_j^{n_j}$ and class at most n_j. Therefore their order is at most $p_j^{t n_j^2}$. Hence there exists k in K such that the π-part of $(G : M_k)$ is maximal. But then $M_k/M_k \cap N_i$ is a π'-group for each i in I. Since it is isomorphic to a linear group of degree n it is soluble (9.19) of derived length at most $f(n)$ by 3.7. Consequently M_k is soluble. ☐

Exercise 10.1. Use the Feit-Thompson Theorem to prove that if R is a finitely generated integral domain and G is a subgroup of $\mathrm{GL}(n, R)$ for which there exists an integer t such that $(H : H^2) \leq t$ for every subgroup H of G of finite index, then G is soluble-by-finite.

10.7. Lemma. *If $\mathfrak{B}$ is a variety (of groups) and G is a $\mathfrak{B}$-subgroup of $\mathrm{GL}(n, F)$ then $\mathscr{A}_F(G)$ is also a $\mathfrak{B}$-group.*

In 5.11 we have already seen special cases of this lemma.

Proof. Let G be any subgroup of $\mathrm{GL}(n, F)$, V any set of words and put $\bar{G} = \mathscr{A}_F(G)$ and $\overline{V(G)} = \mathscr{A}_F(V(G))$. It suffices to prove that $V(\bar{G}) \subseteq \overline{V(G)}$ since if we take V to be the set of laws of $\mathfrak{B}$ then $V(G) = \{1\}$ will imply that $V(\bar{G}) = \{1\}$. Here if X is any group

$$V(X) = \langle w(g_1, \ldots, g_r): w \in V, g_1, \ldots, g_r \in X \rangle.$$

Let $w = w(x_1, \ldots, x_r) \in V$. Suppose that for any $\bar{g}_1, \bar{g}_2, \ldots, \bar{g}_{i-1}$ in $\bar{G}$ and for any $g_i, g_{i+1}, \ldots, g_r$ in G we have $w(\bar{g}_1, \ldots, \overline{g_{i-1}}, g_i, \ldots, g_r) \in \overline{V(G)}$. The mapping

$$\phi: x \mapsto w(\bar{g}_1, \ldots, \overline{g_{i-1}}, x, g_{i+1}, \ldots, g_r)$$

is a continuous mapping of $\mathrm{GL}(n, F)$ mapping G into the closed set $\overline{V(G)}$ by 5.1. Thus $\bar{G}$ lies in the inverse image of $\overline{V(G)}$ under ϕ and consequently $w(\bar{g}_1, \ldots, \bar{g}_i, g_{i+1}, \ldots, g_r) \in \overline{V(G)}$ for all $\bar{g}_1, \ldots, \bar{g}_i$ in $\bar{G}$ and $g_{i+1}, \ldots, g_r$ in G.

A simple induction shows that $w(\bar{G}) \subseteq \overline{V(G)}$ for every w in V and so $V(\bar{G}) = \langle w(\bar{G}): w \in V \rangle \subseteq \overline{V(G)}$. □

10.8. Lemma. *Let $\mathfrak{B}$ be a variety, or the class of soluble groups, or the class of nilpotent groups, and let G be a subgroup of $\mathrm{GL}(n, F)$ such that every finitely generated subgroup of G is a finite extension of a $\mathfrak{B}$-group. Then G is an extension of a $\mathfrak{B}$-group by a periodic linear group over F.*

It can easily be proved (cf. the proof of [69c] Lemma 1) using the dimension of a linear group (see Chapter 14) that 10.8 holds for any class $\mathfrak{B}$ that is the union of varieties.

Proof. Suppose first that $\mathfrak{B}$ is a variety. Let H be any finitely generated subgroup of G. H contains a normal $\mathfrak{B}$-subgroup V of finite index. Then $H \cap \mathscr{A}_F(V)$ is a closed normal $\mathfrak{B}$-subgroup of H of finite index (5.9, 10.7) and in particular contains H^0. Let S_H denote the set of all closed normal $\mathfrak{B}$-subgroups of H of finite index; S_H is a finite non-empty set.

If K is any finitly generated subgroup of G containing H the map $\lambda_H^K: T \mapsto H \cap T$ maps S_K into S_H. It is easily seen that

$$(S_H, \lambda_H^K: H \subseteq K, \text{ finitely generated subgroups of } G)$$

is an inverse system of non-empty finite sets, so its inverse limit is not empty ([31a] I.K.I). Let $(L_H) \in \varprojlim S_H$ where $L_H \in S_H$ and put $L = \bigcup_H L_H$.

L is a normal subgroup of G and since $(H:L_H)$ is finite for every H, G/H is periodic. Also L is locally a $\mathfrak{B}$-group and hence is a $\mathfrak{B}$-group. Therefore $\bar{L} = G \cap \mathscr{A}_F(L)$ is a closed normal $\mathfrak{B}$-subgroup of G and thus $G/\bar{L}$ is isomorphic to a periodic linear group over F by 6.4.

Every soluble subgroup of $\mathrm{GL}(n, F)$ has derived length at most $m = f(n)$ by 3.7. Thus in the case of soluble groups we may apply the first part to the variety of soluble groups of derived length at most m. Now let

$\mathfrak{B}$ be the class of nilpotent groups. A nilpotent subgroup N of $GL(n, F)$ contains a triangularizable normal subgroup T of finite index. By 4.13, T is nilpotent class at most $c = \frac{1}{2}n(n-1)+1$. (By Exercise 1.3 and 7.11, T is in fact nilpotent of class at most n.) Hence every finitely generated subgroup of G contains a normal subgroup of finite index that lies in the variety of nilpotent groups of class at most c. The result again follows from the first case. ☐

A group G is said to have *Prüfer rank* r if every finitely generated subgroup of G can be generated by r elements, and if r is the least integer with this property. (This is a much stronger condition, although equivalent for abelian groups, than G having finite rank r, which merely requires that every finitely generated subgroup of G is contained in an r-generator subgroup of G and r minimal subject to that.)

10.9. Theorem (Platonov [45b]). *Let G be a subgroup of $GL(n, F)$ of finite Prüfer rank r. Then G is soluble-by-finite, and if char $F = q > 0$ then G contains an abelian normal subgroup of finite index bounded in terms of r, n and q.*

It is an immediate corollary of 10.9 that any linear group of finite Prüfer rank is countable. This had previously been pointed out by Mal'cev (1948). If char $F = 0$ in 10.9, G need not be abelian-by-finite since every polycyclic group is isomorphic to a linear group over $\mathbb{Z}$.

Proof. If char $F = 0$ then every finitely generated subgroup of G is soluble-by-finite by 10.6 and so G contains a closed soluble normal subgroup S such that G/S is periodic (10.8). G/S is isomorphic to a linear group over F and therefore G/S is abelian-by-finite by 9.4. This completes the char $F = 0$ case.

Now suppose that char $F = q > 0$ and let H be any finitely generated subgroup of G. Then H is residually finite linear of degree n over a field of characteristic q by 4.2. But a Sylow q-subgroup of one of these finite images of H has Prüfer rank r and thus its order is at most q^{nr}. By the Brauer-Feit Theorem (9.6) there exists a function f of three variable such that H residually contains an abelian normal subgroup of finite index at most $f(nr, n, q)$. H can be generated by r elements and hence contains at most $(f(nr, n, q)!)^r$ normal subgroups of index at most $f(nr, n, q)$. The intersection of these subgroups is residually abelian and hence abelian. The usual inverse limit argument completes the proof. ☐

Even if char $F = 0$ in 10.9 there exists a bound on the index of the maximal soluble normal subgroup of G, but this bound is independent of the rank and is a phenomenon of soluble-by-finite groups. The proof of this result that we give uses quite a lot about algebraic groups. We include it, although we shall never use it, since the basic step in the proof namely 10.10, seems to us to be of independent interest.

Let $\{1\} = N_0 \subseteq \cdots \subseteq N_r \subseteq G$ be a series of normal subgroups of the group G such that $(G:N_r)$ is finite and such that for each i, N_i/N_{i-1} is either locally finite, or torsion-free radicable and abelian. Then G contains a finite subgroup H such that $G = HN_r$. For by induction on r there exists a subgroup K of G such that K/N_1 is finite and $G = KN_r$. If N_1 is locally finite then so is K and K contains a finite subgroup H satisfying $K = HN_1$. Clearly $G = HN_r$.

Suppose that N_1 is torsion-free, radicable and abelian. Then any of the standard proofs of Schur's Splitting Theorem guarantees the existence of a subgroup H of K such that $K = HN_1$ and $H \cap N_1 = \{1\}$; for example, if $t_1, t_2, \ldots, t_s$ is any transversal of N_1 to K and $g \in K$ then for each i, $t_i g = t_{i\sigma} a_i$ for some $a_i \in N_1$ and $\sigma \in S_s$; put $H = \{g \in K : a_1 a_2 \ldots a_s = 1\}$. Clearly H is finite and $G = HN_r$. ▯

10.10. Lemma (Platonov [44]). *If F is algebraically closed and G is a closed subgroup of* $\mathrm{GL}(n, F)$ *then* $G = HG^0$ *for some finite subgroup H of G.*

Proof. The maximal closed connected diagonalizable subgroups T of G^0 are all conjugate (see 14.23 or [3a] 11.3). Hence if $N = N_G(T)$ then $G = NG^0$. By 1.12, N^0 centralizes T and therefore N^0 is nilpotent (see p. 217 or [3a] 11.7, it is of course just a Cartan subgroup of G^0). Therefore $N^0 = T \times U$ for some closed connected unipotent subgroup U of G ([3a] 10.6; it also follows from 7.3 and 7.11). Since F is algebraically closed T is radicable (14.21 or [3a] §8.5, Prop.). Further if char $F = 0$, U is radicable (this is very easy, use 6.4 or see [3a] 10.9) and so its upper central factors are radicable ([47] 6.41 or [24a] 4.8). If char F is positive then U is locally finite. By the remarks above there exists a finite subgroup H of N such that $N = HN^0$. Clearly $G = HG^0$. ▯

10.11. Corollary (Platonov [45a]). *If char $F = 0$ and G is a soluble-by-finite subgroup of* $\mathrm{GL}(n, F)$ *then G contains a soluble normal subgroup of finite index at most $\beta(n)$ where β is any function satisfying Jordan's Theorem* (9.2).

Proof. By 5.11 or 10.7, $\mathscr{A}_{\bar F}(G)^0 = \mathscr{A}_{\bar F}(G^0)$ is soluble. Also by 10.10 there exists a finite subgroup H of $\mathrm{GL}(n, \bar F)$ such that $H \cdot \mathscr{A}_{\bar F}(G)^0 = \mathscr{A}_{\bar F}(G)$. By Jordan's Theorem H contains an abelian normal subgroup of index at most $\beta(n)$ and 10.11 follows. ▯

Exercise 10.2. Use the techniques of Chapter 3 to prove that there exists a function $\gamma(n)$ of n only such that a soluble-by-finite subgroup of $\mathrm{GL}(n, F)$ where char $F = 0$ has a soluble normal subgroup of finite index at most $\gamma(n)$.

We now consider 2-generator conditions on linear groups.

10.12. Lemma. *Let R be a finitely generated integral domain and G a subgroup of $GL(n, R)$ such that every 2-generator subgroup of G is soluble-by-finite. Then G is soluble-by-finite.*

Proof. G contains a normal subgroup H of finite index such that H is residually nilpotent (4.7). It follows from Learner's Lemma ([47] 6.25) that a residually nilpotent group is soluble if it is soluble-by-finite. Hence every 2-generator subgroup of H is soluble. It follows from J.G. Thompson's classification of the (finite) minimal simple groups that every minimal simple group can be generated by 2-elements. Therefore every finite homomorphic image of H is soluble and consequently H is soluble (4.3). □

10.13. Theorem. *Let G be a subgroup of $GL(n, F)$ such that every 2-generator subgroup of G is soluble-by-finite. Then G is soluble-by-periodic, and if char $F = 0$, G is actually soluble-by-finite.*

In [45a] Platonov gives the characteristic zero case of 10.13. His proof avoids the classification of the minimal simple groups, but uses instead something approaching the classification of semisimple algebraic groups. We shall indicate Platonov's proof a little later. The restriction of the characteristic of F to conclude that G is soluble-by-finite is clearly necessary since there exist infinite simple periodic linear groups.

Proof. By 10.12 every finitely generated subgroup of G is soluble-by-finite. Hence G contains a closed soluble normal subgroup S such that G/S is periodic (10.8). By 9.4, G/S is abelian-by-finite if char $F = 0$ and the theorem follows. □

10.14. Theorem ([69c]). *If G is a subgroup of $GL(n, F)$ such that every 2-generator subgroup of G is nilpotent-by-finite, then G is nilpotent-by-periodic. If $G \subseteq GL(n, R)$ for some finitely generated subring R of F then G is nilpotent-by-finite.*

In 10.14 the group G need not be nilpoent-by-finite (though it will for example if $F = \mathbb{Q}$ since periodic linear groups over $\mathbb{Q}$ are necessarily finite, see 9.33). A counter example to this provided by the group
$$\begin{pmatrix} 1 & 0 \\ 1 & 1 \end{pmatrix}, \begin{pmatrix} \zeta & 0 \\ 0 & 1 \end{pmatrix} : \zeta \text{ a root of unity in } \mathbb{C} > \subseteq GL(2, \mathbb{C}).$$

The proof of 10.14 below is due to D. Segal. Two other proofs are given in [69c] and a fourth proof we indicate below by exercises.

Proof. By 10.13, G contains a soluble normal subgroup S such that G/S is periodic. By the Lie-Kolchin Theorem (5.8) S^0 is triangularizable, and also S^0 is normal in G. Write U for the maximal unipotent subgroup of S^0. Then U is normal in G and G/U is abelian-by-periodic.

Let V denote the n-row vector space over F regarded as G-module in the usual way, and put

$$W = C_V(U) = \{v \in V: \ vx = v \text{ for every } x \in U\}.$$

We induct on n. W is a non-trivial FG-submodule of V, so by induction and the above paragraph we may assume that G/A is nilpotent-by-periodic where

$$A = C_U(V/W) = \{x \in U: \ \bar{v}x = \bar{v} \text{ for all } v \in V/W\}.$$

As in the proof of 4.12 there exists a G-monomorphism ϕ of A into $H = \mathrm{Hom}_F(V/W, W)$. Let $a \in A$ and $g \in G$. Then by hypothesis $\langle a, g \rangle$ is nilpotent-by-finite. Since $A \cap \langle a, g \rangle$ is an abelian normal subgroup of $\langle a, g \rangle$ it follows that $\langle a, g^{r(a,g)} \rangle$ is nilpotent for some positive integer $r(a, g)$. Let $r(g)$ denote the product of the $r(a, g)$ taken over some fixed basis of the F-subspace B of H spanned by $A\phi$. Then $g^{r(g)}$ acts unipotently on B.

By the Lie-Kolchin Theorem again S contains a normal subgroup T of G with finite index in S such that T acts triangularizably on B. The set T_1 of elements of T that act unipotently on B is a normal subgroup of G. For each $g \in T$ we have that $g^{r(g)}$ lies in T_1 and hence G/T_1 is periodic. By 1.21, A lies in some finite term of the upper central series of T_1, and G/A is nilpotent-by-periodic. Thus G is nilpotent-by-periodic.

Suppose now that $G \subseteq \mathrm{GL}(n, R)$ where R is a finitely generated integral domain. By 10.12 the group G^0 is soluble. Thus 4.10 implies that $G^0/(G^0)_u$ is a finitely generated abelian group. Hence G contains a nilpotent normal subgroup N such that G/N is periodic and a finite extension of a finitely generated abelian group. Clearly G/N is a finite group. $\square$

Although we have used 10.12 in the proof of 10.14, this latter result does not really depend on Thompson's classification of the minimal simple groups. For in this case we have only to conclude that *a finite group G is soluble if for some prime p every 2-generator subgroup of G is an extension of a nilpotent group by a p-group* (in 10.12 we can take H to be residually a finite p-group for some prime p). Let P be any p-subgroup of G, x an element of P and y a p'-element of $N_G(P)$. Then $\langle x, y \rangle/\eta_1(\langle x, y \rangle)$ is a p-group and $x \in P \cap \langle x, y \rangle$ which is normal in $\langle x, y \rangle$. Therefore $\langle x, y \rangle$ is nilpotent and hence $N_G(P)/C_G(P)$ is a p-group. By Frobenius Theorem ([21] Theorem 14.4.7 or [26a] IV.5.8) G contains a normal p-complement H say. If h and k are any two elements of H of coprime order then $\langle h, k \rangle$ is nilpotent and $[h, k] = 1$. Thus H is nilpotent and G is soluble. $\square$

Exercise 10.3. If G is a linear group each of whose 2-generator subgroups is soluble (resp. nilpotent) prove that G is soluble (resp. hypercentral).

This exercise is an easy consequence of 4.3, 4.16 (or 10.5), 8.2i) and the finite case. The second part also follows from 8.15. It is also true that a linear group each of whose 2-generator subgroups is supersoluble has to be hypercyclic; this will be proved in the next chapter, see 11.23. The supersoluble analogue of 10.14 is a triviality since every supersoluble group is nilpotent-by-finite.

Exercise 10.4. Use Exercises 6.2 and 10.3 to give another proof of 10.14.

Hint: reduce to the case where G is finitely generated by 10.8. Your solution in the characteristic zero case should be very short. In the positive characteristic case use 10.12.

We now consider varietal properties of linear groups. The following two results are the most significant to date. The first can be obtained as a simple corollary of Chevalley's classification of the semisimple algebraic groups over an algebraically closed field [8], but see Chapter 14 for a summary.

10.15. Theorem (Platonov [45]). *A linear group is either soluble-by-finite or generates the variety of all groups.*

If G is a linear group over a field of characteristic zero, whose 2-generator subgroups are soluble-by-finite then by 10.11 the 2-generator subgroups of G generate a variety that is not the variety of all groups. Moreover this variety contains every 2-generator group in the variety generated by G, so G does not generate the variety of all groups and consequently G is soluble-by-finite by 10.15. This is Platonov's approach to 10.13. 10.12 may also be proved in this way.

Proof. Let G be a subgroup $GL(n, F)$ that is not soluble-by-finite. We prove that G generates the variety of all groups. Clearly we may assume that F is as large as we need, so suppose that F is algebraically closed of transcendence degree at least two over its prime field. By 10.7, G and $\bar{G} = \mathcal{A}_F(G)$ generate the same variety and $\bar{G}$ is also not soluble-by-finite. Hence by [8] 23.02, Proposition, $\bar{G}^0/S$ contains a subgroup isomorphic to $SL(2, F)$, where S is the maximal soluble connected normal subgroup of $\bar{G}$. But by 2.8 and Exercise 2.2 (or by the proof of 2.13) $SL(2, F)$ contains non-cyclic free subgroups. Therefore $\bar{G}$, and hence G also, generates the variety of all groups. $\square$

We shall indicate another proof of 10.15 in just a moment. The following theorem of Tits is of major importance.

10.16. Theorem (Tits [57a]). *A finitely generated linear group either is soluble-by-finite or contains a non-cyclic free subgroup.*

10.17. Corollary (Tits [57a]). *If G is a linear group over a field of characteristic $p \geq 0$ that does not contain any non-cyclic free subgroups, then G*

contains a soluble normal subgroup S such that G/S is finite if $p=0$ and locally finite if $p>0$.

This corollary follows immediately from 10.16, 10.13 and 4.9. ☐

A second proof of 10.15 may be based on 10.17 as follows. Suppose that G is a subgroup of $GL(n, F)$ that is not soluble-by-finite and does not generate the variety of all groups. We can clearly assume that F is algebraically closed of positive transcendence degree. $\bar{G} = \mathscr{A}_F(G)$ also is not soluble-by-finite and does not contain a non-cyclic free subgroup by 10.7. Consequently 10.17, 6.4 and 14.12 imply that for some integer m, $GL(m, F)$ contains an infinite periodic closed subgroup S that is not soluble-by-finite. Now no such subgroup of $GL(m, F)$ exists by 11.5(2) of [3a] (since a torus of S is isomorphic to a direct product of copies of the non-periodic group F^* and thus is trivial) and this contradiction completes a proof of 10.15. ☐

The following corollary is a consequence of 10.16 and well-known results of Hirsch characterizing soluble groups with the maximal condition on subgroups. It is a companion piece to 9.8.

10.18. Corollary (Tits [57a]). *A linear group satisfying the maximal condition on subgroups is polycyclic-by-finite.* ☐

Conversely every polycyclic-by-finite group satisfies the maximal condition on subgroups and by 2.5 and 2.3 has a faithful representation of finite degree over the integers.

We give only an outline of Tits' proof of 10.16. We do this in approximately the reverse order that it appears in Tits' paper [57a]. On the whole Tits prefers to work in the projective rather than in the general linear group. Since the kernel of $\phi: GL(n, F) \to PGL(n, F)$ is abelian and the elements x and y of $GL(n, F)$ generate a free group of rank 2 if and only if $x\phi$ and $y\phi$ do, this does not materially affect the problem one way or the other. In this sketch we keep to our usual practice and work in $GL(n, F)$.

Let G be a finitely generated subgroup of $SL(n, F)$ that is not soluble-by-finite. (Clearly there is no loss of generality in taking G to be special linear, and there is also no generality to be gained by replacing the finite generation of G by the finite generation of the subring of F generated by the entries of the elements of G, see the proof of 10.12.) We may assume that F is a finitely generated field, and that G is connected.

The first step is to embed F into an evaluated field in a useful way. To do this we wish to have in G a d-element g of infinite order. If $\operatorname{char} F > 0$ then the existence of such an element g is very easy and follows from 4.8 and the fact that G is finitely generated and infinite. If $\operatorname{char} F = 0$ things are not quite so straightforward. However it follows fairly easily

146

from [3a] 12.1b and 13.17 Corollary 2c that such an element g does exist if $\mathscr{A}_F(G)$ is infinite and contains no non-trivial soluble connected normal subgroups, that is if $\mathscr{A}_F(G)$ is semisimple, and the reduction of our problem to the case when the latter condition is satisfied is very easy (use 6.4). This element g has an eigenvalue λ that is not a root of unity. By Lemma 4.1 of [57a] there exists a locally compact evaluated field F_ω with valuation ω containing F as a subfield and such that $\omega(\lambda) \neq 1$.

Denote the n-row vector space over F_ω by V_ω. Let x be an element of $GL(n, F_\omega)$ and $f(T) = \prod_{i=1}^{n} (T - \mu_i)$ its characteristic polynomial, where the μ_i all lie in some finite extension field F_x of F_ω. Since F_ω is locally compact ω extends uniquely to a valuation (still denoted by ω) of F_x. Set

$$I = \{i: \ \omega(\mu_i) = \max\{\omega(\mu_j): \ 1 \leq j \leq n\}\}$$

and put

$$f_1(T) = \prod_{i \in I} (T - \mu_i), \qquad f_2(T) = \prod_{i \notin I} (T - \mu_i).$$

Then $f_1(T)$ and $f_2(T)$ both lie in $F_\omega[T]$. Define $A(x)$ to be the kernel in V_ω of $f_1(x)$ and $A'(x)$ to be the kernel in V_ω of $f_2(x)$. It follows that $V_\omega = A(x) \oplus A'(x)$. There is an alternative definition of $A(x)$ and $A'(x)$. $A(x)$ is the intersection of V_ω with the subspace of the n-row vector space over F_x spanned by the eigenspaces of x_d corresponding to eigenvalues μ of x_d for which $\omega(\mu)$ is maximal and $A'(x)$ is the analogous subspace of V_ω obtained from the μ for which $\omega(\mu)$ is not maximal.

The free subgroups of G arise from 3.12 of [57a]. The following is only a special case of this proposition. Let x and y be d-elements of $GL(n, F_\omega)$ such that $A(x)$, $A(x^{-1})$, $A(y)$, $A(y^{-1})$ are one-dimensional subspaces and such that

$$\left(A(x) \cup A(x^{-1})\right) \cap \left(A'(y) \cup A'(y^{-1})\right) = \{0\}$$

and

$$\left(A(y) \cup A(y^{-1})\right) \cap \left(A'(x) \cup A'(x^{-1})\right) = \{0\}. \tag{$*$}$$

Then there exists a positive integer m such that x^m and y^m freely generate $\langle x^m, y^m \rangle$. To illustrate the point of this proposition we state the following special case (which the reader should compare with Exercise 2.2 on p. 31). If x and y are d-elements of $SL(2, \mathbb{C})$ such that no eigenvalue of x or y lies on the unit circle and such that x and y have no common eigenvector, then there exists an integer m for which $\langle x^m, y^m \rangle$ is a free group of rank 2.

The main step in proving the existence in G of elements x and y satisfying the above is [57a] Proposition 3.11, which states the following.

Let G be an irreducible connected (in the Zariski topology) subgroup of $GL(n, F_\omega)$ such that G contains at least one d-element g for which $\dim A(g) = 1$. Then

$$X = \{x \in G: \dim A(x) = \dim A(x^{-1}) = 1\}$$

is (Zariski) dense in G.

We return to the group G of the second paragraph containing the d-element g with an eigenvalue λ for which $\omega(\lambda) \neq 1$. Let $d = \dim A(g)$. To apply the proposition of the preceeding paragraph we need $d = 1$. We have to modify the representation of G somewhat to achieve this. Denote by ρ the induced representation of G on the homogeneous component of the exterior algebra of V_ω of degree d, cf. the proof of 6.3. Here $\det g = 1$ and $\omega(\lambda) \neq 1$, so $d < n$ and the kernel of ρ consists of (finitely many) scalar matrices. Hence $G\rho$ is not soluble-by-finite, and $\dim A(g\rho)$ is easily seen to be one. We also need irreducibility, and there clearly does exist an irreducible constituent ρ_i of ρ such that $\dim A(g\rho_i)$ is also one. Provided we have already reduced the problem to the case where $\mathscr{A}_F(G)$ is semisimple it follows that $G\rho_i$ is not soluble-by-finite. Alternatively we may make an initial reduction so as to suppose that the connected group G contains no closed connected normal subgroups N such that G/N is soluble and non-trival, and again deduce that $G\rho_i$ is not soluble-by-finite. Clearly G contains a free subgroup of rank 2 whenever $G\rho_i$ does, so we may assume that G is irreducible and that $d = 1$. The proposition of the preceeding paragraph is now applicable, and hence X is dense in G.

For simplicity let us suppose for the moment that $n = 2$. Then X consists only of d-elements. Choose $x \in X$. By passing to a suitable finite extension of F_ω we may assume that x is diagonal. If G does not contain a free subgroup of rank 2 then we cannot find a y in X to satisfy $(*)$. Hence x and y have a common eigenvalue for every y in X. Let L (resp. U) denote the intersection of G with the full lower (resp. upper) triangular group. Then $X \subseteq L \cup U$. The connectedness of G implies that G is irreducible as topological space (see 14.3). Consequently either $G = L$ or $G = U$; whence G is soluble. This is false by assumption, so G contains a free subgroup of rank 2.

In general, conceivably X contains non-d-elements. However assuming again that we have reduced to the case where $\mathscr{A}_F(G)$ is semisimple it follows from [3a] 12.1 b and 13.17 Corollary 2c that G contains a non-empty open set containing only d-elements. The assumption now that G does not contain a free subgroup of rank 2 leads to the conclusion that G is not absolutely irreducible. However by replacing F_ω by a finite extension of itself and G by a suitable one of its irreducible constituents we can reduce to the case where G is absolutely irreducible and again we obtain the desired contradiction. ☐

Let $w = w(x_1, x_2, \ldots, x_r)$ be any word; so w is an element of the free group on $\{x_1, x_2, \ldots, x_r\}$. If G is any group then w is *finitely valued on* G if $\{w(g_1, g_2, \ldots, g_r): g_1, g_2, \ldots, g_r \in G\}$ is a finite set. As before the *verbal subgroup* of G generated by w is

$$w(G) = \langle w(g_1, g_2, \ldots, g_r): g_1, g_2, \ldots, g_r \in G \rangle.$$

The *marginal subgroup generated by* w is

$$w^*(G) = \{x \in G: w(g_1, \ldots, g_{i-1}, g_i x, g_{i+1}, \ldots, g_r) = w(g_1, \ldots, g_r)$$

$$\text{for } i = 1, 2, \ldots, r \text{ and all } g_1, g_2, \ldots, g_r \text{ in } G\}.$$

It is easy to see that $w^*(G)$ is in fact a subgroup of G. See [24a] 8.4 for the elementary properties of these subgroups. P. Hall's three basic questions concerning the sizes of $w(G)$ and $w^*(G)$ have a positive solution for linear groups.

10.19. Theorem (Merzljakov [36a]). *Let G be a subgroup of $GL(n, F)$ and w any word. Then*

i) *if w is finitely valued on G then $w^*(G)$ has finite index in G, and*

ii) *if $w^*(G)$ has finite index m in G then $w(G)$ is a finite group of order dividing some power of m.*

As an immediate corollary of 10.19 we have:

10.20. Corollary (Merzljakov [36a]). *If G is a linear group and w is any word then the following are equivalent.*

i) *w is finitely valued on G.*

ii) *$w(G)$ is finite.*

iii) *$\big(G: w^*(G)\big)$ is finite.* ⊔

Proof of 10.19. i) If $g_1, g_2, \ldots, g_r \in G$ the mapping

$$\theta: x \mapsto w(g_1, \ldots, g_{i-1}, g_i x, g_{i+1}, \ldots, g_r)$$

is continuous by 5.1 and so $(G^0)\theta$ is a connected set. But it is also a finite set since w is finitely valued on G. Hence $(G^0)\theta$ consists of a single element. This element is clearly $w(g_1, \ldots, g_r)$. It follows that $G^0 \subseteq w^*(G)$ and so $w^*(G)$ has finite index in G. (It is easily seen that for any linear group G, $w^*(G)$ is a closed subgroup and so has finite index in G if and only if it contains G^0. Thus we have not really proved more than is stated in 10.19 i).)

ii) Suppose firstly that G is any finitely generated group such that $(G: w^*(G)) = m$ is finite, and suppose that for some prime p not dividing m there exists a normal subgroup H of G of finite index that is residually a finite p-group. For $i = 1, 2, \ldots$ put $H_i = H^{p^i} \cdot \gamma^i H$. For each i, H_i is a normal subgroup of G of finite index. Put $G_i = G/H_i$ and denote by ϕ_i the natural projection of G onto G_i.

Now $w^*(G)\phi_i \subseteq w^*(G_i)$, so $(G_i: w^*(G_i))$ divides m. There exists an m-subgroup (i.e. a π-subgroup where π is the set of prime divisors of m), K_i of G_i such that $G_i = K_i \cdot w^*(G_i)$, (cf. [22] E 1*). Clearly $w(G_i) \subseteq K_i$ and so $|w(G_i)|$ is an m-number. But p is prime to m. Therefore $|w(G)|$ divides k, the largest m-number to divide $(G:H)$. Now $w(G)\phi_i = w(G_i)$ for each i and $\bigcap_i \ker \phi_i = \{1\}$ by hypothesis. Hence $w(G)$ has finite exponent dividing k. By [24a] 8.4 we have $[w(G), w^*(G)] = \{1\}$ and $w^*(G)$ has finite index in G. Thus $w(G)$ is abelian-by-finite. Also if $x_1, x_2, \dots, x_m$ is any transversal of $w^*(G)$ to G then

$$w(G) = \langle w(x_{i_1}, \dots, x_{i_r}): 1 \leq i_1, \dots, i_r \leq m \rangle,$$

that is $w(G)$ is finitely generated. It follows easily that $w(G)$ is a finite group with order dividing a power of m.

Now let G be as in part ii) of the theorem. If $x_1, x_2, \dots, x_m$ is any transversal of $w^*(G)$ to G and $G_1 = \langle x_1, x_2, \dots, x_m \rangle$ then it is easy to see that $G = G_1 w^*(G)$, $w(G) = w(G_1)$ and $w^*(G_1) = G_1 \cap w^*(G)$. Hence we may assume that $G = G_1$ is finitely generated. If char $F = 0$ or if char $F = q > 0$ where q does not divide m then by 4.7 there exists a prime p not dividing m such that G contains a normal subgroup of finite index that is residually a finite p-group. In these cases the result follows from the above.

Suppose now that char $F = q > 0$ and that q divides m. Clearly we may assume that w is not the empty word. Now $ww^*(G) = \{1\}$ by [24a] 8.4, so the variety of groups generated by $w^*(G)$ is not the variety of all groups. By 10.15 or 10.16, $w^*(G)$, and hence G also, is soluble-by-finite. Let U be the maximal unipotent subgroup of G^0. G/U is finitely generated and abelian-by-finite, and so for every prime p, G/U contains a normal subgroup of finite index that is residually a finite p-group. By the above $w(G/U)$ is a finite m-group. But U has exponent dividing q^n and q divides m. Thus $w(G)$ is an m-group (of finite exponent). As before $w(G)$ is finitely generated and consequently $w(G)$ is finite of order dividing a power of m. $\square$

We conclude this chapter with a complete characterization of linear wreath products, which says that the linear wreath products we constructed in Chapter 2 are essentially the lot. This was done for soluble groups by Ju. E. Vapne in [58a]; he obtained the full result in [58b] using 10.15 to reduce it to the soluble case. Independently the full characteristic zero case was obtained in [69d] using only elementary techniques, and the full characteristic p case in [69e], using Tits' Theorem to reduce to the soluble case. It would be nice to have an elementary proof of this latter case. The full characterization reads as follows.

10.21. Theorem. *The standard wreath product $G = A \wr H$ of the groups A and H, where $A \neq \{1\}$, has a faithful representation of finite degree over some field of characteristic $p (\geq 0)$ if and only if*

either   A has a faithful representation of finite degree over some field of characteristic p and H is finite,

or   A is an abelian group that is torsion-free if $p = 0$ and a p-group of finite exponent otherwise, and H is a finite extension of a torsion-free abelian group.

Linear complete wreath products turn out to be totally uninteresting: every linear complete wreath product is in fact a restricted wreath product, and therefore falls under the mantle of 10.21. The following very easy result, which it is convenient to prove before 10.21, is taken from [69 d].

10.22. Proposition. *The complete wreath product $\bar{G} = A \bar{\wr} G$ of the groups A and H, where $A \neq \{1\}$, has a faithful representation of finite degree over the field F if and only if H is finite and A has such a representation.*

We shall regard the base group B (resp. $\bar{B}$) of $A \wr H$ (resp. $A \bar{\wr} H$) as the set of all mappings of finite support (resp. all mappings) of H into A with pointwise multiplication, and made into an H-module by

$$\phi^y \colon \; x \mapsto (y\,x)\,\phi, \qquad \text{where} \;\; x, y \in H \;\; \text{and} \;\; \phi \in B \;(\text{resp. } \bar{B}).$$

Proof. Suppose that $\bar{G} = A \bar{\wr} H$ is isomorphic to a subgroup of $\mathrm{GL}(n, F)$. It is easily checked that the centralizer of any subset of F_n in F_n is a subspace. Since any chain of subspaces of F_n has length at most n^2, so any chain of centralizers in $\mathrm{GL}(n, F)$, and hence also in $\bar{G}$, has length at most n^2.

Let $H_1 \subset H_2 \subset \cdots \subset H_r$ be a strictly ascending chain of subgroups of H. If ϕ is any element of the base group $\bar{B}$ then $\phi \in C_{\bar{B}}(H_i)$ if and only if $x \phi = x \phi^y = (y\,x)\,\phi$ for all $x \in H$ and $y \in H_i$. Thus $C_{\bar{B}}(H_i)$ is the set of maps of H into A that are constant on the right cosets of H_i in H. But then

$$C_{\bar{B}}(H_1) \supset C_{\bar{B}}(H_2) \supset \cdots \supset C_{\bar{B}}(H_r)$$

and so $r \leq n^2 + 1$. Thus H satisfies both the maximal and the minimal condition on subgroups, and so is finitely generated and periodic. By 4.9, H is finite.

Conversely suppose that A is isomorphic to a subgroup of $\mathrm{GL}(n, F)$ and that H is finite of order h say. Clearly the base group is isomorphic to a subgroup of $\mathrm{GL}(h\,n, F)$ and so $A \wr H$ is isomorphic to a subgroup of $\mathrm{GL}(h^2 n, F)$ by 2.3. (In fact $A \wr H$ is even isomorphic to a subgroup of $\mathrm{GL}(h\,n, F)$.) $\square$

We shall need the following lemma during the proof of 10.21 to enable us to avoid using 10.15 or 10.16, at least in the characteristic zero case. It is a very special case of [69 d] Lemma 3.

10.23. Lemma. *Let G be a linear group over a field of characteristic zero, and suppose that there exists an integer m such that every primitive (i.e. having a faithful irreducible module) image of $\mathbb{Z}H$ for any finite factor H of G, has dimension at most m^2 over its centre. Then there exists an integer e such that G^e is abelian.*

Proof. Put $e = (\max_{i \le m} \beta(i))!$ where β is the function of 9.2. It suffices to prove that $[x^e, y^e] = 1$ for every pair of elements x and y of G. Now by 4.2, $\langle x, y \rangle$ is residually finite. Hence it suffices to prove that H^e is abelian for every finite factor H of G.

Choose a prime p not dividing the order of H. Then $\mathbb{F}_p H$ is a direct sum of central simple algebras of dimension at most m^2 over their centres. Tensoring over $\mathbb{F}_p$ by its algebraic closure $\overline{\mathbb{F}}_p$ it follows that $\overline{\mathbb{F}}_p H$ is a direct sum of matrix algebras over $\overline{\mathbb{F}}_p$ of degree at most m. Applying 9.4 to the images of H in these matrix algebras, it follows that H^e is abelian. $\square$

Proof of 10.21. It follows from 2.15, 2.16 and 10.22 that the conditions are sufficient for G to have a representation of the required type. Hence assume that $G = A \wr H$ is (abusing notation somewhat) a subgroup of $GL(n, F)$ where char $F = p$. Further if H is finite the conditions are trivially satisfied, some assume otherwise.

As in the proof of 10.22 chains of centralizers of G have length at most n^2. The base group B of G is isomorphic to the direct product of infinitely many copies of A. If for $i = 1, 2, \dots$ A_i is a non-abelian group then $D = \overset{\infty}{\underset{i=1}{\times}} A_i$ does not satisfy the minimal condition on centralizers since for $j = 1, 2, \dots$

$$C_D(\langle A_1, A_2, \dots, A_j \rangle) = \left(\overset{j}{\underset{i=1}{\times}} \zeta_1(A_i) \right) \times \left(\underset{i>j}{\times} A_i \right).$$

It follows that A is abelian.

If $P \subset Q$ are finite subgroups of H it follows as in the proof of 10.22 that $C_B(P)$ is the set of mappings of H into A with finite support that are constant on the right cosets of P. Since P and Q are finite we obtain $C_B(P) \supset C_B(Q)$ and hence that chains of finite subgroups of H have length at most n^2. Since periodic linear groups are locally finite it follows that every periodic subgroup of H is finite. Therefore H contains an element h of infinite order.

Suppose that a is an element of A or order q a prime. Then

$$\langle a, h \rangle \cong C_q \wr C_\infty$$

and the base group of this wreath product is an infinite elementary abelian q-group. By 2.2 we have $p=q$. Suppose now that a is an element of A of infinite order. $\langle a, h \rangle$ is soluble so by 3.6, $\langle a, h \rangle$ contains a triangular-izable normal subgroup of index dividing $t = \mu(n)!$. Thus

$$\langle a^t, h^t \rangle \cong \mathbf{C}_\infty \wr \mathbf{C}_\infty$$

is triangularizable and so its derived group is unipotent. But the derived group of $\mathbf{C}_\infty \wr \mathbf{C}_\infty$ is torsion-free. Therefore $p=0$. We have now shown that A is torsion-free if $p=0$ and a p-group if $p>0$. Since if $p>0$ every p-subgroup of $\mathrm{GL}(n, F)$ has finite exponent, we now have all the required properties of A.

Now assume that $p=0$. Since B is abelian B_u is the unique maximal unipotent subgroup of B. If $k \in H$ then $\langle B, k \rangle$ is soluble, so $\langle B^t, k^t \rangle$ is triangularizable and $[B^t, k^t] \subseteq B_u$. If for $i=1, 2$ we have $[B^t, k_i] \subseteq B_u$, then

$$[B^t, k_1 k_2^{-1}] \subseteq \langle [B^t, k_2]^{-k_2^{-1}}, [B^t, k_1]^{k_2^{-1}} \rangle \subseteq B_u,$$

since B_u is normal in G. Therefore $[B^t, H^t]$ is unipotent. If $a \in A^t \smallsetminus \{1\}$ — and a exists since A is torsion-free — and if K is any finitely generated subgroup of H^t then the normal closure of a in $\langle a, K \rangle$ is isomorphic as K-module to $\mathbb{Z}K$ and $[a, K]$ is isomorphic as K-module to the augmenta-tion ideal of K. By 4.26, H^t satisfies the hypotheses of 10.23. Hence H^t is an extension of an abelian group by a group of finite exponent.

We now have that H contains an abelian closed normal subgroup L such that H/L has finite exponent. But H/L is isomorphic to a linear group over F by 6.4 and so 1.23 and 1.6 imply that H/L is finite. Finally L has a torsion-free subgroup of finite index since the periodic subgroups of H are all finite (see [32] Vol. 1, p. 179). Therefore H contains a torsion-free abelian normal subgroup of finite index.

Finally, we consider the case $p>0$. Any chain of finite p-subgroups of H has length at most n^2. Hence a Sylow p-subgroup of any finite subgroup P of H has order at most p^{n^2} and so by 9.6, P contains an abelian normal subgroup with index at most $r = f(n^2, n, p)$.

Let K be any finitely generated subgroup of H and E any finitely generated subgroup of K. Now A is a p-group, so if a is an element of A or order p then $\langle a \rangle^E$ is isomorphic as E-module to $\mathbb{F}_p E$. Thus every irreducible $\mathbb{F}_p E$-module has finite order by 4.26, so clearly E cannot be a non-cyclic free group. Hence K can contain no non-cyclic free subgroups and so K^0 is soluble by 10.16.

K^0 is residually finite (4.2), triangularizable (5.8) and $(K^0)_u$ is finite by the above. Hence K^0 is abelian. But then $(K^0)^{p^n}$ is diagonalizable and K^0 is finitely generated. Hence K^0 is diagonalizable. If $C = C_K(K^0)$ then $(K:C) \leq n!$ by 1.12. The centre of C has finite index in C so by a theorem of Schur ([29] p. 59 or [50] 15.1.13) C' is a finite group. Thus $\eta_1(C')$ has index at most r in C'. Put $L = C_C(C'/\eta_1(C'))$. Then $(K:L) \leq (n!)(r!)$ and L is soluble. It follows that H contains a soluble normal subgroup of finite index (at most $(n!)(r!)$).

We have now shown that $G^0 = T$ is soluble. Hence T is triangularizable and thus T' is nilpotent. Suppose that $H \cap T'$ is infinite. Then

$$C_B(H \cap T') = \{1\}.$$

Since both B and T' are normal in G it follows that $B \cap T' \neq \{1\}$. But then $B \cap \zeta_1(T') \neq \{1\}$ and so $C_B(H \cap T') \neq \{1\}$. This contradiction shows that $H \cap T'$ is finite. Now $H^0 \subseteq H \cap T$, so $(H^0)' \subseteq H \cap T'$ is also finite. But by 5.7, $(H^0)'$ is connected and so $(H^0)' = \{1\}$. Therefore H^0 is abelian and the result follows as in the $p = 0$ case. $\square$

11. Supersoluble and Locally Supersoluble Linear Groups

This chapter consists mainly of an account of papers [69 a] and [69 b], although the order of our development will be somewhat different. The motivation of much of this work came from the following result, a generalization of 1.14.

11.1. Proposition. *Let $\mathfrak{X}$ be a class of groups, closed under taking subgroups and quotient groups such that every finite $\mathfrak{X}$-group is supersoluble. Suppose that F is algebraically closed and that G is a completely reducible subgroup of $\mathrm{GL}(n, F)$ containing a normal subgroup N such that G/N is an $\mathfrak{X}$-group. If either*

   i) *N is abelian or*

   ii) *N is periodic, soluble and all its Sylow subgroups (w.r.t. primes) are abelian,*

then G is monomial.

The question naturally arises from 11.1 as to what is the structure of linear $\mathfrak{X}$-groups. Specifically are they locally supersoluble, or even hypercyclic? The answer we shall see is yes if char $F \neq 0$ but no otherwise, and in the latter case the question is still undecided. The problem resolves itself into two parts. First we descuss the finitely generated case and then we attack the general case by means of localizing techniques.

Proof. Every factor of G is an $\mathfrak{X}$-group modulo a normal subgroup satisfying either i) or ii) above. Thus by 1.11 and 9.13 it suffices to assume that G is also irreducible and primitive and to prove that $n=1$. By 1.13, $\zeta_1(G)$ is the unique maximal abelian normal subgroup of G. Further G is soluble by 4.3, so G contains an abelian normal subgroup of finite index (3.5). Therefore $(G: \zeta_1(G))$ is finite.

   i) Here N is abelian so $N \subseteq \zeta_1(G)$. Thus $G/\zeta_1(G)$ is a finite homomorphic image of the $\mathfrak{X}$-group G/N, and so is supersoluble. If $G \neq \zeta_1(G)$ there exists x in $G \setminus \zeta_1(G)$ such that $A = \langle x, \zeta_1(G) \rangle$ is normal in G. Trivially A is abelian, so we have a contradiction. Hence G is abelian and $n=1$ by 1.3.

   ii) $\eta(N)$ is abelian since it is the direct product of its Sylow subgroups. Hence $\eta(N) \subseteq \zeta_1(G)$. If $N \neq \eta(N)$ there exists a characteristic subgroup A of N such that $A/\eta(N)$ is abelian and non-trivial. But then A is nilpotent, so $A = \eta(N)$. This contradiction proves that N is abelian and we can apply case i). $\square$

We start our discussion in a more general way by studying hypercyclic series in arbitrary linear groups. Let G be any group and $N \subseteq M$ normal subgroups of G. We say that M/N is a G-hypercyclic factor of G if there exists an ascending series of normal subgroups of G running from N to M with cyclic factors. If γ is an ordinal number M/N has G-paraheight γ if there exists an ascending series

$$N = N_0 \subseteq N_1 \subseteq \cdots \subseteq N_\alpha \subseteq \cdots \subseteq N_\gamma = M$$

of normal subgroups of G such that for each ordinal $\alpha < \gamma$, $N_{\alpha+1}/N_\alpha$ is abelian with each of its subgroups normal in G/N_α, and if no such series exists of length less than γ.

Denote by $\lambda(G)$ the product of all the G-hypercyclic normal subgroups of G. Clearly $\lambda(G)$ is a G-hypercyclic normal subgroup of G. We call the G-paraheight of $\lambda(G)$ simply the paraheight of G. It is the hypercyclic analogue of central height. Our first objective is to develop, for linear groups G, a theory of $\lambda(G)$ similar to that developed in Chapter 8 for $\zeta(G)$. Note that there is no real hypercyclic analogue of the upper central series, or of the commutator operation, so this theory is necessarily somewhat sparser.

It is very convenient to regard the abelian factors of the various normal series of a group G as G-modules. If G is an arbitrary group and A a G-module, we call A a hypercyclic G-module if A is H-hypercyclic as a normal subgroup of the split extension H of A by G. Further the paraheight of A as G-module is just the H-paraheight of A as a normal subgroup of H.

11.2. Lemma. *Let G be a finite group and A a hypercyclic G-module. Then A has paraheight at most ω, and if A is torsion-free as $\mathbb{Z}$-module A has paraheight at most $(G:G^2)$.*

Proof. Suppose that A is torsion-free as $\mathbb{Z}$-module. Embed A into $V = \mathbb{Q} \otimes_{\mathbb{Z}} A$ by identifying A and $1 \otimes A$. V is a $\mathbb{Q}G$-module containing an ascending series of $\mathbb{Q}G$-submodules whose factors have $\mathbb{Q}$-dimension at most 1. Also V is completely reducible as $\mathbb{Q}G$-module (1.5 with $H = \{1\}$) and so V is a direct sum of irreducible $\mathbb{Q}G$-modules of $\mathbb{Q}$-dimension 1.

Let $V_1, V_2, \ldots, V_r$ be the non-trivial homogeneous components of V as $\mathbb{Q}G$-module and put $A_i = A \cap V_i$; $A_i \neq \{0\}$. If $g \in G$ there exists a rational number s such that $ag = sa$ for every a in A_i (even V_i). Since A is hypercyclic there exists a non-zero element b of A_i such that $\mathbb{Z}b$ is a G-submodule, that is $bg = nb$ for some integer n. A is torsion-free as

$\mathbb{Z}$-module, so $s = n = \pm 1$. Consequently every $\mathbb{Z}$-submodule of A_i is a G-submodule and A has paraheight at most r. Further G^2 acts trivially on A and thus $r \leq (G : G^2)$.

We now turn to the general case. A contains a free $\mathbb{Z}$-submodule E such that A/E is periodic. Then $E_1 = \bigcap_{g \in G} E g$ is a G-submodule of A such that E_1 is $\mathbb{Z}$-free and A/E_1 is periodic (since G is finite). By the $\mathbb{Z}$-torsion-free case it suffices to assume that A is periodic. Then $A = \bigoplus_p A_p$ where A_p is the p-primary component of A. If B is a $\mathbb{Z}$-periodic G-module such that every $\mathbb{Z}$-submodule of every primary component of B is G-invariant, then every $\mathbb{Z}$-submodule of B is G-invariant. Hence the paraheight of A is equal to the upper bound of the paraheights of the A_p. Thus we may assume that $A = A_p$ for some prime p.

Let $A(1)$ be the G-submodule of A generated by all the irreducible G-submodules of A of order p and inductively define

$$A(i+1)/A(i) = (A/A(i))(1).$$

Put

$$A(\omega) = \bigcup_{i=1}^{\infty} A(i).$$

$A(i+1)/A(i)$ is elementary abelian and completely reducible as $\mathbb{F}_p G$-module. Up to isomorphism there exist only a finite number, r say, of irreducible G-modules of order p. Let $A(i+1, j)/A(i)$ for $j = 1, 2, \dots, r$, denote the homogeneous components of $A(i+1)/A(i)$. G acts on each $A(i+1, j)/A(i)$ as a group of scalars, so each of these G-modules have paraheight at most 1. Hence $A(i+1)/A(i)$ has paraheight at most r and $A(\omega)$ has paraheight at most ω.

It remains to show that $A = A(\omega)$. If $a \in A$, then $a \mathbb{Z} G$ is a finite G-submodule of A since G is finite and A is $\mathbb{Z}$-periodic. Further A is hyper-cyclic so there exists a finite series

$$\{0\} = B_0 \subset B_1 \subset \cdots \subset B_t = a \mathbb{Z} G$$

of G-submodules such that $(B_i : B_{i-1}) = p$ for each i. Clearly $B_i \subseteq A(i)$ for each i, and thus $a \mathbb{Z} G \subseteq A(\omega)$. Therefore $A = A(\omega)$ and 11.2 is proved. $\square$

11.3. Lemma. *Let V be a vector space of dimension n over F, G a subgroup of $\mathrm{Aut}_F(V)$ and W an FG-submodule of V of F-dimension d. If A is a G-hypercyclic normal subgroup of G that stabilizes the series $\{0\} \subseteq W \subseteq V$, then A has G-paraheight at most $d(n-d)$.*

The proof of this lemma is very similar to the proof of 4.12.

Proof. If $a \in A$ let a' denote the linear mapping of V/W into W given by $v + W \mapsto v(a-1)$. The map $\phi : a \mapsto a'$ is a $\mathbb{Z}$-monomorphism of A into the additive group of $\mathrm{Hom}_F(V/W, W)$. The latter is an FG-module, the G-action being given by

$$f^g : v + W \mapsto (v\,g^{-1} + W) f g$$

where $f \in \mathrm{Hom}_F(V/W, W)$ and $g \in G$. A is abelian and so is a G-module via conjugation. A simple check shows that ϕ is a G-module homomorphism.

Let B be the F-subspace of $\mathrm{Hom}_F(V/W, W)$ spanned by $A\phi$. B is an FG-module and since A is G-hypercyclic, B contains an ascending series of FG-submodules whose factors have F-dimension at most 1. B has F-dimension at most $d(n-d)$. Thus there exists a series of FG-submodules

$$\{0\} = B_0 \subseteq B_1 \subset \cdots \subset B_r = B$$

where each B_{i+1}/B_i has F-dimension 1 and $r \le d(n-d)$.

Since $A\phi$ is G-hypercyclic and spans B there exists c in $B_{i+1}/B_i \smallsetminus \{0\}$ such that $\mathbb{Z}c$ is a G-module. Now c generates B_{i+1}/B_i as F-module and if $g \in G$ there exists an integer n such that $c\,g = n\,c$. Hence $(\alpha c)\,g = n(\alpha c)$ for all α in F and so every $\mathbb{Z}$-submodule of B_{i+1}/B_i is G-invariant. Therefore B has paraheight at most r. Since $A \cong_G A\phi \subseteq B$, the result follows. $\quad\square$

Remark. The B_i are F-submodules of B. Hence if char $F = 0$ each factor B_{i+1}/B_i is $\mathbb{Z}$-torsion-free. Thus each factor $A\phi \cap B_{i+1}/A\phi \cap B_i$ is also $\mathbb{Z}$-torsion-free.

11.4. Corollary. *Let G be a subgroup of $\mathrm{GL}(n, F)$ and U a unipotent G-hypercyclic normal subgroup of G. Then U has G-paraheight at most $\frac{1}{2} n(n-1)$.*

Proof. Let V be the n-row vector space over F regarded as a G-module in the usual way. U is unitriangularizable over F by 1.21 and so there exists a non-trivial subspace of V upon which U acts trivially. Put $W = C_V(U)$. Then $d = \dim_F W \ge 1$ and W is G-invariant.

By induction on n, $U C_G(V/W)/C_G(V/W)$ has G-paraheight at most $\frac{1}{2}(n-d)(n-d-1)$. By 11.3, $U \cap C_G(V/W)$ has G-paraheight at most $d(n-d)$. Therefore U has G-paraheight at most $\frac{1}{2}(n-d)(n+d-1)$, which does not exceed $\frac{1}{2} n(n-1)$ since $1 \le d \le n$. $\quad\square$

11.5. Lemma. *If F is algebraically closed and G is an irreducible subgroup of $\mathrm{GL}(n, F)$ then $\lambda(G)$ contains a diagonalizable subgroup A, normal in G, and such that $\lambda(G)/A$ is isomorphic to a subgroup of $\mathbf{S}_n$, the symmetric group on n symbols.*

Proof. Let V be the n-row vector space over F regarded as a G-module in the usual way and suppose that $\{V_1, V_2, \ldots, V_r\}$ is a minimal system of imprimitivity of V as FG-module. If $H = N_G(V_i)$ then H acts primitively and irreducibly on V_i by 1.10. Let ϕ denote the induced homomorphism of H into $\mathrm{Aut}_F(V_i)$ and put $Z = (H \cap \lambda(G)) \phi \cap \zeta_1(H\phi)$.

$H \cap \lambda(G)$ is an H-hypercyclic normal subgroup of H.

If $(H \cap \lambda(G)) \phi \neq Z$ there exists an element h of $(H \cap \lambda(G)) \phi \smallsetminus Z$ such that $\langle h, Z \rangle$ is normal in $H\phi$. But clearly $\langle h, Z \rangle$ is abelian and so lies in the centre of $H\phi$ by 1.13. This contradiction shows that $H \cap \lambda(G)$ acts on V_i as a group of scalars. Let $A = \lambda(G) \cap \bigcap_{i=1}^{r} N_G(V_i)$. Then A is diagonalizable normal subgroup of G and $\lambda(G)/A$ is isomorphic to a subgroup of $\mathbf{S}_r$ since the elements of G permute the V_i. $\square$

Exercise 11.1. F is an algebraically closed field, G is a subgroup of $\mathrm{GL}(n, F)$ and N is a completely reducible G-hypercyclic normal subgroup of G. Prove that N contains a diagonalizable-normal subgroup A of G such that N/A is isomorphic to a subgroup of $\mathbf{S}_n$.

Both 8.4 and 11.5 are effectively special cases of this exercise. To solve the rider prove first a slight extension of 1.10 by showing that S_i is completely reducible whenever G is completely reducible.

11.6. Theorem. *If G is a subgroup of $\mathrm{GL}(n, F)$ then G has paraheight at most $\omega + [\log_2 n!]$. If R is a finitely generated integral domain and G is a subgroup of $\mathrm{GL}(n, R)$ then G has finite paraheight.*

Proof. Clearly we may assume that F is algebraically closed. There exists x in $\mathrm{GL}(n, F)$ and irreducible representations $\rho_1, \rho_2, \ldots, \rho_r$ of G such that

$$g^x = \begin{pmatrix} g\rho_1 & & 0 \\ & \ddots & \\ * & & g\rho_r \end{pmatrix} \qquad \text{for all } g \text{ in } G.$$

Let $U = \bigcap_{i=1}^{r} \ker \rho_i$ and put

$$H = \{\mathrm{diag}\,(g\rho_1, \ldots, g\rho_r) \colon g \in G\} \subseteq \mathrm{GL}(n, F).$$

U is a unipotent normal subgroup of G and G/U is isomorphic to the completely reducible subgroup H of $\mathrm{GL}(n, F)$.

By 11.5, $\lambda(H)$ contains a diagonalizable subgroup A, normal in H such that $\lambda(H)/A$ is isomorphic to a subgroup of $\mathbf{S}_n$. Clearly $\lambda(H)/A$ has H-paraheight at most $[\log_2 n!]$. Now $C_H(A)$ has finite index in H by 1.12. Hence 11.2 implies that A has H-paraheight at most ω and thus H has paraheight at most $\omega + [\log n!]$.

Since G/U and H are isomorphic, $\lambda(G)/(U \cap \lambda(G))$ has G-paraheight at most $\omega + [\log_2 n!]$ and by 11.4, $U \cap \lambda(G)$ has G-paraheight at most $\frac{1}{2} n(n-1)$. Thus G has paraheight at most $\omega + [\log_2 n!]$. Suppose now that $G \subseteq \mathrm{GL}(n, R)$. By 4.10, A is finitely generated. Therefore by 11.2, H, and hence G, has finite paraheight. $\square$

Our next objective is an analogue for $\lambda(G)$ of 8.6 iii). It will necessarily look a little different due to the lack of a "commutator" operation. First a preliminary lemma.

11.7. Lemma. *If p is a prime, G a finite p'-group and A a hypercyclic G-module that is a p-group as a $\mathbb{Z}$-module, then A has paraheight at most*

$$(G:G'G^{p-1}) \leq (G:G').$$

The existence of a bound independent of p is of importance to us.

Proof. Let B be a G-module of order p^2, C a submodule of B of order p and suppose that B/C and C are not isomorphic as G-modules. Then B splits over C as G-modules. For suppose B is $\mathbb{Z}$-generated by a single element b. Then pb generates C and $b C \mapsto pb$ determines a G-isomorphism of B/C onto C. No such isomorphism exists and consequently B has exponent p. But then the extension splits by Maschke's Theorem (1.6).

Let $V_1, V_2, \ldots, V_r$ be a complete set of representatives of the isomorphism classes of the G-modules of order p. Since the automorphism group of a cyclic group of order p is cyclic of order $p-1$, $r \leq (G:G'G^{p-1})$. We consider first the case where A is finite. There exists a series of G-submodules of A running from $\{0\}$ to A with factors of order p. By the paragraph above we may interchange two adjacent factors if they are not G-isomorphic. Hence A contains G-submodules $A_1, A_2, \ldots, A_r$ such that $|A| = \prod_{i=1}^{r} |A_i|$, and such that the G-composition factors of each A_i are isomorphic to V_i. It is easy to see that $A = \bigoplus_{i=1}^{r} A_i$.

We prove that A_i has paraheight at most 1. Let $g \in G$. Then g induces an automorphism of order s say on V_i. There exists an integer k such that the endomorphism $x \mapsto kx$ of A_i is an automorphism of order s that induces upon the G-composition factors of A_i the same automorphism as g. ($\mathbb{Z}|p^n$ contains a subgroup of its group of units of order $p-1$ that maps naturally *onto* the group of units of $\mathbb{Z}|p$). We prove by induction on the order of A_i that $xg = kx$ for every x in A_i. Apart from the trivial case $A_i = \{0\}$, A_i contains a G-submodule B of index p and an element a such that $A_i = \langle a, B \rangle$. By induction $x g = k x$ for every x in B and by the choice of k,

$$a g = k a + b \tag{$*$}$$

for some b in B. If $t = |g|$, repeated application of $(*)$ yields that

$$a = a g^t = k^t a + t k^{t-1} b.$$

Now s divides t, so $k^t a = a$, and t and k are prime to p. Hence $b = 0$ and $x g = k x$ for every x in A_i. Therefore A_i has paraheight at most 1.

We now drop the assumption that A is finite. If X is any finite subset of A, X generates a finite G-submodule (X) of A. By the above

$$(X) = \bigoplus_{i=1}^{r} (X)_i,$$

where the G-composition factors of $(X)_i$ are all isomorphic to V_i and $(X)_i$ has paraheight at most 1. It follows that $A = \bigoplus_{i=1}^{r} A_i$ for some G-submodules A_i of A where all the G-composition factors of each A_i are isomorphic to V_i and A_i has paraheight at most 1. Therefore A has paraheight at most r. ☐

11.8. Theorem. *If G is a linear group of degree n then $\lambda(G)$ contains a diagonalizable periodic normal subgroup A of G such that G/A has paraheight at most $n! + \frac{1}{2} n(n-1) + \log_2 n! + 1$ and a diagonalizable normal $n!$-subgroup A_1 of G such that G/A_1 has paraheight at most*

$$2(n!) + \tfrac{1}{2} n(n-1) + \log_2 n! + 1.$$

Proof. By 11.5, $\lambda(G)$ contains a triangularizable normal subgroup T of G such that $(\lambda(G):T)$ divides $n!$. Put $K = \mu(T \cap G')$ and $\bar{G} = KG$. $\lambda(G) \cap G'$ is locally nilpotent since the automorphism group of a cyclic group is abelian. Thus K is triangularizable, locally nilpotent (hence nilpotent) and $K = K_u \times K_d$ (see 7.12 and 7.14). Since

$$K = K_u(T \cap G') = K_d(T \cap G'),$$

K is $\bar{G}$-hypercyclic.

K_u has $\bar{G}$-paraheight at most $\frac{1}{2} n(n-1)$ by 11.4. K_d is diagonalizable since K is triangularizable. Hence $(\bar{G}:C_{\bar{G}}(K_d))$ divides $n!$ by 1.12. Let B denotes the periodic part of K_d. By 11.2, K_d/B has $\bar{G}$-paraheight at most $n!$. Therefore $G/B \cap G$ has paraheight at most $n! + \frac{1}{2} n(n-1) + \log_2 n! + 1$.

Denote by B_1 the subgroup of B consisting of all the $n!$-elements of B. By 11.7, B/B_1 has $\bar{G}$-paraheight at most $(\bar{G}:\bar{G}'C_{\bar{G}}(B)) \le n!$. Thus $G/(B_1 \cap G)$ has paraheight at most $2(n!) + \frac{1}{2} n(n-1) + \log_2 n! + 1$. ☐

11.9. Corollary. *Let G be a linear group of degree n and suppose that there exists an integer k such that the diagonalizable $n!$-subgroups of G have order at most k. Then G and each of its subgroups have paraheight at most*

$$2(n!) + \tfrac{1}{2} n(n-1) + \log_2 k n! + 1.\quad ☐$$

Trivially a torsion-free linear group of degree n satisfies the hypothesis of 11.9 and by 4.8 so does a finitely generated linear group of degree n. That a subgroup of a finitely generated linear group has finite paraheight we already knew from 11.6. 11.9 adds that the paraheights of all these subgroups are bounded, the bound depending on the group.

Further examples of groups satisfying the hypothesis of 11.9 are $GL(n, \mathbb{Q})$ and $GL(n, \mathbb{Q}_p)$ for any prime p, where $\mathbb{Q}_p$ denotes the (complete) field of p-adic numbers, see 9.33. It also follows at once from 11.8 that a linear group containing no C_{p^∞}-subgroups for any prime p has finite para-height, cf. 8.8. Incidentally, unlike the hypercentral case, there is no advantage in proving the whole of 11.6 during the proof of 11.8 since this method leads to a slightly larger bound.

Our next collection of results concern finitely generated linear groups. Notice that if $\mathfrak{X}$ is the class of groups of 11.1 then every $\mathfrak{X}$-group has the property that each of its finite homomorphic images is super-soluble. Our next main theorem will concern finitely generated linear groups all of whose finite images are supersoluble, but first we need some lemmas.

11.10. Lemma (R. Baer). *A finitely generated hypercyclic group G is supersoluble.*

Proof. Let $\{G_\alpha: \alpha \leq \gamma\}$ be a hypercyclic series of G and suppose that G is not supersoluble. There exists at least α such that G/G_α is supersoluble. Clearly α is a limit ordinal. Since G/G_α is supersoluble it is finitely presentable ([47] 2.24, Corollary), and therefore G_α is finitely G-generated ([47] 2.24). But then $G_\alpha = G_\beta$ for some $\beta < \alpha$, contradicting the minimality of α. Thus G is supersoluble. $\square$

11.11. Lemma (R. Baer [2a]). *If G is a polycyclic group all of whose finite homomorphic images are supersoluble, then G is supersoluble.*

Proof. Suppose that G is not supersoluble. Since G satisfies the maximal condition on subgroups G contains a normal subgroup N that is maximal subject to the condition that G/N is not supersoluble. Hence we may suppose that every proper homomorphic image of G is supersoluble. Since G is residually finite every finite normal subgroup of G is G-hyper-cyclic. Therefore G contains no non-trivial finite normal subgroups.

If H is a normal subgroup of finite index in G' then $K = \bigcap_{g \in G} H^g$ also has finite index in G', see [50] 7.1.6). Consequently $K \neq \{1\}$ and G/K is supersoluble. Therefore G'/K is nilpotent and in particular G'/H is nilpotent. By a result of K. A. Hirsch ([50] 7.1.12, or couple 2.5 with 4.16) G' is nilpotent. We prove by induction on the minimal number of generators of G/G' that $\zeta_1(G')$ contains a non-trivial cyclic normal subgroup of G. This will imply that G is supersoluble, a contradiction that will complete the proof of the lemma.

Suppose that L is a subgroup G containing G', x an element of G such that $G = \langle L, x \rangle$ and a an element of $\zeta_1(G')$ such that $\langle a \rangle$ is normal in L. If $A = \langle a^G \rangle$, A is a free abelian subgroup of $\zeta_1(G')$, since G contains no non-trivial finite normal subgroups, of rank n say. A determines a homomorphism ϕ of G into $\mathrm{GL}(n, \mathbb{Z})$.

If p is any prime G/A^p is supersoluble, that is $x\phi$ is triangularizable modulo p. Hence the characteristic polynomial $\det(X1_n - x\phi)$ is a monic polynomial with integer coefficients that splits completely modulo p for every prime p. Therefore $\det(X1_n - x\phi)$ has a root, s say, in $\mathbb{Z}$ by [6b] p. 229, Corollary 2. But then $x\phi$ has an eigenvector over $\mathbb{Z}$ with eigenvalue s; that is there exists $c \neq 1$ in A such that $c^x = c^s$.

If $h \in L$ there exists an integer r such that $a^h = a^r$. Now $[x^i, h] \in G'$, and so centralizes $\zeta_1(G')$. Hence

$$a^{x^i h} = a^{h x^i [x^i, h]} = (a^{x^i})^r.$$

Clearly $A = \langle a^{x^i} : i \in \mathbb{Z} \rangle$, consequently $c^h = c^r$. Thus $\langle c \rangle$ is a non-trivial cyclic normal subgroup of G lying in $\zeta_1(G')$ as G satisfies the maximal condition. (A different proof is given by Baer in [2a], but unfortunately it too uses non-elementary results from algebraic number theory.) $\square$

11.12. Lemma. *If G is a subgroup of $\mathrm{GL}(n, F)$ and U is a unipotent G-hypercyclic normal subgroup of G then U is contained in*

$$\zeta_{\frac{1}{2}n(n-1)}(UG'G^{p-1}), \quad \text{if } \operatorname{char} F = p > 0,$$
$$\zeta_{\frac{1}{2}n(n-1)}(UG^2), \quad \text{if } \operatorname{char} F = 0.$$

A useful special case of 11.12 is that $U \subseteq \zeta_{\frac{1}{2}n(n-1)}(UG')$, whatever the characteristic of F.

Proof. Let $\operatorname{char} F = p > 0$. Then U is a nilpotent p-group. Hence there exists an ascending series

$$\{1\} = U_0 \subseteq U_1 \subseteq \cdots \subseteq U_\alpha \subseteq \cdots \subseteq U_\gamma = U$$

of normal subgroups of G such that each factor $U_{\alpha+1}/U_\alpha$ is cyclic of order p and central in U. But then $UG'G^{p-1} \subseteq \bigcap_\alpha C_G(U_{\alpha+1}/U_\alpha)$ and so

$$U \subseteq \zeta_{\frac{1}{2}n(n-1)}(UG'G^{p-1})$$

by 4.13.

Now let $\operatorname{char} F = 0$. It follows from the proofs of 11.3 and 11.4 (see remark before 11.4) that there exists a series $\{1\} = U_0 \subseteq U_1 \subseteq \cdots \subseteq U_r = U$ of normal subgroups of G such that $r \leq \frac{1}{2}n(n-1)$ and such that for each i every subgroup of U_{i+1}/U_i is an abelian torsion-free normal subgroup of G/U_i. Hence if $x \in U_{i+1}$, $\langle x, U_i \rangle$ is normal in G and G^2 centralizes $\langle x, U_i \rangle / U_i$. Certainly U is nilpotent, so by 4.13, $U \subseteq \zeta_{\frac{1}{2}n(n-1)}(UG^2)$. $\square$

The only case of 11.12 that we use is where U is finitely generated. In this situation, because of the following easy result, the characteristic zero case of 11.12 is as elementary as the positive characteristic case.

11.13. *If G is a group and A is a hypercyclic G-module that is free of finite rank as $\mathbb{Z}$-module then A contains a finite series of G-submodules whose factors are infinite cyclic ($\mathbb{Z}$-modules).*

Proof. There exists a series $\{0\} = A_0 \subseteq A_1 \subseteq \cdots \subseteq A_r = A$ of G-submodules of A whose factors are cyclic as $\mathbb{Z}$-module. Embed A in $B = \mathbb{Q} \otimes_{\mathbb{Z}} A$ by identifying A and $1 \otimes A$. Then

$$\{0\} = A_0 \subseteq \mathbb{Q}A_1 \subseteq \cdots \subseteq \mathbb{Q}A_r = B$$

is a series of $\mathbb{Q}G$-submodules of B whose factors have $\mathbb{Q}$-dimension at most 1. Therefore

$$\{0\} = A_0 \subseteq A \cap \mathbb{Q}A_1 \subseteq \cdots \subseteq A \cap \mathbb{Q}A_r = A$$

is a series of G-submodules of A whose factors are isomorphic to subgroups of $\mathbb{Q}^+$. Since every finitely generated subgroup of $\mathbb{Q}^+$ is cyclic the point is proved. $\quad\square$

We now come to our main theorem on finitely generated linear groups with all their finite images supersoluble.

11.14. Theorem. *Let R be a finitely generated integral domain of characteristic $p \geq 0$ and G a subgroup of $\mathrm{GL}(n, R)$ such that every finite homomorphic image of G is supersoluble.*

    i) *If $p > 0$ then G is hypercyclic (of finite paraheight by 11.6).*

    ii) *If $p = 0$ then G' is nilpotent and G contains a unipotent normal subgroup U such that G/U is supersoluble and abelian-by-finite.*

    iii) *G is supersoluble if and only if G is polycyclic.*

It follows from i) (and either 8.9 or 11.6) that G' is also nilpotent if $p > 0$. A hypercyclic group with finite paraheight is called *parasoluble*. Such a group is soluble and nilpotent-by-abelian, and a group is supersoluble if and only if it is finitely generated and parasoluble (11.10). Perhaps all parasoluble groups should be called supersoluble.

Proof. By 4.2 there exists a set $\mathcal{H}$ of normal subgroups of G such that $\bigcap_{H \in \mathcal{H}} H = \{1\}$, such that for each H in $\mathcal{H}$, G/H is isomorphic to a finite linear group of degree n (over a field of characteristic p if $p > 0$), and such that the image of any unipotent subgroup of G in any of these representations is still unipotent. By supposition G/H is soluble for every H in $\mathcal{H}$ and so is soluble of derived length at most $f(n)$ by 3.7. Thus G is soluble and therefore contains a closed triangularizable normal subgroup T of finite index by 5.8. (Triangularizable here means triangularizable over the algebraic closure F of the quotient field of R.) Let U denote the maximal unipotent subgroup of T. T/U is a finitely generated abelian group (4.10). Hence by 11.11, G/U is supersoluble.

i) If $H \in \mathscr{H}$ then the image of UH/H in the given faithful representation of G/H is also unipotent and so 11.13 implies that $[U, {}_{\frac{1}{2}n(n-1)}UG'G^{p-1}] \subseteq H$. Since this is for every H in $\mathscr{H}$ we have $U \subseteq \zeta_{\frac{1}{2}n(n-1)}(UG'G^{p-1})$. Put $U_i = U \cap \zeta_i(UG'G^{p-1})$. Then U_i is a closed normal subgroup of G (for U is closed in G) and U_i/U_{i-1} is central in $UG'G^{p-1}$. Now $UG'G^{p-1}$ has finite index in G; thus if $x \in U_1$, then x has only finitely many conjugates in G and $\langle x^G \rangle$ is a finite normal subgroup of G. Since G is residually finite there exists a normal subgroup N of G of finite index such that $\langle x^G \rangle \cap N = \{1\}$. Therefore $\langle x^G \rangle$ is G-hypercyclic. This holds for every x in U_1, so U_1 is G-hypercyclic.

If we knew that G/U_i were residually finite then this same argument would show that U_{i+1}/U_i is G-hypercyclic. Thus since G/U is supersoluble, the proof of part i) would be complete. U_i is a closed normal subgroup of G, so there exists a rational homomorphism ρ of G with kernel U_i into $\mathrm{GL}(m, F)$ for some m, see 6.4. If S denotes the subring of F generated by R and the coefficients of a set of rational functions determining ρ, then S is a finitely generated integral domain and $G\rho \subseteq \mathrm{GL}(m, S)$. Therefore $G\rho \cong G/U_i$ is residually finite and the proof of part i) is complete.

ii) It remains here to show that G' is nilpotent. By 11.13 we have $[U, {}_{\frac{1}{2}n(n-1)}UG'] \subseteq H$ for every H in $\mathscr{H}$, so $U \subseteq \zeta_{\frac{1}{2}n(n-1)}(UG')$. Also $UG'/U = (G/U)'$ is nilpotent since G/U is supersoluble. Therefore UG' is nilpotent.

iii) This is an immediate consequence of 11.11. □

It follows from 11.14 i) and 11.10 that in the non-zero characteristic case at least, linear $\mathfrak{X}$-groups are locally supersoluble, where $\mathfrak{X}$ is the class of 11.1. Since the class of locally supersoluble groups clearly satisfies the hypothesis on $\mathfrak{X}$, this reduces the problem, in this case, to the study of locally supersoluble linear groups. The characteristic zero case is not so simple as the following easy example shows.

11.15. Example. $\mathrm{GL}(2, \mathbb{Q})$ *contains a 2-generator metabelian subgroup G that is not supersoluble but all of whose finite factors are supersoluble (in fact metacyclic). Thus every finite image of every subgroup of G is supersoluble, yet G is not supersoluble. Further G is not nilpotent-by-finite.*

Proof. Let A denote the additive group of all rational numbers of the form $r/2^s$ where r and s are integers, and x the automorphism of A given by $a \mapsto 2a$. Put H equal to the split extension of A by $\langle x \rangle$. Note that $H = \langle 1_{\mathbb{Q}}, x \rangle$. A has rank 1, so every finite factor of A is cyclic. Hence every finite factor of H is metacyclic. If $a \in A \setminus \{0\}$, then $x a x^{-1} = \frac{1}{2}u \notin \mathbb{Z} a$, so A contains no non-trivial cyclic normal subgroups of H. Consequently H is not supersoluble.

It is easily seen that H is isomorphic to

$$G = \left\langle \begin{pmatrix} 1 & 0 \\ 1 & 1 \end{pmatrix}, \begin{pmatrix} 2 & 0 \\ 0 & 1 \end{pmatrix} \right\rangle \subseteq \text{GL}(2, \mathbb{Q}),$$

the maximal unipotent subgroup of G being isomorphic to A. If G is nilpotent-by-finite then $\langle A, x^r \rangle$ is nilpotent for some integer r and yet it is clearly not even hypercyclic. ⬜

Question 17. *What further can be said about a finitely generated linear group G over a field of characteristic zero if all its finite homomorphic images are supersoluble? In particular is G "hyper-rank-1-abelian"?*

11.16. Corollary. *If R is a finitely generated integral domain of positive characteristic and G is a subgroup of $\text{GL}(n, R)$ then the following are equivalent:*

   i) *G is hypercyclic with finite paraheight (i.e. G is parasoluble).*
   ii) *$G/\phi(G)$ is hypercyclic.*
   iii) *Every maximal subgroup of G has finite index a prime.*
   iv) *Every maximal subgroup of G of finite index has index a prime.*

The finite case of 11.16 is a result of B. Huppert.

Proof. Trivially i) implies ii) and iii) implies iv). Let M be a maximal subgroup of a hypercyclic group C. We prove that the index of M in C is a prime, this will show that ii) implies iii). We may assume that $\bigcap_{g \in C} M^g = \{1\}$. C contains a non-trivial cyclic normal subgroup A. $AM = C$ and $A \cap M$ is normal in C. The latter yields $A \cap M = \{1\}$. If B is any proper subgroup of A then B is also normal in C and $BM = C$. Thus $A = B(A \cap M) = B$ and A is cyclic of prime order. Finally suppose that iv) holds. By VI.9.5. of [26a] every finite homomorphic image of G is super-soluble. Thus G is hypercyclic of finite paraheight by 11.14. ⬜

Always i) implies ii), ii) implies iii) and iii) implies iv). It is easy to see that the subgroup G of $\text{GL}(2, \mathbb{Q})$ constructed in 11.15 satisfies $\phi(G) = \{1\}$ and that each of its maximal subgroups have finite index a prime.

11.17. Proposition. *If R is a finitely generated integral domain and G is a subgroup of $\text{GL}(n, R)$ then*

   i) *if every maximal subgroup of G of finite index has index a prime, then every maximal subgroup of G has finite index; and*

   ii) *if G is finitely generated and $G/\phi(G)$ is hypercyclic then G is supersoluble.*

In view of 11.16 it is only the characteristic zero case of 11.17 that is new. These results leave one question outstanding, which presumably has a positive answer.

Question 18. *If R is a finitely generated integral domain and G is a sub-group of $GL(n, R)$ such that $G/\phi(G)$ is hypercyclic, is G hypercyclic?*

Proof of 11.17. i) By Huppert's Theorem every finite image of G is super-soluble. Hence by 11.14, G is soluble so the result is a corollary of 4.23.

ii) $G/\phi(G)$ is supersoluble by 11.10 and so is nilpotent-by-finite. Hence G has a nilpotent subgroup N of finite index (4.17). N is also finitely generated (since G is) and so G is polycyclic. Further every finite image of G is supersoluble by Huppert's result. Therefore 11.11 implies that G is supersoluble. □

While we are dealing with examples of finitely generated linear groups one further remark might be worth making. Any supersoluble group is nilpotent-by-finite, so any connected supersoluble linear group is necessarily nilpotent. It is also triangularizable by the Lie-Kolchin Theorem (5.8). But *a triangularizable supersoluble linear group need not be nilpotent unless the characteristic of the ground field is* 2. For

$$\left\langle \begin{pmatrix} 1 & 0 \\ 1 & 1 \end{pmatrix}, \begin{pmatrix} -1 & 0 \\ 0 & 1 \end{pmatrix} \right\rangle$$

is dihedral of order depending on the characteristic and thus is super-soluble, but not nilpotent if the characteristic is not 2. If T is a triangular-izable locally supersoluble linear group over a field of characteristic 2 then T is locally nilpotent by 11.12 and hence is nilpotent. □

A result that is closely related to those of the preceding discussion is the following proposition (it also has affinities with Exercise 4.9, 12.8, 12.9 and 12.10).

11.18. Proposition. *Let R be a finitely generated integral domain and G a subgroup of $GL(n, R)$.*

i) *If every maximal subgroup of G of finite index has index a prime or the square of a prime then G is soluble.*

ii) *If every maximal subgroup of G of finite index is hypercyclic then G is soluble.*

iii) *If every maximal subgroup of G of finite index is soluble then either G is soluble or G is nilpotent-by-finite.*

The finite case of i) is due to P. Hall and of ii) is due to B. Huppert.

Proof. In order to prove that G is soluble it suffices to consider a normal subgroup N of G of finite index and to prove that G/N is soluble (cf. the proofs of 4.3 and 12.14). For part i) this is a theorem of P. Hall ([26a] VI.9.4) and for part ii) of B. Huppert ([26a] VI.9.6). Consider part iii). If G is not soluble it has a normal subgroup N of finite index such that G/N is not soluble. But then every maximal subgroup of G of finite index must contain N. Hence if H is any normal subgroup of G of finite index then $NH/H \subseteq \phi(G/H)$, which is nilpotent. Therefore N is nilpotent by 10.5 and the result follows. (This final step can also be accomplished with the techniques of Chapter 4, cf. Exercise 4.3.) ☐

For the remainder of this chapter we discuss locally supersoluble linear groups.

11.19. Lemma. *If G is locally supersoluble and an extension of a hypercentral group by a finitely generated group then G is hypercyclic.*

Proof. The given properties of G are preserved by homomorphisms. Thus it suffices to prove that G contains a non-trivial cyclic normal subgroup. G contains subgroups H and K such that H is a hypercentral group and is normal in G, K is finitely generated and $G = HK$. If a is any non-trivial element of the centre of H then $\langle a, K \rangle$ is supersoluble. Thus $\langle a^K \rangle$ contains a non-trivial cyclic subgroup C normalized by K. Since $\langle a^K \rangle \subseteq \zeta_1(H)$, C is normal in G. ☐

11.20. Lemma. *A locally supersoluble linear group G is locally-nilpotent by periodic-abelian.*

Proof. This may be proved directly by a standard inverse limit argument similar to that used in the proof of 10.8. Alternatively by 10.8, G contains a nilpotent normal subgroup N such that G/N is periodic. $N \subseteq \eta(G)$, so $G/\eta(G)$ is periodic. If X is any finite subset of G' there exists a finitely generated subgroup Y of G such that $X \subseteq Y'$. Y is supersoluble, so Y' is nilpotent. Therefore G' is locally nilpotent and $G/\eta(G)$ is abelian. ☐

11.21. Theorem. *A locally supersoluble linear group is nilpotent-by-finite, locally-nilpotent by finite-abelian, and hypercyclic.*

Proof. Let G be a locally supersoluble subgroup of $GL(n, F)$ and put

$$q = \begin{cases} 2, & \text{if } \operatorname{char} F = 0, \\ \operatorname{char} F - 1, & \text{otherwise}. \end{cases}$$

G is soluble and so contains a triangularizable normal subgroup T of finite index. By 11.12, for every finitely generated subgroup H of G, $T' \cap H \subseteq \zeta_{\frac{1}{2}n(n-1)}((T' \cap H)H'H^q)$. Therefore $T' \subseteq \zeta_{\frac{1}{2}n(n-1)}(G'G^q)$ and consequently $T'T^q$ is nilpotent. The closure C of $T'T^q$ in T is also a nilpotent normal subgroup of T (5.9 and 5.11) and by 6.4, T/C is isomorphic to an abelian linear group over F of exponent dividing q. By 2.2 we have that T/C is finite and so $\eta(T)$ has finite index in T. But $\eta(T) \subseteq \eta(G)$, so $\eta(G)$ has finite index in G. The theorem now follows from 11.20, 11.19 and the fact that every locally nilpotent linear group is hypercentral and nilpotent-by-finite (8.2). ☐

Our next objective is a slight generalization of 11.21, see 11.23 below, but first we give a simple lemma.

11.22. Lemma. *Let N be a normal subgroup of the group G such that N is a hypercentral group and G/N is a finitely generated abelian group. If every 2-generator subgroup of G is supersoluble then G is hypercyclic.*

Proof. It clearly suffices to prove that $\zeta_1(N)$ contains a non-trivial cyclic normal subgroup of G. This we prove by induction on the minimal number of generators of G/N. Suppose that L is a subgroup of G containing N, x an element of G such that $G = \langle x, L \rangle$ and a a non-trivial element of $\zeta_1(N)$ such that $\langle a \rangle$ is normal in L. $R = \langle a, x \rangle$ is supersoluble, so there exists a non-trivial element b of $\langle a^R \rangle$ such that $\langle b \rangle$ is normal in R. Now

$$\langle a^R \rangle = \langle a^{x^i} : i \in \mathbb{Z} \rangle \subseteq \zeta_1(N).$$

If $h \in L$ there exists an integer s such that $a^h = a^s$. Since G' centralizes $\langle a^R \rangle$, $c^h = c^s$ for every c in $\langle a^R \rangle$. Therefore $\langle b \rangle$ is normal in G. ◻

11.23. Corollary. *A linear group is hypercyclic if each of its 2-generator subgroups is supersoluble.*

In [6a] R. Carter, B. Fischer and T. Hawkes prove that a finite group is supersoluble if each of its 2-generator subgroups is supersoluble. The very nice proof given in [6a] uses formation theory, a direct proof is given in [69a] 3.5.

Proof. Suppose that every 2-generator subgroup of the linear group G is supersoluble. If H is any finitely generated subgroup of G then every finite homomorphic image of H is supersoluble. By 11.14 and 11.22, H is hypercyclic. Therefore G is locally supersoluble (11.10) and so is hypercyclic (11.21). (It is possible to base a proof of 11.23 on 10.14 and 11.21 instead of on 11.14 and 11.21.) ◻

A consequence of 11.21 is that a linear group G is locally supersoluble if and only if $G = \lambda(G)$. As one would expect the paraheights of hypercyclic linear groups are somewhat more restricted than the paraheights of linear groups in general. For example there exist linear groups with paraheight ω. $C_{2^\infty} \wr C_2$ has paraheight 2 but we shall see below that $C_{3^\infty} \wr C_3$ has paraheight $\omega + 1$. It follows in a similar way to 8.7 that the subgroup

$$\left\langle \begin{pmatrix} 0 & 1 & 0 & 0 & 0 \\ 0 & 0 & 1 & 0 & 0 \\ 1 & 0 & 0 & 0 & 0 \\ 0 & 0 & 0 & 1 & 0 \\ 0 & 0 & 0 & 2 & 1 \end{pmatrix}, \begin{pmatrix} 0 & 1 & 0 & 0 & 0 \\ 0 & 0 & 1 & 0 & 0 \\ 1 & 0 & 0 & 0 & 0 \\ 0 & 0 & 0 & 1 & 2 \\ 0 & 0 & 0 & 0 & 1 \end{pmatrix}, \begin{pmatrix} \varepsilon_i & 0 & 0 & 0 & 0 \\ 0 & 1 & 0 & 0 & 0 \\ 0 & 0 & 1 & 0 & 0 \\ 0 & 0 & 0 & 1 & 0 \\ 0 & 0 & 0 & 0 & 1 \end{pmatrix} : i = 1, 2, \ldots \right\rangle$$

of GL(5, $\mathbb{C}$) has paraheight ω, where ε_i is a primitive 3^i-th root of unity. In contrast to this we have:

11.24. Proposition. *There exist no hypercyclic linear groups with paraheight ω.*

Proof. Let G be a hypercyclic linear group of degree n and paraheight ω and suppose that n is the least integer for which such a group exists. By 11.4, G is absolutely irreducible. Hence G contains an abelian normal subgroup A of finite index by 3.5. Also G contains a series

$$\{1\} = G_0 \subseteq G_1 \subseteq \cdots \subseteq G_i \subseteq \cdots \subseteq G_\omega = G$$

of length ω of normal subgroups such that for each i, every subgroup of G_{i+1}/G_i is abelian and normal in G. Since $(G:A)$ is finite there exists a finite r such that $G = AG_r$. But then G/G_r is abelian, so G has paraheight at most $r+1$. This contradiction proves the proposition. $\Box$

Denote by $\sigma(n, p)$ the upper bound of all the paraheights of hyper-cyclic linear groups of degree n over some field of characteristic p (possibly $p=0$). By 11.6, $\sigma(n, p)$ is less than $\omega 2$ and clearly $\sigma(n, p) \leq \sigma(n+1, p)$ for all n and p. It therefore follows from 11.25 below that $\sigma(n, p)$ is never a limit ordinal and therefore is actually the paraheight of some hypercyclic linear group of degree n over a field of characteristic p. Let $s(n)$ denote the upper bound of the paraheights of the supersoluble subgroups of the symmetric group S_n on n symbols.

11.25. Theorem.

$$\sigma(n, p) \leq \omega + s(n) \leq \omega + [\log_2 n!].$$

$\sigma(1, p) = 1$	for all $p \geq 0$.
$\sigma(2, p) = 2$	for all $p \geq 0$.
$\sigma(3, 3) = 4$; $\sigma(3, p) = \omega + 2$	for all $p \neq 3$.
$\sigma(4, p) = \omega + 2$	for all $p \geq 0$.

Question 19. *What is $\sigma(n, p)$ in general?*

The corresponding function for the central height of hypercentral linear groups is completely determined by 8.2 and 8.3. As a start to answering this question it seems a reasonable conjecture that the following question has a positive solution.

Question 20. *Does $\sigma(n, 0) = \omega + s(n)$ for all $n \geq 3$? If not does there exist an integer r such that $\sigma(n, 0) = \omega + s(n)$ for all $n \geq r$?*

Proof of 11.25. The proof of 11.6 yields at once that

$$\sigma(n, p) \leq \omega + s(n) \leq \omega + [\log_2 n!].$$

Clearly $\sigma(1, p) = 1$ for all p. Throughout what follows F denotes an algebraically closed field of characteristic p and G denotes a hypercyclic subgroup of $\mathrm{GL}(n, F)$.

Case $n = 2$. If G is reducible then G is triangularizable. By 11.4, G' has G-paraheight at most one, so G has paraheight at most two. If G is irreducible then G is monomial. We show that the full monomial subgroup M of $\mathrm{GL}(2, F)$ is hypercyclic with paraheight two. Let $D = D(2, F)$, $B = D \cap \mathrm{SL}(2, F)$ and $c = \begin{pmatrix} 0 & 1 \\ 1 & 0 \end{pmatrix}$. Then $M = \langle c, D \rangle$, B is normal in M, M/B is abelian and $x^c = x^{-1}$ for every x in B. Thus M is hypercyclic with paraheight at most two. M is not abelian so M has paraheight two.

Case $n = 3$, $p \neq 3$. Clearly $s(3) = 2$, so $\sigma(3, p) \leq \omega + 2$. It is easy to check that $\mathbf{C}_3 \wr \mathbf{S}_3$ (where $\mathbf{S}_3$ is regarded as a permutation group in the natural way) is supersoluble. Let $H = \mathbf{C}_{3\infty} \wr \mathbf{S}_3$. If B is the base group of H put $B_i = \{b \in B : b^{3^i} = 1\}$. For each i, B_{i+1}/B is isomorphic as $\mathbf{S}_3$-module to the base group of $\mathbf{C}_3 \wr \mathbf{S}_3$ and so is $\mathbf{S}_3$-hypercyclic with finite paraheight. Thus B is $\mathbf{S}_3$-hypercyclic with $\mathbf{S}_3$-paraheight ω and H is hypercyclic with paraheight at most $\omega + 2$.

Suppose that H has paraheight at most $\omega + 1$ and let a be a 3-cycle in $\mathbf{S}_3$. Then $a \in H'$, so there exists a finite series of normal subgroups of $\langle a, B \rangle$, say

$$\{1\} = A_0 \subseteq A_1 \subseteq \cdots \subseteq A_r \subseteq \langle a, B \rangle$$

such that each factor A_i/A_{i-1}, is either cyclic of order 3, or a $\mathbf{C}_{3\infty}$-group, and $a \in A_r$. Then $\langle a, B \rangle/A_r$ is abelian. Since the 3-adic integers contain no primitive cube root of unity, a centralizes each of the factors in the above series and so $\langle a, B \rangle$ is nilpotent. But $\langle a, B \rangle \cong \mathbf{C}_{3\infty} \wr \mathbf{C}_3$ which is not nilpotent (e.g. [47] 2.32). This contradiction proves that H has paraheight $\omega + 2$. (It also effectively shows that $\mathbf{C}_{3\infty} \wr \mathbf{C}_3$ has paraheight $\omega + 1$.)

Since F is algebraically closed and char $F \neq 3$, the full monomial subgroup of $\mathrm{GL}(3, F)$ contains a subgroup isomorphic to H. Therefore $\sigma(3, p) = \omega + 2$ if $p \neq 3$.

Case $n = 4$. It is easily seen that $s(4) = 2$, so $\sigma(4, p) \leq \omega + 2$ for all p. But $\sigma(3, p) \leq \sigma(4, p)$. Thus $\sigma(4, p) = \omega + 2$ for all $p \neq 3$.

Let P be a Sylow 2-subgroup of $\mathbf{S}_4$ regarded as a permutation group in the obvious way. Then $\mathbf{C}_{2\infty} \wr P$ is a locally finite 2-group satisfying the minimal condition on subgroups, and therefore is hypercentral. In particular $\mathbf{C}_{2\infty} \wr P$ is hypercyclic. The 2-adic integers contains no primitive fourth root of unity and $C_{2\infty} \wr C_2$ is not nilpotent. A very similar argument to that employed in the case $n=3$ proves that $C_{2\infty} \wr C_4$ has paraheight $\omega+1$ and that $C_{2\infty} \wr P$ has paraheight $\omega+2$. Since F is algebraically closed, if $p \neq 2$, then the full monomial subgroup of $GL(4, F)$ contains a subgroup isomorphic to $C_{2\infty} \wr P$ and so $\sigma(4, 3)=\omega+2$.

Case $n=p=3$. Now G is a hypercyclic subgroup of $GL(3, F)$ where F is an algebraically closed field of characteristic 3. Suppose that the paraheight of G exceeds 4 and seek a contradiction. By 11.3 and the case $n=2$, G is irreducible and hence is monomial. By replacing G by a conjugate if necessary we may assume that G is actually in monomial form. Now $F^*G=\{\alpha g: \alpha \in F^*, g \in G\}$ is also hypercyclic, so we may also assume that $F^*1_3 \subseteq G$.

Let D denote the (full) diagonal subgroup of G. If $D=F^*1_3$ then G has paraheight at most 3, so this is not the case. Hence there exists an element

$$a=\text{diag}(\alpha, \beta, \gamma) \in D \smallsetminus F^*1_3$$

such that $\langle a, F^*1_3 \rangle$ is normal in G. G is irreducible so 3 divides $(G:D)$. Suppose that $(G:D)=6$ (i.e. that $G/D \cong S_3$). Then there exists integers i, j and elements δ, δ_1 of F^* such that

$$(\gamma, \alpha, \beta)=(\alpha^i \delta, \beta^i \delta, \gamma^i \delta)$$

and

$$(\beta, \alpha, \gamma)=(\alpha^j \delta_1, \beta^j \delta_1, \gamma^j \delta_1).$$

Solving these six equation we find that $(\alpha, \beta, \gamma)=(\alpha, \alpha \xi, \alpha \xi^{-i})$ where

$$\xi^{i^2+i+1}=\xi^{j+1}=\xi^{ij-i-1}=1.$$

Now a is not a scalar, so $\xi \neq 1$. Hence there exists a prime q dividing $i^2+i+1, j+1$ and $ij-i-1$. Taking congruences modulo q we get $j \equiv -1$, so the third expression yields that $2i \equiv -1$. Therefore $4i^2 \equiv 1$. But from the first expression $4i^2 \equiv -4i-4 \equiv -2$. Thus $q=3$. Consequently some power of ξ is a primitive cube root of 1 in F and since char $F=3$ this is impossible. We have now shown that $(G:D)=3$.

As $F^*1_3 \subseteq G$ there exists λ, μ in F^* such that

$$x=\begin{pmatrix} 0 & \lambda & 0 \\ 0 & 0 & \mu \\ \lambda^{-1}\mu^{-1} & 0 & 0 \end{pmatrix}$$

lies in G. If $X=\langle x \rangle$ then $G=DX$, and $D \cap X=\{1\}$. Let T be the torsion subgroup of D/F^*1_3 and T_q the q-primary component of T. Suppose that $T_q \neq \{1\}$. Then $q \neq$ char $F=3$. If the field of q elements does not contain a primitive cube root of unity then there exists a non-trivial cyclic subgroup of T_q centralized by x. A trivial check (e.g. take $i=1$ in the previous calculation) shows that this is impossible. Hence the field of q elements contains a primitive cube root of unity and hence so does the ring of q-adic integers.

Let $A_q = \langle \alpha_j : \alpha_j^q = \alpha_{j-1}, \alpha_0 = 1, j = 1, 2, \ldots \rangle$ be the $\mathbf{C}_{q^\infty}$-subgroup of F^*. A_q has an automorphism c of order 3. Then $\alpha_j^c = \alpha_j^{i_j}$ for some integer i_j. Let

$$S_q = \langle F^* 1_3, \operatorname{diag}(\alpha_j, \alpha_k, 1) : j, k = 1, 2, \ldots \rangle / F^* 1_3;$$

clearly $T_q \subseteq S_q$. Put

$$B_q = \langle F^* 1_3, \operatorname{diag}(\alpha_j, \alpha_j^c, \alpha_j^{c^2}) : j = 1, 2, \ldots \rangle / F^* 1_3.$$

Both S_q and B_q are invariant under conjugation by x since

$$\left(\operatorname{diag}(\alpha_j, \alpha_j^c, \alpha_j^{c^2}) \right)^x = \left(\operatorname{diag}(\alpha_j, \alpha_j^c, \alpha_j^{c^2}) \right)^{i_j}$$

and

$$\left(\operatorname{diag}(\alpha_j, \alpha_k, 1) \right)^x = \alpha_k \left(\operatorname{diag}(\alpha_k^{-1}, \alpha_j \alpha_k^{-1}, 1) \right).$$

Also B_q and S_q/B_q are both $\mathbf{C}_{q^\infty}$-groups. Therefore S_q hs paraheight at most 2 as X-module. This is for all primes q with $T_q \neq \{1\}$, so T has G-paraheight at most 2.

Now $2 \nmid (G : D)$ so by 11.13, $(G/F^* 1_3)/T$ is locally nilpotent and hence is isomorphic to the abelian group $X \times (D/F^* 1_3)/T$. Therefore G has paraheight at most 4. Consequently $\sigma(3, 3) \leq 4$ and it remains only to construct an example to show that this bound is actually attained.

Let $a = \operatorname{diag}(-1, 1, 1)$ and $b = (1, 1, -1)$. If

$$x = \begin{pmatrix} 1 & 0 & 0 \\ \xi & 1 & 0 \\ \eta & \zeta & 1 \end{pmatrix} \quad \text{then } x^a = \begin{pmatrix} 1 & 0 & 0 \\ -\xi & 1 & 0 \\ -\eta & \zeta & 1 \end{pmatrix} \quad \text{and } x^b = \begin{pmatrix} 1 & 0 & 0 \\ \xi & 1 & 0 \\ -\eta & -\zeta & 1 \end{pmatrix},$$

where ξ, η and ζ lie in F. Let $U = \operatorname{Tr}_1(3, F)$ and put $R = \langle a, b, U \rangle \subseteq \operatorname{GL}(3, F)$. If $U_1 = \zeta_1(U) = \{x \in U : \xi = \zeta = 0\}$ and $U_2 = \{x \in U : \xi = 0\}$ then

$$\{1\} = U_0 \subset U_1 \subset U_2 \subset U \subset R$$

is a normal series of R whose factors are R-hypercyclic with R-paraheight 1.

It is easy to check that $R' = U$ and that if A is an abelian normal subgroup of R lying in U with R-paraheight 1, then A is contained in U_1. U/U_1 does not have R-paraheight 1 since for example $x^a \notin \langle x, U_1 \rangle$ if $\xi = \zeta = 1$. Therefore R has paraheight 4 and $\sigma(3, 3) = 4$. (In fact if F is any field at all with $\operatorname{char} F \neq 2$ then R has paraheight 4: if $\operatorname{char} F = 2$ then R has paraheight 2). $\square$

12. A Localizing Technique and Applications

Although we have derived a number of theorems on linear groups from properties of finitely generated linear groups by standard group-theoretic techniques we have seldom used the linear structure of the matrix ring to accomplish this. The object of this chapter is to describe a general method for extending theorems from finitely generated linear groups to more general linear groups that relies heavily on the linearity. Although the fundamental results hold in any linear group the technique has only enjoyed any significant success, so far at least, in the theory of periodic linear groups.

The first few pages of this chapter contain an account of this general method together with one or two simple applications to show how it is used. From then on the chapter is devoted exclusively to periodic linear groups. We extend a number of well-known theorems on finite groups to periodic linear groups concerning p-complements and p-quotient groups. We then use these to prove that a periodic linear group is soluble if it contains a locally nilpotent maximal subgroup satisfying a few relatively mild restrictions.

We recall a definition from Chapter 9. Let G be a subgroup of $GL(n, F)$ and S a subset of G. Then

$$\mathscr{L}_G(S) = \left\langle \sum_{i=1}^{n^2} \alpha_i s_i \, ; \, \alpha_i \in F, \, s_i \in S, \, \sum \alpha_i s_i \in G \right\rangle.$$

If S is triangularizable, then so is $\mathscr{L}_G(S)$ and hence $\mathscr{L}_G(S)$ contains a unique maximal unipotent subgroup $\mathscr{L}_G(S)_u$.

12.1. Let G be a subgroup of $GL(n, F)$, P a maximal unipotent subgroup of G and Q a subgroup of P. Let $h_1, h_2, \ldots, h_s$ be elements of $\zeta_1(P)$ that span $\zeta_1(P)$ linearly (over F). Extend this set to $h_1, h_2, \ldots, h_t$, $t \geq s$, where $h_1, h_2, \ldots, h_t$ are elements of P spanning P linearly. Let $k_1, k_2, \ldots, k_t$ be elements of Q spanning Q linearly (we can take the same t since $Q \subseteq P$; this is a notational simplification only). Suppose that S is any subset of G and put

$$\Gamma = \langle S, h_i, k_i \colon i = 1, 2, \ldots, t \rangle.$$

Denote by Π a maximal unipotent subgroup of Γ containing $P \cap \Gamma$ and set $K = Q \cap \Pi$. We claim that the following eight statements are true.

1. $P = \mathscr{L}_G(\Pi)_u$.
2. $\Pi = P \cap \Gamma$.
3. $N_\Gamma(\Pi) = N_G(P) \cap \Gamma$ and $C_\Gamma(\Pi) = C_G(P) \cap \Gamma$.
4. $\zeta_1(P) = \mathscr{L}_G(\zeta_1(P))_u$.
5. $\zeta_1(\Pi) = \zeta_1(P) \cap \Gamma$.
6. $C_\Gamma(\zeta_1(\Pi)) = C_G(\zeta_1(P)) \cap \Gamma$.
7. $Q \subseteq \mathscr{L}_G(K)_u \subseteq P$.
8. *If Q is weakly closed in P with respect to G then K is weakly closed in Π with respect to Γ and $N_\Gamma(K) = N_G(Q) \cap \Gamma$.*

If G is a group, K a subgroup of G and H a subgroup of K, then H is weakly closed in K with respect to G if $g \in G$ and $H^g \subseteq K$ together imply that $H = H^g$. For example, in the set-up above it is easy to see that P is weakly closed in P with respect to G.

12.1 says that the embedding of Π in Γ is the same, or at least very similar, to the embedding of P in G. Notice the effect of the arbitrary subset S of G. It means that 1. to 8. above not only hold for Γ, but also for any subgroup of G containing Γ.

Proof. By 1.21, Π is unitriangularizable, so $\mathscr{L}_G(\Pi)_u$ is a unipotent subgroup of G. Clearly

$$P \subseteq \mathscr{L}_G(h_1, \ldots, h_t)_u \subseteq \mathscr{L}_G(\Pi)_u \subseteq G,$$

and P is a maximal unipotent subgroup of G. Therefore $P = \mathscr{L}_G(\Pi)_u$. But then $\Pi \subseteq P$, so $\Pi = P \cap \Gamma$. It will follow from 8. and the choice $P = Q$ that $N_G(P) \cap \Gamma = N_\Gamma(\Pi)$. It follows easily from 1. and 2. that $C_G(\Pi) = C_G(P)$ and $C_\Gamma(\Pi) = C_G(P) \cap \Gamma$.

Clearly

$$\zeta_1(\Pi) \subseteq C_G(P) \cap \Pi = \zeta_1(P) \cap \Gamma \subseteq \zeta_1(P \cap \Gamma) = \zeta_1(\Pi),$$

using 2., so $\zeta_1(\Pi) = \zeta_1(P) \cap \Gamma$. Further

$$\mathscr{L}_G(\zeta_1(\Pi))_u \subseteq \mathscr{L}_G(\Pi)_u \cap C_G(\Pi) = P \cap C_G(P) = \zeta_1(P),$$

using 1. and the above. Also $\zeta_1(P) \subseteq \mathscr{L}_G(h_1, \ldots, h_s)_u \subseteq \mathscr{L}_G(\zeta_1(\Pi))_u$. This proves 4. and 5., and 6. is an immediate consequence of them.

It is clear that $Q \subseteq \mathscr{L}_G(K)_u \subseteq \mathscr{L}_G(\Pi)_u = P$. Suppose that Q is weakly closed in P with respect to G. Let $g \in \Gamma$ be such that $g K g^{-1} \subseteq \Pi$. Then

$$Q \subseteq \mathscr{L}_G(K)_u \subseteq \mathscr{L}_G(\Pi^g)_u = (\mathscr{L}_G(\Pi)_u)^g = P^g.$$

By weak closure $Q = Q^g$. Thus $K^g = (Q \cap \Gamma)^g = Q^g \cap \Gamma = K$ since $g \in \Gamma$. That is K is weakly closed in Π with respect to Γ. Trivially $N_G(Q) \cap \Gamma \subseteq N_\Gamma(K)$. If $g \in N_\Gamma(K)$ then $g K g^{-1} \subseteq \Pi$, and the above argument shows that $Q = Q^g$. Therefore $N_\Gamma(K) = N_G(Q) \cap \Gamma$. □

Simply in order to illustrate the way in which 12.1 is used we prove the following easy result.

12.2. *Let G be a subgroup of* $GL(n, F)$ *such that every finite subset of G lies in a subgroup of G whose maximal unipotent subgroups are all conjugate. Then the maximal unipotent subgroups of G are all conjugate.*

Proof. Let P_1 and P_2 be maximal unipotent subgroup of G. For $i=1, 2$, there exists a finitely generated subgroup Γ_i of G satisfying 12.1 with respect to P_i. By hypothesis there exists a subgroup Γ of G containing $\langle \Gamma_1, \Gamma_2 \rangle$ such that the maximal unipotent subgroups of Γ are all conjugate. Now in 12.1, S is any subset of G, so 12.1 holds not only for Γ_i but also for any subgroup of G containing Γ_i, for example Γ. In particular $P_i \cap \Gamma$ is a maximal unipotent subgroup of Γ for $i=1, 2$, and $P_i = \mathscr{L}_G(P_i \cap \Gamma)_u$. By the choice of Γ there exists an element $g \in \Gamma \subseteq G$ such that $(P_1 \cap \Gamma)^g = (P_2 \cap \Gamma)$. Thus
$$P_1^g = (\mathscr{L}_G(P_1 \cap \Gamma)_u)^g = \mathscr{L}_G((P_1 \cap \Gamma)^g)_u = \mathscr{L}_G(P_2 \cap \Gamma)_u = P_2,$$
and 12.2 is proved. $\square$

If G is a closed subgroup of $GL(n, F)$ where F is algebraically closed, then the maximal unipotent subgroups of G are all conjugate: if char $F = 0$ then any unipotent subgroup of $GL(n, F)$ is connected, so this result is just 11.3.2 of [3a]. If the characteristic of F is positive then it follows from 4.6 of [44] [8]. Call a subgroup G of $GL(n, F)$ *locally algebraic* if every finite subset of G is contained in an algebraic (i.e. closed in $GL(n, F)$) subgroup of G. Then a corollary of 12.2 is that *the maximal unipotent subgroups of a locally algebraic linear group over an algebraically closed field are all conjugate.* Clearly, algebraic linear groups are locally algebraic and periodic linear groups are locally algebraic. In view of the following exercise locally algebraic linear groups are a not very interesting generalization of these two groups.

Exercise 12.1. If G is a subgroup of $GL(n, F)$ prove that G is locally algebraic if and only if G contains a connected algebraic normal subgroup N such that G/N is isomorphic to a periodic linear group. (Hint: use the dimension theory of Chapter 14.)

If G is a subgroup of $GL(n, F)$ and S is a diagonalizable subset of G, then $\mathscr{L}_G(S)$ is also diagonalizable and in particular abelian. Consequently for any prime p, $\mathscr{L}_G(S)$ contains a unique maximal p-subgroup, which we denote by $\mathscr{L}_G(S)_p$.

Let G be a p-subgroup of $GL(n, F)$ where $p \neq$ char F, and A any abelian subgroup of G. The elements of A of order less than or equal to p form an elementary abelian subgroup whose rank we denote by $d(A)$.

---

[8] If G is finite it is simply Sylow's Theorem.

It follows from 2.2 that $d(A) \leq n$. Let $m(G)$ denote the maximal value taken by $d(A)$ as A ranges over all the abelian subgroups of G. $J(G)$ denotes the subgroup of G generated by all the abelian subgroups A of G for which $d(A) = m(G)$. If G is a finite group then $J(G)$ is simply the (old) Thompson subgroup of G. One can formally make the same definitions if $p = \operatorname{char} F$ but since $m(G)$ may now be an infinite cardinal the concepts do not seem to be very useful. Similarly the variant of the definition of $J(G)$ in which one replaces $d(A)$ by $|A|$ does not seem to be relevant for infinite groups.

12.3. Let G be a subgroup of $\operatorname{GL}(n, F)$, P a maximal p-subgroup of G for some prime p other than the characteristic of F and Q a subgroup of P. Put $m = n!$. Now P satisfies the minimal condition on subgroups by 2.6. Let P_1 (resp. $\zeta_1(P)_1$, $J(P)_1$, Q_1) denote the minimal subgroup of P (resp. $\zeta_1(P)$, $J(P)$, Q) of finite index. P_1, $\zeta_1(P)_1$, $J(P)_1$ and Q_1 are divisible abelian groups. Hence there exists elements,

$$h_1, \ldots, h_n \in P_1,$$
$$b_1, \ldots, b_n \in \zeta_1(P)_1,$$
$$v_1, \ldots, v_n \in J(P)_1,$$

and

$$k_1, \ldots, k_n \in Q_1$$

such that

$$h_1^m, \ldots, h_n^m \text{ span } P_1 \qquad \text{linearly}$$
$$b_1^m, \ldots, b_n^m \text{ span } \zeta_1(P)_1 \quad \text{linearly}$$
$$v_1^m, \ldots, v_n^m \text{ span } J(P)_1 \quad \text{linearly}$$

and

$$k_1^m, \ldots, k_n^m \text{ span } Q_1 \qquad \text{linearly.}$$

Let X be a transversal of P_1 to P, D of $\zeta_1(P)_1$ to $\zeta_1(P)$, W of $J(P)_1$ to $J(P)$ and Y of Q_1 to Q. There exists a finite number of finite abelian subgroups $A_1, \ldots, A_c$ of P such that $d(A_i) = m(P)$ for each i and

$$\langle v_1, \ldots, v_n, W \rangle \subseteq \langle A_1, \ldots, A_c \rangle = V$$

say. V is clearly a finite subgroup of $J(P)$.

Let S be any subset of G and put

$$\Gamma = \langle S, X, D, V, Y, h_i, b_i, k_i : i = 1, 2, \ldots, n \rangle.$$

Suppose that Π is any maximal p-subgroup of Γ containing $P \cap \Gamma$. Π is monomial (over the algebraic closure of F) by 1.6 and 1.14, and so Π^m diagonalizable. Put $K = Q \cap \Gamma$. We claim that the following ten statements are true.

1. $P = \Pi\big(\mathcal{L}_G(\Pi^m)_p\big)$.
2. $\Pi = P \cap \Gamma$.
3. $N_\Gamma(\Pi) = N_G(P) \cap \Gamma$ and $C_\Gamma(\Pi) = C_G(P) \cap \Gamma$.
4. $\zeta_1(P) = \zeta_1(\Pi)\big(\mathcal{L}_G(\zeta_1(P)^m)_p\big)$.
5. $\zeta_1(\Pi) = \zeta_1(P) \cap \Gamma$.
6. $C_\Gamma(\zeta_1(\Pi)) = C_G(\zeta_1(P)) \cap \Gamma$.
7. $Q \subseteq K\big(\mathcal{L}_G(K^m)_p\big) \subseteq P$.
8. If Q is weakly closed in P with respect to G, then K is weakly closed in Π with respect to Γ and $N_\Gamma(K) = N_G(Q) \cap \Gamma$.
9. $J(\Pi) \subseteq J(P) \subseteq J(\Pi)\big(\mathcal{L}_G(J(\Pi)^m)_p\big) \subseteq P$.
10. $N_G(J(\Pi)) \subseteq N_G(J(P))$.

Proof. Π normalizes Π^m, so Π normalizes $\mathcal{L}_G(\Pi^m)_p$. Hence $\Pi\big(\mathcal{L}_G(\Pi^m)_p\big)$ is a p-subgroup of G. Also $h_1, \ldots, h_n$ lie in Π so $h_1^m, \ldots, h_n^m$ lie in Π^m, and $X \subseteq \Pi$. Therefore $P \subseteq \Pi\big(\mathcal{L}_G(\Pi^m)_p\big) \subseteq G$, and so $P = \Pi\big(\mathcal{L}_G(\Pi^m)_p\big)$. But then $\Pi \subseteq P$, so $\Pi = P \cap \Gamma$. Again it will follow from 8. that $N_\Gamma(\Pi) = N_G(P) \cap \Gamma$ and it is an immediate consequence of 1. and 2. that $C_G(\Pi) = C_G(P)$ and $C_\Gamma(\Pi) = C_G(P) \cap \Gamma$.

By 2. we have

$$\zeta_1(\Pi) \subseteq C_G(P) \cap \Pi \subseteq \zeta_1(P) \cap \Gamma \subseteq \zeta_1(P \cap \Gamma) = \zeta_1(\Pi).$$

Thus $\zeta_1(\Pi) = \zeta_1(P) \cap \Gamma$, proving 5. Since $b_i \in \zeta_1(\Pi)$ by 5., $b_i^m \in \zeta_1(\Pi)^m$. Hence

$$\zeta_1(P) \subseteq \zeta_1(\Pi)\big(\mathcal{L}_G(\zeta_1(\Pi)^m)_p\big) \subseteq C_G(P) \cap \Pi\big(\mathcal{L}_G(\Pi^m)_p\big) = \zeta_1(P),$$

by 5. and 1. This proves 4., and 6. follows at once from 4. and 5.

It is clear that $Q \subseteq K\big(\mathcal{L}_G(K^m)_p\big) \subseteq P$. Suppose that Q is weakly closed in P with respect to G and let g be an element of Γ satisfying $gKg^{-1} \subseteq \Pi$. Then $K \subseteq \Pi^g$, $K^m \subseteq (\Pi^m)^g$ and

$$Q \subseteq K\big(\mathcal{L}_G(K^m)_p\big) \subseteq \Pi^g\big(\mathcal{L}_G((\Pi^m)^g)_p\big) = P^g$$

by 1. By weak closure $Q = Q^g$ and so $K^g = (Q \cap \Gamma)^g = Q^g \cap \Gamma = K$, since $g \in \Gamma$. Therefore K is weakly closed in Π with respect to Γ. Trivially

$$N_G(Q) \cap \Gamma \subseteq N_\Gamma(K).$$

Let $g \in N_\Gamma(K)$. Then $gKg^{-1} \subseteq \Pi$, so $Q = Q^g$ as above. Therefore

$$N_\Gamma(K) = N_G(Q) \cap \Gamma.$$

Trivially $m(\Pi) \leq m(P)$, and since $V \subseteq \Pi$ and $m(V) = m(P)$ we have that $m(\Pi) = m(P)$. Therefore $V \subseteq J(\Pi) \subseteq J(P)$ and so

$$J(P) \subseteq V\big(\mathcal{L}_G(V^m)_p\big) \subseteq J(\Pi)\big(\mathcal{L}_G(J(\Pi)^m)_p\big) \subseteq P.$$

This proves 9. Let $a \in N_G(J(\Pi))$. Then

$$J(P)^a \subseteq J(\Pi)^a \left(\mathscr{L}_G((J(\Pi)^m)^a)_p\right) = J(\Pi)\left(\mathscr{L}_G(J(\Pi)^m)_p\right) \subseteq P.$$

This clearly implies that $J(P)^a \subseteq J(P)$. But a^{-1} also lies in $N_G(J(\Pi))$, so $J(P)^{a^{-1}} \subseteq J(P)$ in the same way. Thus $J(P)^a = J(P)$ and 10. is proved. □

The following exercise may be proved in a similar way to 12.2 for the case $p \neq \operatorname{char} F$, using 12.3 in place of 12.1. If $p = \operatorname{char} F$ use 12.2 itself.

Exercise 12.2. If p is a prime and G is a subgroup of $\mathrm{GL}(n, F)$ such that every finite subset of G lies in a subgroup of G whose maximal p-subgroups are all conjugate, prove that the maximal p-subgroups of G are all conjugate.

9.10 is a special case of this exercise. The proof of 9.10 that we gave is essentially the same as the proof of Exercise 12.2 that we have hinted at above. By 4.6 of [44] for every prime p the maximal p-subgroups of an algebraic linear group over an algebraically closed field are all conjugate. Hence Exercise 12.2 implies that *for every prime p the maximal p-subgroups of a locally algebraic linear group over an algebraically closed field are all conjugate.*

As a further simple illustration of 12.1 and 12.3 in action, we give a second proof of the following result (see 9.13).

12.4. *Let G be a periodic subgroup of $\mathrm{GL}(n, F)$, N a normal subgroup of G such that G/N is a p-group for some prime p, and P a Sylow p-subgroup of G. Then $PN = G$.*

It is an immediate consequence of 12.4 that if G is a periodic linear group then the Sylow subgroups of a homomorphic image of G are exactly the images of the Sylow subgroups of G.

Proof. Let $x \in G$ and suppose that Γ and Π are the groups constructed in 12.1 or 12.3 (depending on the characteristic of F), where $S = \{x\}$ and P is as in the data. Γ is a finite group, Π is a Sylow p-subgroup of Γ and $\Gamma/(N \cap \Gamma)$ is a p-group. Therefore $\Gamma = \Pi(N \cap \Gamma) \subseteq PN$. But $x \in \Gamma$, and x is any element of G, so $G = PN$. □

It is possible to carry out analogous processes to those used in the proofs of 12.1 and 12.3 for other soluble subgroups P of G, using the existence of triangularizable normal subgroups of finite bounded index in P, see 3.6. This was essentially done for soluble maximal π-subgroups during the proof of 9.15. Also the groups Γ of 12.1 and 12.3 may be chosen to have further properties (see §2 of [68]).

12.1 and 12.3 can be used to lift a whole host of p-complement and p-quotient group theorems from finite groups to periodic linear groups ([68]). We give here only a selection. Our choice is restricted to those results that we shall need in order to discuss periodic linear groups with locally nilpotent maximal subgroups.

If G is a periodic group and π is a set of primes, a normal π-complement of G is a normal π'-subgroup L of G such that $G = PL$ for every maximal π-subgroup P of G. It is easy to see that a normal π-complement of G is a Hall π'-subgroup of G and it follows from 12.4 that if p is a prime, a normal Hall p'-subgroup of a periodic linear group G is necessarily a normal p-complement of G. (This is actually true for any set π of primes, see 9.18.)

12.5. Theorem. *Let G be a periodic subgroup of $\mathrm{GL}(n, F)$, p an odd prime with $p \neq \operatorname{char} F$, and P a Sylow p-subgroup of G. If $C_G(\zeta_1(P))$ and $N_G(J(P))$ have normal p-complements, then G also has a normal p-complement.*

The finite analogue of 12.5 is a famous theorem of J.G. Thompson (see [26a] IV. 6.2). Thompson's original theorem has now been generalized in two different directions by G. Glauberman and Thompson himself. However the original version is more than adequate for our purposes, although stronger variations will extend to periodic linear groups in much the same way. Remember that we are referring always to theorems involving the "old" Thompson subgroup.

Proof. We have only to show that the set of p'-elements of G is a subgroup of G. Suppose that x and y are any two p'-elements of G. Let Γ be the subgroup constructed in 12.3 with $S = \{x, y\}$ and P as in the statement of the theorem. Γ is a finite group and $\Pi = P \cap \Gamma$ is a Sylow p-subgroup of G. Also by 12.3.6 and 12.3.10 we have that $C_\Gamma(\zeta_1(\Pi)) \subseteq C_G(\zeta_1(P))$ and $N_\Gamma(J(\Pi)) \subseteq N_G(J(P))$. Therefore $C_\Gamma(\zeta_1(\Pi))$ and $N_\Gamma(J(\Pi))$ both have normal p-complements. Hence by Thompson's Theorem so does Γ. Since x and y are p'-elements of Γ we have that $\langle x, y \rangle$ is a p'-group and the theorem is proved. □

If G is any group and π a set of primes $O^\pi(G)$ denotes the intersection of all the normal subgroups N of G such that G/N is a π-group. If G is periodic then $G/O^\pi(G)$ is a π-group and $O^\pi(G)$ is generated by the set of π'-elements of G. Thus if G is a periodic linear group then G has a normal p-complement if and only if $O^p(G)$ is a p'-group, and then $O^p(G)$ is the normal p-complement of G.

A p-group is called *regular* if for each pair of elements x, y of G we have $(xy)^{-p} x^p y^p = d^p$ for some $d \in \langle x, y \rangle'$. See [21] §12.4 or [26a] §III.10 for a discussion of regular p-groups.

180

12.6. Theorem. *Let G be a periodic subgroup of* $\mathrm{GL}(n, F)$, p *a prime and* P *a Sylow p-subgroup of G. If P is regular and* $N = N_G(P)$, *then*

$$O^p(N) = O^p(G) \cap N.$$

In particular $G/O^p(G) \cong N/O^p(N)$ *and* $O^p(G)$ *is a normal p-complement of G if and only if* $O^p(N)$ *is a normal p-complement of N.*

The finite case of 12.6 is well-known and due to H. Wielandt ([26a] IV.8.1). Again a number of generalizations and variations are known (due to Wielandt and P. Hall). These also extend to periodic linear groups, see [68] 3.12, 3.3 and 3.5.

Proof. Trivially $O^p(N) \subseteq O^p(G) \cap N$. Let $x \in O^p(G) \cap N$. Then $x = x_1 x_2 \ldots x_r$ for some p'-elements $x_1, \ldots, x_r$ of G. Let Γ be the group of 12.1 or 12.3 as appropriate with $S = \{x_1, \ldots, x_r\}$ and P as in the theorem. Then $\Pi = P \cap \Gamma$ is a Sylow p-subgroup of Γ and $N_\Gamma(\Pi) = N \cap \Gamma$. Since $S \subseteq \Gamma$,

$$x \in O^p(\Gamma) \cap N \cap \Gamma.$$

By Wielandt's Theorem $x \in O^p(N \cap \Gamma) \subseteq O^p(N)$. Therefore

$$O^p(N) = O^p(G) \cap N \quad \text{and} \quad G/O^p(G) \cong N/O^p(N).$$

Trivially if $O^p(G)$ is a p'-group, so is $O^p(N)$. If $O^p(N)$ is a p'-group, $P \cap O^p(N) = \{1\}$ and so $P \cap O^p(G) = \{1\}$. Hence $O^p(G)$ is also a p'-group, and 12.6 now follows. □

12.7. Theorem. *Let G be a periodic subgroup of* $\mathrm{GL}(n, F)$, p *a prime and P a Sylow p-subgroup of G. If* $\zeta_1(P)$ *is weakly closed in P with respect to G (i.e. if G is p-normal) and* $H = N_G(\zeta_1(P))$, *then* $O^p(H) = O^p(G) \cap H$. *In particular* $G/O^p(G) \cong H/O^p(H)$ *and* $O^p(G)$ *is a normal p-complement of G if and only if* $O^p(H)$ *is a normal p-complement of H.*

The finite case of 12.7 is a generalization of Grün's Second Theorem due to P. Hall ([21] Theorem 14.4.6). The proof of 12.7 is very similar to the proof of 12.6, using 5. and 8. of 12.1 or 12.3 instead of 3. and the Grün-Hall Theorem in place of Wielandt's Theorem. □

We have now all the results we need. We add two results of a similar nature to the above three as exercises. The interested reader may consult [68] for the proofs of these and related results.

Exercise 12.3. If G is a periodic linear group and P a Sylow p-subgroup of G, prove that G has a normal p-complement if and only if $N_G(\zeta_1(P))$ has a normal p-complement and G is p-normal. ([68] 3.8.)

Exercise 12.4. If G is a periodic linear group and P is an abelian Sylow p-subgroup of G prove that $P \cap \zeta_1(N_G(P))$ is a complement of $O^p(G)$ and hence that G has a normal p-complement if $C_G(P) = N_G(P)$. ([68] 3.10.)

The final part of this exercise is just the extension to periodic linear groups of Burnside's Transfer Theorem and can quickly be proved independently of the first part of the exercise.

We now come to our second main topic; periodic linear groups containing a locally nilpotent maximal subgroup. The main theorem is the following.

12.8. Theorem ([68]). *Let G be a periodic subgroup of $GL(n, F)$ and H a locally nilpotent, maximal subgroup of G. If* i) *the Sylow 2-subgroup of H is nilpotent of class at most 2, and if* ii) char $F > 2$ *implies that the maximal unipotent subgroup of H is finite or regular, then G is soluble.*

A Sylow 2-subgroup of $PSL(2, 17)$ is dihedral of order 16 (and so has class 3) and is a maximal subgroup of $PSL(2, 17)$ ([26a] II.8.10, II.8.27). Thus condition i) is necessary even for finite groups. Condition ii) is presumably redundant.

Proof. Put $N = \bigcap_{g \in G} H^g$ and denote by $\bar{N}$ the closure of N in G. By 5.11, $\bar{N}$ is soluble, so if $N \neq \bar{N}$ then $G = H\bar{N}$ is soluble. Suppose therefore that $N = \bar{N}$. G/N is isomorphic to a linear group over F by 6.4, so we may assume that $N = \{1\}$.

Let $\pi = \pi(H)$. If $p \in \pi$ then the Sylow p-subgroup P of H is non-trivial and $H \subseteq N_G(P)$. Since H is maximal and $N = \{1\}$ we have that $H = N_G(P)$. If P_1 is a p-subgroup of G properly containing P then $P \subset N_{P_1}(P)$, as P_1 is a hypercentral group. Hence no such P_1 exists and P is also a Sylow p-subgroup of G.

We prove now that G contains a normal p-complement L_p. There are three separate cases to consider.

Case. $p = 2$. Since $\zeta_1(P) \neq \{1\}$, $N_G(\zeta_1(P)) = H$ has a normal p-complement. Thus by 12.7 it suffices to show that $\zeta_1(P)$ is weakly closed in P with respect to G. Let x be an element of G such that $x \zeta_1(P) x^{-1} \subseteq P$. Then $\zeta_1(P) \subseteq P^x$, so $\zeta_1(P^x)$ is contained in $C_G(\zeta_1(P)) = H$. P is the unique Sylow p-subgroup of H and thus $\zeta_1(P^x) \subseteq P$. Therefore

$$Z = \langle \zeta_1(P), \zeta_1(P^x) \rangle \subseteq P \cap P^x$$

P is nilpotent of class 2, so $P \cup P^x \subseteq N_G(Z)$ and clearly $N_G(Z) = H$. Hence $P^x \subseteq H$, which implies that $P = P^x$. Consequently $\zeta_1(P)^x = \zeta_1(P^x) = \zeta_1(P)$ and $\zeta_1(P)$ is weakly closed in P w.r.t. G.

Case. $p > 2$ and $p \neq$ char F. $N_G(J(P)) = C_G(\zeta_1(P)) = H$ since $J(P)$ and $\zeta_1(P)$ are both non-trivial and $N = \{1\}$. H trivially has a normal p-complement and therefore G has a normal p-complement by 12.5.

Case. $p > 2$ and $p =$ char F. In this case P is regular or finite. If P is regular then G has a normal p-complement by 12.6. If P is finite then every finite subgroup of G containing P has a normal p-complement by Thompson's J-Theorem (the finite case of 12.5, namely [26a] IV.6.2) and so G contains a normal p-complement.

We have now shown that for each prime $p \in \pi$, G contains a normal p-complement L_p. By 12.4, $G = PL_p$, so G/L_p is soluble of derived length at most $f(n)$ by 3.7. Put $L = \bigcap_{p \in \pi} L_p$. Clearly L is a normal Hall π'-subgroup of G and G/L is soluble. (Alternatively $G = HL$ by 9.18 and 9.26 and therefore G/L is soluble.)

H is a hypercentral group so its centre contains an element x of prime order (if $H = \{1\}$ then G is cyclic and there is nothing to prove). Since $N = \{1\}$ we have that $C_G(x) = H$ and that x acts fixed-point freely on L. If T is any finite subgroup of L then $K = \langle x, T \rangle$ is a finite group and x induces a fixed-point-free automorphism of prime order on the normal subgroup $K \cap L$ of K. By Thompson's Fixed-Point-Free Automorphism Theorem ([26a] V.8.14) $K \cap L$ is nilpotent. Hence L is locally nilpotent and G is soluble.

(This final paragraph may be replaced by the following. If $L \neq \{1\}$ let Q be a non-trivial Sylow q-subgroup of L for some prime q. Then $G = L \cdot N_G(Q)$ and $N_L(Q)$ is a normal Hall π'-subgroup of $N_G(Q)$. By 9.18, $N_G(Q) = S \cdot N_G(Q)$ for some maximal π-subgroup S of G. By 9.26 there exists an element g of G such that $S^g = H$. But then $\langle H, Q^g \rangle \subseteq N_G(Q^g)$ and $Q^g \nsubseteq H$ since $q \notin \pi$. Thus Q is normal in G and so also in L. This applies to every non-trivial Sylow subgroup of L; therefore L is locally nilpotent and G is soluble.) □

It is easy to see that in the above theorem $G^0 \nsubseteq H$ implies that $G \in \mathfrak{N}\mathfrak{A}(\mathfrak{F} \cap \mathfrak{N})$, and $G^0 \subseteq H$ implies that $G \in (L\mathfrak{N} \cap \mathfrak{S})(\mathfrak{F} \cap \mathfrak{A})(\mathfrak{F} \cap \mathfrak{N})$. However G need not lie in $\mathfrak{N}(L\mathfrak{N})(L\mathfrak{N})$ as the following example shows.

Let H be the non-central extension of a cyclic group of order 7 by a cyclic group of order 3. There exists a non-trivial irreducible $\mathbb{F}_3 H$-module V say, and clearly H must act faithfully upon V. Denote by S the split extension of V by H. V is the Fitting subgroup of S and S is not nilpotent-by-nilpotent. Put $G = C_{3^\infty} \wr S$ (the standard wreath product). G is isomorphic to a subgroup of $\mathrm{GL}(|S|, \mathbb{C})$ for example. A Sylow 3-subgroup of G has index 7 and hence is a locally nilpotent, maximal subgroup of G. If B is the base group of G then BV is locally nilpotent and $\eta(G) = BV$. Moreover BV is not nilpotent since $C_{3^\infty} \wr C_3$ is not, B is abelian and BV/B is a chief factor of G (because V is H-irreducible). Therefore B is the Fitting subgroup of G and $G \notin \mathfrak{N}(L\mathfrak{N})(L\mathfrak{N})$. □

$C_{2^\infty} \wr S_4$, where S_4 is taken with its natural permutation representation of degree 4, would also be a counterexample but its Sylow 2-subgroups do not satisfy condition i) of 12.8. Further G need not lie in $(L\mathfrak{N})\mathfrak{F}$; for if F is any infinite locally finite field then $D(2, F)$ is an abelian maximal subgroup of $\text{Tr}(2, F)$. The latter group is easily seen not to be locally-nilpotent-by-finite.

12.9. Theorem. *Let G be a periodic subgroup of $\text{GL}(n, F)$ containing an abelian maximal closed subgroup A. Then G is centre-by-metabelian.*

12.10. Corollary. *Let G be a periodic subgroup of $\text{GL}(n, F)$ containing an abelian maximal subgroup A. Then G is centre-by-metabelian.*

For if $\bar{A}$ denotes the closure of A in G then either $\bar{A} = G$, whence G is abelian, or $A = \bar{A}$ and 12.9 applies. $\square$

Proof of 12.9. If A is normal in G then G/A is a periodic CZ-group containing no proper closed subgroups, and hence is cyclic of prime order. Thus G is metabelian. Suppose that A is not normal in G. By 5.4, $N_G(A)$ is closed in G and so $A = N_G(A)$. Let $x \in G \smallsetminus A$. Trivially $Z = \zeta_1(G) \subseteq A \cap A^x$. But $C_G(A \cap A^x)$ is a closed subgroup of G by 5.4 containing $A \cup A^x$ and hence must be G itself. Therefore $Z = A \cap A^x$ for every x in $G \smallsetminus A$ (which says that G/Z is a Frobenius group, but will not quote properties of locally finite Frobenius groups at the reader).

Since A is not normal in G we have $Z \neq A \neq G$. Let $a \in A \smallsetminus Z$ and $b \in G \smallsetminus A$. Then every finite subgroup of G/Z containing $\langle a, b, Z \rangle / Z$ is a Frobenius group with an abelian Frobenius complement, and so is soluble ([26a] V.8.7). Hence G is locally soluble and consequently G is soluble. Clearly $Z = \bigcap_{g \in G} A^g$. Since G is soluble there exists a non-trivial abelian normal subgroup B/Z of G/Z. $B \nsubseteq A$, so the closure of AB in G is G itself. AB is centre-by-metabelian and hence G is also by 5.10. $\square$

In 12.9, G need not be metabelian nor nilpotent, even if G is finite: for if Q denotes the quaternion group (of order 8), Q has an automorphism ϕ of order 3. Let G be the split extension of Q by $\langle \phi \rangle$. Clearly G is neither metabelian nor nilpotent and yet $\langle Q', \phi \rangle$ is an abelian maximal subgroup of G.

For other results concerning linear groups with soluble maximal subgroups see Exercise 4.9 and 11.18.

Many familiar results on normal π-complements in finite groups can also be lifted to periodic linear groups using 12.1 and 12.3. We conclude this chapter by proving the extension to periodic linear groups of a well-known theorem of H. Wielandt ([26a] IV.7.3). See §4 of [68] for further results of this kind.

12.11. Theorem. *Let G be a periodic subgroup of $\text{GL}(n, F)$ and H a subgroup but not a Sylow subgroup of G such that $N_G(P) = H$ for every non-trivial Sylow subgroup P of H. Then H has a normal complement in G.*

Proof. Let P be a non-trivial Sylow p-subgroup of H. Since linear p-groups are hypercentral and $N_G(P)=H$, it follows that P is a Sylow p-subgroup of G. By hypothesis there exists a non-trivial Sylow q-subgroup Q of H for some prime $q \neq p$. Q is also a Sylow q-subgroup of G. Let S be any finite set of p'-elements of G and put $\pi = \pi(H)$.

By 12.1 and 12.3 (cf. the proof of 12.2) there exists a finite subgroup Γ of G containing S such that $\Pi = P \cap \Gamma$ and $K = Q \cap \Gamma$ are non-trivial Sylow subgroups of Γ and

$$N_\Gamma(K) = N_G(Q) \cap \Gamma = H \cap \Gamma.$$

Let x and y be elements of Π and g an element of Γ such that $x^g = y$. Then both K and K^g are Sylow q-subgroups of $C_\Gamma(y)$, so there exists an element h of $C_\Gamma(y)$ such that $K = K^{gh}$. Thus $gh \in N_\Gamma(K) = H \cap \Gamma$. Since H is locally nilpotent there exists an element g_1 of Π and a p'-element g_2 of $H \cap \Gamma$ such that $gh = g_1 g_2$. Then

$$y = y^h = x^{gh} = x^{g_1 g_2} = x^{g_1},$$

so whenever elements of Π are conjugate in Γ they are conjugate in Π. Therefore by [26a] IV.4.9, Γ contains a normal p-complement and so $\langle S \rangle$ is a p'-subgroup of Γ. This is for any such S, so for every $p \in \pi$, G contains a normal p-complement L_p.

$L = \bigcap_{p \in \pi} L_p$ is clearly a normal Hall π'-subgroup of G and trivially $H \cap L = \{1\}$. Also G/L is a locally nilpotent π-group and so is the direct product of its Sylow subgroups. By 12.4, P covers the Sylow p-subgroup of G/L and so H does also. Therefore $HL = G$. (Alternatively use 9.18 and 9.26.) ☐

Note that in fact L is a normal π-complement of G and H is a Hall π-subgroup of G.

13. Module Automorphism Groups over Commutative Rings

We shall now widen our field of vision by considering subgroups of the automorphism groups of finitely generated modules over commutative rings. We begin with a few elementary ring-theoretic lemmas that will enable us to reduce the problem to the linear case.

13.1. Lemma. *Let R be a commutative ring, M a non-zero R-module and $\mathfrak{p}$ a maximal annihilator (of a non-zero element of M). Then $\mathfrak{p}$ is a prime ideal.*

Proof. It is trivial to confirm that $\mathfrak{p}$ is an ideal. Let $\alpha, \beta \in R$ with $\alpha\beta \in \mathfrak{p}$. There exists $a \in M \setminus \{0\}$ such that $\mathfrak{p} = \{\gamma \in R : a\gamma = 0\}$. If $\alpha \notin \mathfrak{p}$ then $a\alpha \neq 0$. But $a\alpha(\mathfrak{p}, \beta) = 0$ so $(\mathfrak{p}, \beta)$ is contained in the annihilator of $a\alpha$. The maximality of $\mathfrak{p}$ implies $\beta \in \mathfrak{p}$.   □

13.2. Lemma. *Let R be a Noetherian ring and M a finitely generated R-module. Then M contains a finite series $\{0\} = M_0 \subseteq M_1 \subseteq \cdots \subseteq M_n = M$ of fully invariant submodules such that for each $i = 1, 2, \ldots, n$, R contains a prime ideal $\mathfrak{p}_i$ satisfying $M_i \mathfrak{p}_i \subseteq M_{i-1}$ and such that the induced $R/\mathfrak{p}_i$-module structure on M_i/M_{i-1} is torsion-free.*

Proof. Clearly we can suppose that $M \neq \{0\}$. Let $\mathfrak{p}_1$ be maximal among all the annihilators of the non-zero elements of M; $\mathfrak{p}_1$ exists since R is Noetherian. Put $M_1 = \{a \in M : a\mathfrak{p}_1 = \{0\}\}$. M_1 is a fully invariant submodule of M. Suppose that $\alpha \in R$ and $b \in M_1 \setminus \{0\}$ are such that $b\alpha = 0$. Then $(\mathfrak{p}_1, \alpha)$ annihilates b and so $\alpha \in \mathfrak{p}_1$. Therefore M_1 is torsion-free as $R/\mathfrak{p}_1$-module. If $M_1 \neq M$ we can define $\mathfrak{p}_2$ and $M_2/M_1 \subseteq M/M_1$ in the same way, and if $M_2 \neq M$ we repeat the process again. By the Noetherian condition this process stops after a finite number of steps. It is easy to see that the M_i are fully invariant.   □

A group G is called *quasi-linear* if it is isomorphic to a subgroup of a direct product of a finite number of linear groups. Every subgroup of a quasi-linear group is quasi-linear. By 2.2 a direct product of an infinite number of copies of the cyclic group of order 6 is quasi-linear but not linear.

Let R be a commutative ring and M a finitely generated R-module. An element g of $\operatorname{Aut}_R(M)$ is called *unipotent* if $(g-1)$ is nilpotent, that is if some power of $(g-1)$ is the zero map. Note that this happens if and only if $\langle g \rangle$ stabilizes a finite series of submodules of M. (If g is unipotent put $M_i = M(g-1)^i$ etc.) A subgroup G of $\operatorname{Aut}_R(M)$ is called *unipotent* if every element of G is unipotent.

13.3. Theorem. *Let R be a Noetherian ring and M a finitely generated R-module. Then $\operatorname{Aut}_R(M)$ contains a normal subgroup U stabilizing a finite series of submodules of M such that $\operatorname{Aut}_R(M)/U$ is quasi-linear. In particular U is unipotent and nilpotent (as an abstract group).*

Proof. Let M_i and $\mathfrak{p}_i$, $i = 1, 2, \ldots, n$ be as in 13.2, let F_i denote the quotient field of $R/\mathfrak{p}_i$ and put $V_i = (M_i/M_{i-1}) \otimes_{R/\mathfrak{p}_i} F_i$. Since M_i/M_{i-1} is torsion-free as $R/\mathfrak{p}_i$-module M_i/M_{i-1} is embedded in V_i, and this embedding determines a ring monomorphism of $\operatorname{End}_{R/\mathfrak{p}_i}(M_i/M_{i-1})$ into $\operatorname{End}_{F_i}(V_i)$. Thus we have a ring homomorphism η_i of $\operatorname{End}_R(M)$ into $\operatorname{End}_{F_i}(V_i)$. Let

$$U = \bigcap_{i=1}^{n} C_{\operatorname{Aut}_R(M)}(M_i/M_{i-1}) = \bigcap_{i=1}^{n}(1 + \ker \eta_i) \cap \operatorname{Aut}_R(M).$$

Then $\operatorname{Aut}_R(M)/U$ is isomorphic to a subdirect product of the $\operatorname{Aut}_R(M)\eta_i$ and 13.3 is proved. $\quad\square$

An immediate corollary of the proof of 13.3 is that *if M is irreducible as $\operatorname{End}_R(M)$-module then $\operatorname{Aut}_R(M)$ is isomorphic to a linear group.*

13.4. Lemma. *Let R be a commutative ring, M a finitely generated R-module and put $G = \operatorname{Aut}_R(M)$. Then G is locally, a group of automorphisms of a finitely generated module over a finitely generated commutative (and hence Noetherian) ring.*

Proof. Let H be any finitely generated subgroup of G. H is generated as semigroup by some finite subset, say $\{g_1, \ldots, g_r\}$; suppose that M is generated by $x_1, \ldots, x_e$. Then $x_i g_k = \sum_{j=1}^{e} x_j \alpha_{ijk}$ for some α_{ijk} in R. Let S denote the subring of R generated by the $e^2 r$ elements α_{ijk} and put $N = \sum_{i=1}^{e} x_i S \subseteq M$. Clearly N is an SH-submodule of M that is faithful as H-module. The result now follows. $\quad\square$

13.3 allows us to reduce the Noetherian case to the field case and 13.4 allows us to reduce most of our problems to the Noetherian case. Occasionally however we need the following theorem (whose existence I learnt from Otto Kegel). The proof is essentially in N. Jacobson's book [26 b]. Note that the theorem does not say that G/N is quasi-linear since here an infinite number of linear factors may be involved.

13.5. Theorem. *Let R be a commutative ring, M a finitely generated R-module and put $G = \operatorname{Aut}_R(M)$. If M can be generated by e elements then G contains a normal subgroup N such that G/N is residually linear of degree $f \leq [\frac{1}{2}(e^2+1)]$ and N is unipotent, locally nilpotent and hyperabelian. If M is free we can take $f = e$ and if in addition the nil radical $\mathfrak{n}$ of R is nilpotent then N is nilpotent.*

Proof. Let $E = \operatorname{End}_R(M)$ and $d = e^2 + 1$. If $x_1, \ldots, x_d$ are elements of a ring T write

$$[x_1, \ldots, x_d] = \sum_{\sigma \in S_d} (-1)^\sigma x_{1\sigma} x_{2\sigma} \ldots x_{d\sigma},$$

where $(-1)^\sigma$ is the sign of the permutation σ. This expression is linear in each x_i and is zero if any two of the x's are equal. Thus $[x_1, \ldots, x_d] = 0$ for all $x_1, \ldots, x_d$ in R_e. E is isomorphic to a quotient of a subring of R_e, so E also satisfies this polynomial identity over $\mathbb{Z}$. (A non-trivial result of Amitsur and Levitzki is that R_e, and hence E, satisfies this polynomial identity with $d = 2e$. Using this result one then arrives at the bound $f \leq e$ for any such M.) A simple induction on the degree shows that if a ring T satisfies this identity, then so does the polynomial ring $T[X]$. Let $\mathfrak{a}$ denote the upper nil radical of E. The proofs of Proposition 1 and Theorem 1 and the first part of the proof of Theorem 2 of [26b] § 10.5 show that there exists a set of fields $\{F_\alpha : \alpha \in I\}$, a positive integer $f \leq [\frac{1}{2}d]$ and ring homomorphisms $\mu_\alpha : E \to (F_\alpha)_f$ such that $\mathfrak{a} = \bigcap_I \ker \mu_\alpha$.

Put $N = G \cap (1 + \mathfrak{a})$. Certainly N is a normal subgroup of G such that G/N is residually linear of degree f. By the proof of [26b] Theorem 10.8.2, E contains an ascending series of ideals

$$\{0\} = \mathfrak{a}_0 \subseteq \mathfrak{a}_1 \subseteq \cdots \subseteq \mathfrak{a}_\beta \subseteq \cdots \subseteq \mathfrak{a}_\gamma = \mathfrak{a}$$

such that $\mathfrak{a}_{\beta+1} \mathfrak{a}_{\beta+1} \subseteq \mathfrak{a}_\beta$ for each β. Let $N_\beta = G \cap (1 + \mathfrak{a}_\beta)$; N_β is a normal subgroup of G. If $x = 1 + a$ and $y = 1 + b$ are elements of $N_{\beta+1}$ then modulo $\mathfrak{a}_\beta$, $x^{-1} \equiv 1 - a$, $y^{-1} \equiv 1 - b$ and $[x, y] \equiv 1$. Therefore $N'_{\beta+1} \subseteq N_\beta$ for each β and N is hyperabelian. Since $\mathfrak{a}$ is nil, N is unipotent. By the proof of [26b] Theorem 10.8.1, $\mathfrak{a}$ is locally nilpotent, from which it follows easily that N is locally nilpotent (or alternatively use 13.4 and 13.6 below).

Suppose now that M is free of rank e. Identify E and the matrix ring R_e. If $\mathfrak{b}$ is any ideal of R let $G(\mathfrak{b}) = \{x \in G : x \equiv 1_e \bmod \mathfrak{b}\}$. Now $\mathfrak{n}$ is the intersection of all the prime ideals of R, hence $N = G(\mathfrak{n}) = \bigcap_{\mathfrak{p} \text{ prime}} G(\mathfrak{p})$. Clearly $G/G(\mathfrak{p})$ is isomorphic to a subgroup of $\operatorname{GL}(e, R/\mathfrak{p})$ which is linear since $\mathfrak{p}$ is prime. Therefore G/N is residually linear of degree e. It is easily seen that $\mathfrak{a} = \mathfrak{n}_e$ is a nil ideal of R_e (its upper nil radical in fact) and that $N = G \cap (1 + \mathfrak{a})$. N is unipotent, locally nilpotent and hyperabelian as above.

If M is free and $\mathfrak{n}$ is nilpotent, say $\mathfrak{n}^r = \{0\}$, then $N = G(\mathfrak{n})$ stabilizes the series $M \supseteq M\mathfrak{n} \supseteq M\mathfrak{n}^2 \supseteq \cdots \supseteq M\mathfrak{n}^r = \{0\}$. Therefore N is nilpotent of class less than r. □

For the remainder of this chapter we shall use the following notation. R is a commutative ring, $\mathfrak{n}$ is the nil radical of R, M is a finitely generated R-module on e generators and G is a subgroup of $\mathrm{Aut}_R(M)$. If R is Noetherian then $\{0\} = M_0 \subseteq M_1 \subseteq \cdots \subseteq M_n = M$ denotes a series of fully invariant submodules of M for which there exist prime ideals $\mathfrak{p}_1, \ldots, \mathfrak{p}_n$ of R such that for each i, $M_i \mathfrak{p}_i \subseteq M_{i-1}$ and M_i/M_{i-1} is torsion-free as $R/\mathfrak{p}_i$-module. F_i denotes the quotient field of $R/\mathfrak{p}_i$, $V_i = (M_i/M_{i-1}) \otimes_{R/\mathfrak{p}_i} F_i$, $e_i = \dim_{F_i} V_i$, η_i is the induced ring homomorphism of $\mathrm{End}_R(M)$ into $\mathrm{End}_{F_i}(V_i)$ and $U = \bigcap_{i=1}^{n} C_G(M_i/M_{i-1})$.

Unipotent Groups

13.6. *If R is Noetherian, then G is unipotent if and only if G stabilizes a finite series of submodules of M. In particular if G is unipotent then G is nilpotent of class at most $e_1 + e_2 + \cdots + e_n$.*

It is clear that if G stabilizes a finite series in M then G is unipotent and nilpotent. Suppose that G is unipotent. Then $G\eta_i$ is a unipotent subgroup of $\mathrm{Aut}_{F_i}(V_i)$ – because η_i is a ring homomorphism – and so stabilizes a finite series of subspaces of V_i of length at most e_i (1.21). Now M_i/M_{i-1} is embedded in V_i and thus G stabilizes a finite series of $R/\mathfrak{p}_i$-submodules (and hence R-submodules) of M_i/M_{i-1} of length at most e_i. The result now follows. □

If R is Noetherian the periodic subgroup of the additive group of M has finite exponent since $N_i = \{x \in M : (i!) x = 0\}$ is an R-submodule of M and $N_i \subseteq N_{i+1}$ for each positive integer i.

13.7. *If R is Noetherian and G is unipotent then the periodic subgroup of G has finite exponent.*

G stabilizes a finite series of submodules $\{0\} = N_0 \subseteq N_1 \subseteq \cdots \subseteq N_r = M$ of M. The proof is by induction on the length of this series. Clearly we can suppose that G is periodic and by induction we can assume that there exists an integer k such that $G^k \subseteq C_G(M/N_1)$. Then G^k is isomorphic to a periodic subgroup of $\mathrm{Hom}_R(M/N_1, N_1)^+$. Since the periodic subgroup of N_1 has finite exponent the periodic subgroup of $\mathrm{Hom}_R(M/N_1, N_1)^+$ has finite exponent and the result follows. □

The exponent can be bounded in terms of the structure of R and M only.

13.8. *If R is Noetherian and W is a unipotent subgroup of G such that We G, then* $W \subseteq \zeta_h(G)$ *where*

$$h = e\left(\sum_{i=1}^{n-1} e_i\right) + \tfrac{1}{2} \max_{1 \leq i \leq n} \{e_i(e_i - 1)\}.$$

If we apply the proofs of 4.12 and 4.13 to the above situation using 13.6 in place of 1.21 we obtain 13.8 except that the finite h so obtained is not bounded in terms of the structure of M, see Gruenberg [19]. However the proof needs only minor modifications to obtain the bound.

We induct on n. By 4.13 we have

$$[W\eta_1,\ _{\frac{1}{2}e_1(e_1-1)}G\,\eta_1] = \{1\}, \quad \text{so} \quad [W,\ _{\frac{1}{2}e_1(e_1-1)}G] \subseteq C_G(M_1).$$

Induction on n yields $[W, {}_k G] \subseteq C_G(M/M_1)$ where

$$k = e\left(\sum_{i=2}^{n-1} e_i\right) + \tfrac{1}{2} \max_{1 < i \leq n} \{e_i(e_i - 1)\}.$$

Hence

$$[W, {}_{h-ee_1}G] \subseteq A = C_G(M/M_1) \cap C_G(M_1).$$

There exists a G-monomorphism ϕ of A into $H_1 = \operatorname{Hom}_R(M/M_1, M_1)$, cf. the proof of 4.12. Now M_1 is torsion-free as $R/\mathfrak{p}_1$-module, and hence so is H_1. Consequently H_1 embeds as RG-module into $F_1 \otimes_{R/\mathfrak{p}_1} H_1$, the latter being a vector space over F_1 of dimension at most $e\,e_1$. Hence $[W \cap A, {}_{ee_1}G] = \{1\}$ as in the proof of 4.12 and the result follows. □

13.9. *If R is Noetherian and W is a unipotent G-hypercyclic normal subgroup of G, then W has paraheight at most*

$$k = e\left(\sum_{i=1}^{n-1} e_i\right) + \tfrac{1}{2}\sum_{i=1}^{n} e_i(e_i - 1).$$

The proof of 13.9 is very similar to that of 13.8, using 11.4 and the latter part of the proof of 11.3 in place of 4.13 and the latter part of the proof of 4.12. The slightly different bound arises as follows. If J is any group and K_i is a normal subgroup of J such that J/K_i has central height k_i and paraheight l_i, $i = 1, 2$, then although $J/(K_1 \cap K_2)$ has central height at most $\max\{k_1, k_2\}$, we can only assert that $J/(K_1 \cap K_2)$ has paraheight at most $l_1 + l_2$. (This problem did not arise during the proof of 11.4 since there we could arrange for one of the l_i to be zero.) □

13.10. *If M is free and R has nilpotent nil radical then G is unipotent if and only if G stabilizes a finite series of submodules of M. In particular, if G is unipotent then G is nilpotent.*

If M is free of rank m, G unipotent and $\mathfrak{p}$ any prime ideal of R, then

$$\underbrace{M(G-1)(G-1)\ldots(G-1)}_{m \text{ times}} \subseteq M\mathfrak{p}.$$

Since this is for all such $\mathfrak{p}$,

$$M(G-1)(G-1)\ldots(G-1) \subseteq \bigcap_{\mathfrak{p}} M\mathfrak{p} = M\left(\bigcap_{\mathfrak{p}} \mathfrak{p}\right) = M\mathfrak{n}$$

since M is free. Therefore $M\mathfrak{n}^i(G-1)\ldots(G-1) \subseteq M\mathfrak{n}^{i+1}$ for each i and the result follows. □

The nilpotency class of G in 13.10 is bounded of course in terms of the rank of M and the exponent of the nilradical of R.

In general a unipotent group need not stabilize a finite series of submodules, and need not be nilpotent. The stability group U_r of the series

$$M_r \supset p\,M_r \supset p^2\,M_r \supset \cdots \supset p^r\,M_r = \{0\}$$

in the free $\mathbb{Z}|p^r$-module M_r of rank two is nilpotent of class $r-1$. The direct product of the U_r for $r = 1, 2, \ldots$ is therefore a non-nilpotent group that is isomorphic to a unipotent group of automorphisms of a free module of rank two over the cartesian product of the $\mathbb{Z}|p^r$. However it does follow from 13.6 and 13.4 that a unipotent group is always locally nilpotent.

Soluble and Locally Soluble Groups

13.11 (Gruenberg [19]). *If R is Noetherian and G is soluble then G contains a normal subgroup T of finite index such that T' is unipotent.*

For by Mal'cev's Theorem (3.6) $G\eta_i$ contains a triangularizable normal subgroup T_i of finite index. Put $T = \left(\bigcap_{i=1}^{n} T_i\,\eta_i^{-1}\right) \cap G$. Certainly T is a normal subgroup of finite index in G. But T_i' acts unipotently on $V_i \supseteq M_i/M_{i-1}$ and $T' \subseteq T_i'\,\eta_i^{-1}$. Therefore T' is unipotent. □

Let $\mathfrak{X}$ be a class of groups such that $Q\mathfrak{X} = \mathfrak{X}$, $\mathfrak{X} \subseteq L(\mathfrak{G} \cap \mathfrak{X})$ and $\mathfrak{X} \cap \mathfrak{F} \subseteq \mathfrak{S}$.

13.12. (Gruenberg [19].) *If R is Noetherian and $G \in \mathfrak{X}$ then G is soluble.*

For by 4.3 each $G\eta_i$ is soluble. Since U is nilpotent G is soluble. □

13.13. *If $G \in \mathfrak{X}$ then G is locally soluble.*

If H is any finitely generated subgroup of G there exists a finitely generated $\mathfrak{X}$-subgroup K of G containing H. By 13.4 and 13.12 K is soluble. Therefore G is locally soluble. □

13.14. *If G is locally soluble then G is hyperabelian-and-locally-nilpotent by soluble, and in particular G is hyperabelian. If in addition M is free of* *rank m and* $\mathfrak{n}^r = \{0\}$ *then G is nilpotent of class less than r by soluble with derived length at most* $f(m)$, *and in particular G is soluble.*

A soluble linear group of degree m has derived length at most $f(m)$ by 3.7. If G is residually soluble with derived length at most s then G is soluble with derived length at most s. 13.14 follows at once from these remarks and 13.5. ☐

13.15. *If G is hyper locally soluble then G is locally soluble.*

This follows at once from 13.13 since the class of hyper locally soluble groups satisfies the conditions on $\mathfrak{X}$. ☐

13.16. *G contains a unique maximal locally soluble normal subgroup.*

By Zorn's Lemma G contains maximal locally soluble normal subgroups. If H and K are locally soluble normal subgroups of G, then HK is locally soluble (by either 13.14 or 13.15) and the result follows. ☐

Needless to say the proceeding four results do not hold for arbitrary groups. For examples see [47] 4.22, 4.24 and its corollary. One can partially generalise 13.11.

13.17. *If M is free and G is locally soluble then G contains a normal subgroup T such that G/T has finite exponent and T' is unipotent (and so locally nilpotent).*

If $\mathfrak{p}$ is a prime ideal of R there exists a normal subgroup $T(\mathfrak{p})$ of G of finite index at most $\mu(n)$ such that $T(\mathfrak{p})'$ acts unipotently on $M/M\mathfrak{p}$ – see 3.6 and 13.5. Put $T = \bigcap_{\mathfrak{p}} T(\mathfrak{p})$. Certainly G/T has finite exponent. If M has rank m,

$$\underbrace{M(T'-1)(T'-1)\ldots(T'-1)}_{m \text{ times}} \subseteq \bigcap_{\mathfrak{p}} M\mathfrak{p} = M\mathfrak{n}.$$

But M is free and $\mathfrak{n}$ is a nil ideal. Therefore T' is unipotent. ☐

The Upper Central Series

We use the notation of the latter part of Chapter 8.

13.18 (Gruenberg [19]). *If R is Noetherian then*
    i) $L(G) = \sigma(G) = \eta(G)$, *which is hypercentral with height less than* $\omega 2$.
    ii) $\bar{L}(G) = \bar{\sigma}(G) = \eta_1(G)$, *which is nilpotent.*
    iii) $R(G) = \rho(G) = \zeta(G)$, *and G has central height less than* $\omega 2$.
    iv) $\bar{R}(G) = \bar{\rho}(G) = \zeta_\omega(G)$.

By 13.12, $\langle L(G), R(G)\rangle$ is soluble (see the proof of 8.15). Hence $L(G) = \sigma(G) = \eta(G)$, $\bar{L}(G) = \bar{\sigma}(G)$, $R(G) = \rho(G)$ and $\bar{R}(G) = \bar{\rho}(G)$.

Clearly $U \subseteq \eta_1(G) \subseteq \eta(G)$, so $U \in \eta(G)$. Therefore by 13.8, $U \subseteq \zeta_h(\eta(G))$ for some finite h. Since G/U is quasi-linear $\eta(G/U)$ is hypercentral with height less than $\omega 2$ and $\bar{\sigma}(G)/U \subseteq \bar{\sigma}(G/U)$ which is nilpotent. Parts i) and ii) now follow.

By 13.8, $\rho(G) \cap U \subseteq \zeta_k(G)$ for some finite k. Since G/U is quasi-linear $\rho(G)/\rho(G) \cap U \cong_G \rho(G)\,U/U$ is a hypercentral factor of G of height less than $\omega 2$ and $\bar{\rho}(G)/\rho(G) \cap U$ is a hypercentral factor of G with height at most ω. The result now follows.   □

13.19. *If M is free and R has nilpotent nil radical then*

   i) $L(G) = \sigma(G) = \eta(G)$;
   ii) $\bar{L}(G) = \bar{\sigma}(G)$;
   iii) $R(G) = \rho(G)$;
   iv) $\bar{R}(G) = \bar{\rho}(G)$.

By 13.14, $\langle L(G), R(G)\rangle$ is soluble and the result follows as above.   □

13.20.

   i) $L(G) = \sigma(G) = \eta(G)$;
   ii) $R(G) = \rho(G)$.

$\langle L(G), R(G)\rangle$ is locally soluble by 13.12. In any soluble group the right Engel elements are left Engel elements and all such lie in the Hirsch-Plotkin radical by [17] Theorem 4. Therefore $\langle L(G), R(G)\rangle$ is locally nilpotent and so $R(G) \subseteq L(G) = \eta(G)$. By [17] Prop. 3 and 13.14 we have $\eta(G) = \sigma(\eta(G)) = \sigma(G)$ and from [17] Lemma 15 and what we have just shown it follows that $R(G) = \rho(G)$.   □

Question 21. *Does $R(G) = \zeta(G)$? In particular if G is locally nilpotent is G necessarily hypercentral?*

13.21. *If R is Noetherian and G is locally nilpotent, then $(G : \zeta_\omega(G))$ and $(G : \eta_1(G))$ are both finite.*

By 13.8 there exists a positive integer h such that $U \subseteq \zeta_h(G)$. The result now follows from 8.6, 13.18 and the trivial fact that $\zeta_\omega(G) \subseteq \eta_1(G)$.   □

Question 22. *If R is Noetherian are $(\zeta(a) : \zeta_\omega(G))$ and $(\eta(G) : \eta_1(G))$ finite ?*

Question 23. *If R is Noetherian and G is locally nilpotent, does there exist a positive integer $c = c(G)$ such that $\zeta_{\alpha+1}(G)/\zeta_\alpha(G)$ is finite for all $\alpha > c$? If the answer is yes, what if G is not assumed to be locally nilpotent?*

13.22 (Gruenberg [20]). *Let L be a group and $A \subseteq H$ normal subgroups of L such that $A \subseteq \zeta_k(L)$ and such that H/A is a locally finite π-group,*

where k is a non-negative integer and π is some set of primes. Then $[H, {}_kL]$
is a locally finite π-group.

The proof uses induction on k, if $k=0$ the result is a triviality. Let $K=[H, {}_{k-1}L]$ and $B=[A, {}_{k-1}L]$; K and B are normal subgroups of L, B is central in L and by induction K/B is a locally finite π-group. If $x \in K$ and $g \in L$ there exists a π-number r such that $x^r \in B$. Hence $\{1\} = [x^r, g] \equiv [x, g]^r$ modulo K' and thus $[K, L]/K'$ is an abelian π-group.

Suppose that $c_1, \ldots, c_s$ are elements of K'. There exist elements $x_1, \ldots, x_t$ of K such that if $X = \langle x_1, \ldots, x_t \rangle$ then $\langle c_1, \ldots, c_s \rangle \subseteq X'$. Now $X/B \cap X$ is a finite π-group (since K/B is a locally finite π-group) and $(B \cap X) \subseteq \zeta_1(X)$. By Schur's Theorem (see [27] p. 59, or [24 a] 8.1) X' is a finite π-group and so K' is a locally finite π-group. 13.22 now follows. $\square$

13.23. *If R is Noetherian there exists a positive integer r such that $[\zeta(G), {}_rG]$ is a π-group for some finite set of primes π, and is an extension of a unipotent group by an abelian group with minimal condition on subgroups.*

By 8.6 iii) there exists a positive integer s such that

$$[\zeta(G), {}_sG]/U \cap [\zeta(G), {}_sG]$$

is an abelian group satisfying the minimal condition on subgroups, and by 13.8 there exists a positive integer h such that $U \cap [\zeta(G), {}_sG] \leq \zeta_h(G)$. Therefore $[\zeta(G), {}_{s+h}G]$ has the desired properties by 13.22. $\square$

Notice that in the above r can be chosen to depend only on R and M and independent of the particular subgroup G of $\text{Aut}_R(M)$ under consideration.

13.24 (Gruenberg [20]). *If R is Noetherian and G is torsion-free then G has finite central height and nilpotent Hirsch-Plotkin radical.*

This is an immediate consequence of 13.23. $\square$

In 13.24, G is even centrally stunted.

Question 24. *If R is Noetherian and G contains no $\mathbf{C}_{p^\infty}$-groups for any prime p, does G have finite central height and nilpotent Hirsch-Plotkin radical? Cf. 8.8.*

Finitely Generated Groups

If G is finitely generated then we can assume that R is a finitely generated commutative ring (13.4). In particular, we are in the Noetherian situation. We shall be slightly more general and just consider R to be finitely generated.

13.25. *Let R be a finitely generated integral domain, F the quotient field of R and $M = (x_1, \ldots, x_e)$ a finitely generated torsion-free R-module. Then $E = \mathrm{End}_R(M)$ is isomorphic to a subring of the matrix ring S_m where $m = \dim_F(M \otimes_R F)$ and S is a finitely generated subring of F containing R.*

Note that in particular $\mathrm{Aut}_R(M)$ is isomorphic to a subgroup of $GL(m, S)$. Thus the results of Chapter 4 can now be stated more generally (replacing finitely generated free R-module where R is a finitely generated integral domain by finitely generated torsion-free R-module where R is as before). This we leave to the reader.

The proof of 13.25 is as follows. There exist elements $y_1, \ldots, y_m$ of M such that $y_1 \otimes 1, \ldots, y_m \otimes 1$ is a basis of $M \otimes_R F$, and elements α_{ij} of F such that

$$ x_i \otimes 1 = \sum_j y_j \otimes \alpha_{ij}. $$

Let S be the subring of F generated by R and the em elements α_{ij}. It is easy to see that

$$ \sum_j (y_j \otimes 1) S = \bigoplus_j (y_j \otimes 1) S = (M \otimes 1) S \subseteq M \otimes_R F. $$

Since M is torsion-free, $M \to M \otimes 1$ is an embedding and

$$ \phi \mapsto (\phi \otimes 1)|_{(M \otimes 1)S} $$

determines a ring monomorphism of $\mathrm{End}_R(M)$ into S_m. ▯

13.26. *If R is finitely generated then $\mathrm{Aut}_R(M)$ is an extension of a unipotent group by a subdirect product of a finite number of subgroups of groups $GL(n_i, S_i)$ where the n_i are positive integers and the S_i are finitely generated integral domains.*

This follows from 13.2 and 13.25. ▯

13.27. *If R is finitely generated then G has finite central height and nilpotent Hirsch-Plotkin radical.*

This follows from 8.8 and 13.26, cf. the proof of 13.23. ▯

In 13.27, G is even centrally stunted.

13.28. *If R is finitely generated then G contains a normal subgroup T of finite index such that every element of T of finite order is unipotent.*

This follows from 4.8, 13.2 and 13.25. ▯

13.29. *If R is finitely generated and G is soluble then G is unipotent by finitely generated abelian by finite.*

This follows from 4.10, 13.2 and 13.25. ☐

The residual finiteness theorems of Chapter 4 do not seem to extend so easily, but we do have the following trivial result.

13.30. *If R is residually finite (as a ring, e.g. if R is finitely generated) and if M is free then* $\text{Aut}_R(M)$ *is a residually finite group.* ☐

13.31. *If R is finitely generated then either G is soluble-by-finite, or G contains a non-cyclic free subgroup.*

If G is not soluble-by-finite, then neither is G/U. In this case G/U contains a non-cyclic free subgroup E/U say, by 10.16. Since E/U is free the extension splits and G contains a subgroup isomorphic to E/U. ☐

13.32. *G satisfies the maximal condition on subgroups if and only if G is polycyclic-by-finite.*

If G satisfies the maximal condition on subgroups, then G is finitely generated. Hence by 13.31, G is soluble-by-finite, and thus G is poly-cyclic-by-finite. The converse is very easy. ☐

13.33. *If R is finitely generated and every maximal subgroup of G of finite index either is of prime or prime square index, or is hypercyclic, then G is soluble.*

This follows at once from 13.26 and 11.18. ☐

13.34. *If R is finitely generated and every 2-generator subgroup of G is soluble-by-finite, then G is soluble-by-finite.*

This follows from 10.13 and 13.26. ☐

Exercise 13.1. If R is finitely generated and if there exists an integer t such that $(H:H^2) \le t$ for every subgroup H of G of finite index, prove that G is soluble-by-finite. (Use Exercise 10.1.)

Periodic Groups

13.35. *If G is periodic then G is locally finite.*

This follows from 13.28 or from 13.3 and 4.9. ☐

13.36. *If R is Noetherian and G is periodic then G is unipotent-by-countable.*

This follows from 13.3 and 9.5. ☐

13.37. *If G is periodic and if every Sylow subgroup of G satisfies the minimal condition on subgroups then G is abelian-by-finite.*

It is not difficult to prove that the Sylow subgroups of any homomorphic image of G also satisfy the minimal condition on subgroups. (See [31a] 3.13.) Hence by 13.5 and 9.7, G contains a series containing $\{1\}$ and G with finite factors. A theorem of M. I. Kargarpolov (cf. [31a] 3.19) now guarantees that G is abelian-by-finite. $\quad\square$

An immediate consequence of 13.37 is the following.

13.38. *If G satisfies the minimal condition on subgroups then G is abelian-by-finite.* $\quad\square$

The conjugacy theorems break down in this more general situation.

13.39. *There exist R, M and G such that G is soluble, is an extension of a 3-group by a 2-group, and has non-conjugate Sylow 2-subgroups.*

Choose any field F and integer n such that $GL(n, F)$ contains a dihedral subgroups S of order six. Let R be the cartesian product of $\aleph_0$ copies of F and let M be the cartesian product of $\aleph_0$ copies of the n-row vector space over F. Then M is a free R-module of rank n in the obvious way. Further $\mathrm{Aut}_R(M)$ contains a subgroup G that is the direct product of copies S_i of S, $i = 1, 2, \ldots$. The group S_i contains two distinct Sylow 2-subgroups P_i and Q_i say. Then

$$P = \langle P_i \colon i = 1, 2, \ldots \rangle \quad \text{and} \quad Q = \langle Q_i \colon i = 1, 2, \ldots \rangle$$

are maximal 2-subgroups of G. P and Q will be conjugate in G if and only if there exists an element g of G such that $P_i^g = Q_i$ for each i. Since G is only the direct product of the S_i, all but a finite number of the components of g are 1. Hence no such g can exist and P and Q are not conjugate in G. $\quad\square$

Question 25. *If R is Noetherian, G periodic and p a prime, are the Sylow p-subgroups of G necessarily all conjugate?*

Chief Factors

13.40. [69d]. *G contains a unique maximal unipotent normal subgroup.*

Clearly unipotence is a local property, so it suffices to prove that the product of any two unipotent normal subgroups V and W of G is also unipotent. Let $v \in V$ and $w \in W$, and suppose that M is generated by its elements $x_1, x_2, \ldots, x_e$. For $i, j = 1, 2, \ldots, e$ there exist elements α_{ij} and β_{ij} of R such that

$$x_i v = \sum_{j=1}^{e} x_j \alpha_{ij} \quad \text{and} \quad x_i w = \sum_{j=1}^{e} x_j \beta_{ij}.$$

Let S denote the subring of R generated by the $2e^2$ elements α_{ij} and β_{ij}, and put $N = \sum_{i=1}^{e} x_i S$ and $H = \{g \in G : Ng = N\}$. Then H is a subgroup of G, $v \in H \cap V -$ since $v^{-1} = \sum_{i=0}^{\infty} (1-v)^i$, $w \in H \cap W$, and $H \cap V$ and $H \cap W$ both act unipotently on N.

S being finitely generated, is Noetherian. By 13.6, $H \cap V$ stabilizes in N a finite series of S-submodules. Put $N_0 = \{0\}$ and define N_i inductively by

$$N_{j+1}/N_j = \{x \in N/N_j : xy = x \text{ for every } y \in H \cap V\} = C_{N/N_j}(H \cap V).$$

Then N_i is an S-submodule of N, and since $H \cap W$ normalizes $H \cap V$, N_i is also an $H \cap W$-submodule of N. Thus $H \cap W$ acts unipotently on each factor N_{i+1}/N_i. But $N = N_r$ for some r. Hence $(H \cap V)(H \cap W)$ acts unipotently on N and $N(vw-1)^t = 0$ for some positive integer t. Consequently

$$M(vw-1)^t = NR(vw-1)^t = N(vw-1)^t R = \{0\}$$

and therefore VW is a unipotent normal subgroup of G.   ☐

This unique maximal unipotent normal subgroup of G we denote by $u(G)$. $\operatorname{Ann}_T(X)$ denotes the annihilator of the T-module X in the ring T.

13.41. [69 d]. *Suppose that $G/u(G)$ is finitely generated. If X/Y is any chief factor of G covered by $u(G)$, (meaning $X \subseteq Y \cdot u(G)$), then X/Y is finite and $\mathbb{Z}G/\operatorname{Ann}_{\mathbb{Z}G}(X/Y)$ is isomorphic to a matrix ring of finite degree, m say, over a finite field. Further if R is Noetherian there exists a bound for m independent of X/Y (but dependent on G).*

The proof of 13.41 is very similar to the proof of 4.26, so we leave it to the reader (see Theorem 4 of [69 d] for the details). It contains only one extra complication; it is now no longer a triviality that $[u(G), X) \subseteq Y$. If H is any finitely generated subgroup of G then $H \cap u(G)$ is unipotent and hence $H \cap u(G)$ is nilpotent by 13.4 and 13.6. Thus the following lemma resolves the difficulty.

Lemma. *Let J be a group and V a locally nilpotent normal subgroup of J such that $H \cap V$ is nilpotent for every finitely generated subgroup H of J. Then $[N, V] = 1$ for every minimal normal subgroup N of J.*

There do exist groups with locally nilpotent and non-abelian chief factors, see [23].

Proof. If N is not contained in V then $N \cap V = \{1\}$ and $[N, V] = \{1\}$. Assume therefore that $N \subseteq V$ and suppose that $[N, V] \neq \{1\}$. Then there exist elements $a \in N$ and $v \in V$ such that $b = [a, v] \neq 1$. Since N is a minimal normal subgroup of J, b generates N as a normal subgroup of J. Hence $a \in \langle b^X \rangle$ for some finite subset X of J. Then $H = \langle a, v, X \rangle$ is a finitely generated subgroup of J. If $A = \langle a^H \rangle$ then

$$b = [a, v] \in [A, H \cap V],$$

which is a normal subgroup of H. Hence

$$a \in \langle b^X \rangle \subseteq [A, H \cap V]$$

and so $A = [A, H \cap V]$. But by assumption $A \subseteq H \cap N \subseteq H \cap V$ and $H \cap V$ is nilpotent. Therefore $A \supset [A, H \cap V]$ and this contradiction proves the point. $\square$

Just as 4.26 implied 4.28, so 13.41 has the following corollary.

13.42. [69 d]. *If R is finitely generated and G is soluble-by-finite, then every chief factor of G is finite and every maximal subgroup of G has finite index.* $\square$

Miscellaneous Results

13.43. *Suppose that every 2-generator subgroup of G is soluble-by-finite. If R is Noetherian then G is soluble-by-locally-finite and if R is finitely generated then G is soluble-by-finite.*

This is an immediate consequence of 10.13, 13.2, 10.12 and 13.26. $\square$

13.44. (D. Segal). *Suppose that every 2-generator subgroup of G is nilpotent-by-finite. If R is Noetherian then G is nilpotent-by-locally-finite and if R is finitely generated then G is nilpotent-by-finite.*

We induct on n. By 10.14, 13.2 and induction we may assume that G/A is nilpotent-by-locally-finite, where

$$A = C_G(M_1) \cap C_G(M/M_1).$$

There exists a G-monomorphism ϕ of A into $H_1 = \mathrm{Hom}_R(M/M_1, M_1)$, cf. the proof of 4.12. Now M_1 is torsion-free as $R/\mathfrak{p}_1$-module, so if F_1 denotes as usual the quotient field of $R/\mathfrak{p}_1$ then H_1 embeds as RG-module into the finite-dimensional vector space $H = F_1 \otimes_{R/\mathfrak{p}_1} H_1$. The proof is completed in a similar way to that of 10.14. $\square$

Exercise 13.2. If every 2-generator subgroup of G is soluble (resp. nilpotent) show that G is locally soluble (resp. locally nilpotent).

13.45. (D. Segal). *If every 2-generator subgroup of G is supersoluble then G is locally supersoluble. If R is Noetherian and G is locally supersoluble, then G is hypercyclic.*

If H is a finitely generated subgroup of G then H is nilpotent-by-finite by 13.44 and 13.4. H is isomorphic to a linear group by 2.5 and 2.3 and hence H is supersoluble by 11.23. Therefore G is locally supersoluble.

Now assume that R is Noetherian. We induct on n. Thus with the notation of the preceeding proof we may suppose that G/A is hypercyclic. As in 13.44 there exists a G-monomorphism ψ of A into the finite-dimensional vector space H over F_1; let B denote the subspace of H spanned by $A\psi$. Denote by ρ the induced homomorphism of G into $\mathrm{Aut}_{F_1}(B)$. Choose a basis $X\psi$ of B in $A\psi$ and a finite subset Y of G such that $Y\rho$ spans $G\rho$ over F_1.

By hypothesis $\langle X, Y \rangle$ is supersoluble. Therefore there exists a series of $F_1\langle Y \rangle$-submodules

$$\{0\} = B_0 \subset B_1 \subset \cdots \subset B_s = B$$

of B such that $\dim_{F_1}(B_i/B_{i-1}) = 1$ for $i = 1, 2, \ldots, s$. Let $g \in G$. By the choice of Y there exist elements $\alpha_y \in F_1$ such that

$$b\,g = \sum_{y \in Y} \alpha_y\, b\, y$$

for each $b \in B$. Hence $B_i g \subseteq B_{i-1}$ for each i. There exists a_i in A such that $a_i \psi \in B_i \smallsetminus B_{i-1}$, and $\langle a_i, g \rangle$ is supersoluble. Hence there exists a_i' in A such that $a_i' \psi \in B_i \smallsetminus B_{i-1}$ and $a_i' \psi g \in n_i a_i' \psi + B_{i-1}$ for some integer n_i. If $b \in B_i \smallsetminus B_{i-1}$ then $b \in \alpha(a_i' \psi) + B_{i-1}$ for some $\alpha \in F_1$, from which it follows that $b g \in n_i b + B_{i-1}$. Therefore B is a hypercyclic G-module and consequently G is hypercyclic. □

The following result follows at once from 13.9, 13.3, 13.26 and 11.6.

13.46. (B. Hill). *If R is Noetherian then G has paraheight less than $\omega 2$ and if R is finitely generated then G has finite paraheight.* □

13.47. *If R is Noetherian then either G is soluble-by-finite or G generates the variety of all groups.*

This result follows at once from 10.15 and 13.3. □

13.48. *If R is Noetherian and G has finite Prüfer rank then G is soluble-by-finite.*

This follows at once from 10.9 and 13.3. □

200

Linear Groups over Non-Commutative Rings

Nothing much seems to be known about the structure of linear groups over non-commutative rings. A.E. Zalesskii in [71a] has investigated the subgroups of $GL(n, D)$ where D is a division ring, but more counter examples seem to exist than theorems.

It would be very useful to have some structural theorems about subgroups of $GL(n, R)$ where R is the integral group-ring of a finitely generated nilpotent group. If one did have such things one could hope, for example, to be able to say something significant about the auto-morphism groups of finitely generated abelian-by-nilpotent groups (prehaps even of $\mathfrak{G} \cap \mathfrak{N}^2$-groups). Such a ring R can be embedded into a division ring, so Zalesskii's results tell us a little about subgroups of $GL(n, R)$. Perhaps this is the direction in which the abstract theory of linear groups ought now to go.

14. Appendix on Algebraic Groups

In this chapter we have set ourselves two tasks. There are a number of simplifications that take place if one can reduce a problem to involve only closed subgroups of $GL(n, F)$. Our first aim is to give an account of these results, and in most cases also their proofs. In a number of places in this book we have skirted round some of these properties of algebraic groups and here and there we have come very close to using them. I hope that this chapter will fill in the background of these a bit more.

The second object of this chapter is to give a description (without proofs of course) of the much deeper and very extensive structure theorems for algebraic groups, especially those that we have referred to or hinted at earlier in the book. Hopefully we shall help the reader with little knowledge of algebraic geometry to get some idea of what an algebraic group looks like. The reader who seriously wishes to learn about algebraic groups should consult Borel's book [3a]. The classical reference for algebraic groups is Chevalley's seminar [8]. See also the monumental work "Groupes Algébriques" by M. Demazure and P. Gabriel (Tome 1, North-Holland Pub. Co., Amsterdam 1970).

For simplicity, throughout this chapter F denotes an algebraically closed field. By an *algebraic linear group over F* we mean a closed subgroup of $GL(n, F)$. We shall often drop the word linear and just use the term algebraic group to mean algebraic linear group. Strictly an algebraic group is a group that carries the structure of an algebraic variety in a sensible way. It can then be shown that an algebraic group is isomorphic (as algebraic group) to what we have called an algebraic linear group if and only if its varietal structure is affine. However, we are interested here in subgroups of the general linear groups, and we therefore confine ourselves to the affine case. (Incidentally this is not a very strong restriction as things turn out since every algebraic group has an affine normal subgroup modulo which the group is abelian-by-finite.)

One word of warning on the literature: to say that a group G is defined over F does not mean that it is a subgroup of $GL(n, F)$, but, at least for perfect F, that it is the zeros of a set of polynomials with coefficients in F. For example $SL(n, \mathbb{C})$ is defined over $\mathbb{Q}$ since it is defined by a polynomial, namely det $X - 1$, whose coefficients lie in $\mathbb{Q}$. We are not interested here in this aspect of algebraic groups and will avoid using the phrase "defined over".

Morphisms

Suppose that G is a subset of $GL(n, F)$ and H a subset of $GL(m, F)$. Call a mapping ϕ of G into H a *polynomial map* if for each connected component C of G there exists m^2 polynomials p_{ij} for $i, j = 1, 2, ..., m$ in the variables $X_{ij}, i, j = 1, 2, ..., n$ and $(\det X)^{-1}$ with coefficients in F such that

$$x \phi = (p_{ij}(x))$$

for each $x \in C$. Here X denotes the $n \times n$ matrix whose entries are the n^2 indeterminates X_{ij}. We call ϕ a *polynomial homomorphism* if G and H are subgroups and if ϕ is a polynomial map and a homomorphism of groups. Such a ϕ is "a morphism in the category of algebraic groups". However we are mainly interested in the properties of algebraic groups regarded as objects in the category of groups, so rather than change categories, or specifying the category every time, we keep the word homomorphism to mean morphism in the category of groups and we have introduced the rather cumbersome term polynomial homomorphism.

At first sight the choice of polynomial homomorphisms as the morphisms in the category of algebraic groups is rather extraordinary and we owe the reader some explanation. The two obvious questions, perhaps, are why polynomials and not, say, rational functions, and if polynomials why not have them just in the X_{ij}. The remainder of this section is devoted to clearing up these points. Note that every polynomial map is continuous by 5.1.

A topological space is said to be *irreducible* if it is not the union of two closed proper subsets.

14.1. Lemma. *Let U denote the m-row vector space over F, V a subset of U and $\mathfrak{a} = \text{Ann}(V)$, the annihilator of V in the polynomial ring*

$$R = F[X_1, X_2, ..., X_m].$$

Then V is irreducible if and only if $\mathfrak{a}$ is a prime ideal of R.

Proof. Suppose that V is irreducible. Always $\mathfrak{a}$ is an ideal, so let f and g be elements of R with $fg \in \mathfrak{a}$. Let A denote the subset of U of all common zeros of the elements of $(f, \mathfrak{a})$ and B of $(g, \mathfrak{a})$. If $x \in V$, then

$$f(x) \cdot g(x) = fg(x) = 0,$$

so either $f(x)=0$ or $g(x)=0$. That is $x \in A \cup B$, so $V \subseteq A \cup B$. But V is irreducible; thus either $V \subseteq A$ or $V \subseteq B$. Consequently either $f \in \mathrm{Ann}(V) = \mathfrak{a}$ or $g \in \mathfrak{a}$, and therefore $\mathfrak{a}$ is prime.

Now suppose that $\mathfrak{a}$ is prime. If $V \subseteq A \cup B$, where A and B are both closed in U, then

$$\mathrm{Ann}(A) \cdot \mathrm{Ann}(B) \subseteq \mathfrak{a}.$$

Hence either $\mathrm{Ann}(A) \subseteq \mathfrak{a}$ or $\mathrm{Ann}(B) \subseteq \mathfrak{a}$. Now since A and B are closed they are equal to the sets of common zeros of their annihilators. Therefore either $V \subseteq A$ or $V \subseteq B$ and consequently V is irreducible. $\square$

Let U, V, R and $\mathfrak{a}$ be as in 14.1 and suppose that V is closed and irreducible. Then $R/\mathfrak{a}$ is an integral domain by 14.1 and so has a quotient field K, say. Let $k \in K$. If $x \in V$ there may exist elements r_x and s_x of R such that

$$k = (r_x + \mathfrak{a})/(s_x + \mathfrak{a}) \quad \text{and} \quad s_x(x) \neq 0.$$

In such a situation the element $x \phi_k = r_x(x) s_x(x)^{-1}$ of F depends only on k and not on the choice of r_x and s_x. We say that k is *defined at* x. The ϕ_k here is a mapping of the domain of definition of k in V into F; it is not difficult to show that it is a rational mapping in the sense that we used the term in Chapter 6. Incidentally the elements of K are what the geometer means by the rational functions on V, even when they are not defined everywhere in V. We keep the above notation in the following lemma.

14.2. Lemma. *If k is defined at every point of the closed irreducible subset V of U, then $k \in R/\mathfrak{a}$ and there exists a polynomial p in R such that $x \phi_k = p(x)$ for every x in V.*

Proof. Let $\mathfrak{b}$ denote the ideal of R generated by all the elements s of R for which there exists an r in R satisfying $k = (r + \mathfrak{a})/(s + \mathfrak{a})$. Suppose that $\mathfrak{b} \neq R$. Then by the Nullstellensatz (and the assumption that F is algebraically closed, [29] Theorem 32) there exists an element u of U with $f(u) = 0$ for every f in $\mathfrak{b}$. Trivially $\mathfrak{a} \subseteq \mathfrak{b}$, so $u \in V$. But by construction k is not defined at u and therefore $\mathfrak{b} = R$.

It follows that there exist elements r_i, s_i and q_i of R for, say, $i = 1, 2, \ldots, t$ such that $k = (r_i + \mathfrak{a})/(s_i + \mathfrak{a})$ for each i and $\sum s_i q_i = 1$. Hence in K

$$k = \sum k(s_i + \mathfrak{a})(q_i + \mathfrak{a}) = \sum (r_i + \mathfrak{a})(q_i + \mathfrak{a}) = (\sum r_i q_i) + \mathfrak{a} \in R/\mathfrak{a}.$$

Finally set $p = \sum r_i q_i$. Clearly $x \phi_k = p(x)$ for every x in V. $\square$

Exercise 14.1. V is a closed subset of the m-row vector space U over the algebraically closed field F, and r and s are elements of

$$R = F[X_1, X_2, \ldots, X_m]$$

such that $s(v) \neq 0$ for every v in V. Prove that R contains an element p such that
$$r(v)/s(v) = p(v)$$
for every v in V.

Hint: consider $(\text{Ann}(V), s)$.

The following lemma we have also recorded in Exercise 5.7. See [27] 8.8 for a somewhat different proof.

14.3. Lemma. *A connected CZ-group G is irreducible as topological space.*

Proof. A simple application of the descending chain condition on closed sets shows that $G = S_1 \cup S_2 \cup \cdots \cup S_r$ for some finite collection of closed irreducible subsets S_i of G. We may clearly assume that $S_i \nsubseteq S_j$ for each $i \neq j$, a condition which implies by the irreducibility of the S_i that
$$S_i \nsubseteq \bigcup_{i \neq j} S_j$$
for each i.

Suppose that $G = S \cup T$ where S and T are closed, S is irreducible and $S \nsubseteq T$. Then from $S \subseteq \bigcup S_i$ we deduce that $S \subseteq S_j$ for some j. Also the irreducible set S_j lies in $S \cup T$ and $S_j \nsubseteq T$ since $S \subseteq S_j$ and $S \nsubseteq T$. Hence $S_j \subseteq S$, which yields $S = S_j$. It now follows that the decomposition $G = \bigcup S_i$ above is unique.

Since the S_i are unique the elements of G permute the set $\{S_1, S_2, \ldots, S_r\}$ by right multiplication. Suppose that $1 \in S_1$; denote the stabilizer of S_1 in G by G_1 and the closure of G_1 in G by $\bar{G}_1$. If $s \in S_1$ then the map $x \mapsto sx$ is a homeomorphism of G that maps G_1 into S_1. Since S_1 is closed we have that $s\bar{G}_1 \subseteq S_1$ for each $s \in S_1$. Thus for $g \in \bar{G}_1$, $S_1 g \subseteq S_1$ and because $S_i \nsubseteq S_1$ for each $i > 1$ it follows that $\bar{G}_1 = G_1$. But G is connected and G_1 has finite index in G, so by 5.3, $G_1 = G$. Therefore
$$G = 1\, G_1 \subseteq S_1\, G_1 = S_1$$
and so $G = S_1$ is irreducible. $\quad\square$

Suppose now that G is a closed subgroup of $\text{SL}(n, F)$ and that ϕ is a mapping of G into the m-row vector space U over F such that for each connected component C of G there exist elements $r_1, r_2, \ldots, r_m$ of $F(X_{11}, X_{12}, \ldots, X_{nn})$ such that
$$x\,\phi = \big(r_1(x), \ldots, r_m(x)\big)$$
for each $x \in C$. Now $\text{SL}(n, F)$ is a closed subset of F_n being the set of zeros of the polynomial $\det X - 1$. Hence G is also a closed subset of F_n. Since G is a group a connected component C of G is homeomorphic to G^0 and consequently is irreducible by 14.3. Put $\mathfrak{a} = \text{Ann}(C)$. Then each r_i determines an element k_i of the quotient field of $F[X_{11}, X_{12}, \ldots, X_{nn}]/\mathfrak{a}$ defined everywhere on C such that in the notation of 14.2, $r_i(x) = x\,\phi_{k_i}$

for every x in C. Therefore there exists by 14.2 a polynomial p_i over F in $X_{11}, X_{12}, \ldots, X_{nn}$ such that $r_i(x) = p_i(x)$ for each x in C. Thus we have shown that ϕ is actually a polynomial map, and so by using rational functions in place of polynomials we do not get, for closed subgroups of the special linear group, a class of maps wider than the class of polynomial maps.

Our aim now is to show this for closed subgroups of $GL(n, F)$. The trouble is that $GL(n, F)$ is not a closed subset of F_n. But it is a principal open subset of F_n, being the complement of the set of zeros of the polynomial $\det X$, and as such it does carry the structure of a closed subset of a vector space of dimension $n^2 + 1$ over F. Rather than saying what this means we do the following.

Let G be a closed subgroup of $GL(n, F)$ and consider the embedding π of $GL(n, F)$ into $SL(n+1, F)$ given by

$$g \, \pi = \mathrm{diag}\big(g, (\det g)^{-1}\big)$$

where $g \in GL(n, F)$. It is easily seen that π is continuous and that the image of any closed subset of $GL(n, F)$ is a closed subset of $SL(n+1, F)$.

Now suppose that ϕ is a mapping of G into U such that for each connected component C of G there exist elements $r_1, r_2, \ldots, r_m$ of $F(X_{11}, X_{12}, \ldots, X_{nn})$ such that $x \phi = (r_1(x), \ldots, r_m(x))$ for each $x \in C$. These same rational functions define the map $(\pi|_C)^{-1} \phi$ of $C \pi$ into U. Hence by the above there exist polynomials $p_1, p_2, \ldots, p_m$ in

$$F[X_{11}, \ldots, X_{nn}, X_{1,n+1}, \ldots, X_{n+1,n+1}]$$

such that

$$x \phi = x \, \pi \, (\pi|_C)^{-1} \phi = \big(p_1(x \, \pi), \ldots, p_m(x \, \pi)\big)$$

for each $x \in C$. But for $i \leq n$ the $(i, n+1)$ and $(n+1, i)$ entries of $x \, \pi$ are zero and the $(n+1, n+1)$ entry of $x \, \pi$ is $(\det x)^{-1}$, so if we let q_i denote the image of p_i in

$$F[X_{11}, X_{12}, \ldots, X_{nn}, (\det X)^{-1}]$$

under the map induced by

$$X_{n+1,n+1} \mapsto (\det X)^{-1}, \quad X_{i,n+1} \mapsto 0, \quad X_{n+1,i} \mapsto 0 \quad \text{for } i \leq n,$$

it follows that $x \phi = (q_1(x), \ldots, q_m(x))$ for every x in C. We have now proved the following theorem.

14.4. Theorem. *Let G be a closed subgroup of $GL(n, F)$ and ϕ a mapping of G into the m-row vector space U over F such that for each connected component C of G there exist elements $r_1, \ldots, r_m$ of $F(X_{11}, X_{12}, \ldots, X_{nn})$ such that $x \phi = (r_1(x), \ldots, r_m(x))$ for each x in C. Then ϕ is a polynomial map.* □

Exercise 14.2. Let G be a closed subgroup of $GL(n, F)$ and ϕ a mapping of G into the m-row vector space U over F. Suppose that G has an open covering $\{O_i : i \in I\}$ such that for each $i \in I$ there exists a rational function r_i in $F(X_{ij} : i, j = 1, 2, \ldots, n)$ satisfying $r_i(x) = \phi(x)$ for every x in O_i. Prove that ϕ is a polynomial map.

Hint: Reduce to considering a closed irreducible subset C of $SL(n+1, F)$ as in the proof of 14.5. If ψ is the map of C induced by ϕ deduce that $\psi = \psi_k$ for some rational function k on C and apply 14.2.

The Dimension of a Linear Group

(In this section the assumption that F is algebraically closed is not used.)

If R is an integral domain and an F-algebra we call the transcendence degree over F of the quotient field of R the dimension of R over F; $\dim_F R$ for short. The following lemma justifies in term dimension.

14.5. Lemma. *Let R be a finitely generated commutative F-algebra and $\mathfrak{p}$ and $\mathfrak{q}$ prime ideals of R such that $\mathfrak{p} \subseteq \mathfrak{q}$. Then $\dim_F R/\mathfrak{p} \geq \dim_F R/\mathfrak{q}$ with equality if and only if $\mathfrak{p} = \mathfrak{q}$.*

Proof. Since the image in $R/\mathfrak{q}$ of an algebraically dependent subset of $R/\mathfrak{p}$ (over F) is still algebraically dependent it follows that

$$\dim_F R/\mathfrak{p} \geq \dim_F R/\mathfrak{q}.$$

Further as R is finitely generated over F both these dimensions are finite. Put $n = \dim_F R/\mathfrak{p}$ and suppose that $\mathfrak{p} \neq \mathfrak{q}$. The lemma can be false only if there exist n elements $x_1, x_2, \ldots, x_n$ of $R/\mathfrak{p}$ whose images $\bar{x}_1, \bar{x}_2, \ldots, \bar{x}_n$ in $R/\mathfrak{q}$ are algebraically independent over F. If y is any non-trivial element of $\mathfrak{q}/\mathfrak{p}$ then $x_1, \ldots, x_n, y$ are algebraically dependent, so there exists a non-zero polynomial f in $X_1, X_2, \ldots, X_n, Y$ with coefficients in F such that $f(x_1, x_2, \ldots, x_n, y) = 0$. Choose f to have the least possible degree in Y, which must be positive since $x_1, \ldots, x_n$ are algebraically independent. Since $y \neq 0$ it follows that Y does not divide f and so $f(X_1, \ldots, X_n, 0)$ is non-zero. But $f(\bar{x}_1, \bar{x}_2, \ldots, \bar{x}_n, 0) = 0$, which contradicts the independence of the $\bar{x}_i$. This completes the proof of the lemma. □

There is another approach to the dimension of a commutative ring R. Define Dim R to be the upper bound of the lengths of all the (finite) chains of prime ideals ($\neq R$) of R. It is clear at once from this definition that if R is a commutative ring with Dim R finite and if $\mathfrak{p}$ and $\mathfrak{q}$ are prime ideals of R with $\mathfrak{p} \subseteq \mathfrak{q}$, then Dim $R/\mathfrak{p}$ and Dim $R/\mathfrak{q}$ are both finite and Dim $R/\mathfrak{p} \geq$ Dim $R/\mathfrak{q}$ with equality if and only if $\mathfrak{p} = \mathfrak{q}$. It can be shown for a polynomial ring P over a field in n indeterminates that Dim $P = n$, and thus Dim R is finite for any finitely generated F-algebra R, cf. [29] pp. 109, 110. If R is also an integral domain then the Noether Normalization Lemma ([36e] p. 4) and Theorem 48 of [29] (the invariance of Dim under integral extensions) show that $\dim_F R = \text{Dim } R$, and so the two concepts of dimension are equivalent (or rather the second is a generalization of the first).

Now let G be any (not necessarily closed) subgroup of $\text{GL}(n, F)$. Then G^0 is irreducible by 14.3 and consequently its annihilator ideal $\mathfrak{a}$ in $R = F[X_{11}, X_{12}, \ldots, X_{nn}]$ is prime by 14.1. Define the dimension $\dim G$ of G by $\dim G = \dim_F(R/\mathfrak{a})$. (More generally if V is an irreducible subset of the m-row vector space over F and if $\mathfrak{a}$ is the annihilator ideal of V in $F[X_1, X_2, \ldots, X_m]$ we can put

$$\dim V = \dim_F(F[X_1, \ldots, X_n]/\mathfrak{a}).$$

In this sense each connected component of G has a dimension and it is very easy to see that all these dimensions coincide with what we have called $\dim G$.)

We have written $\dim G$ and not $\dim_F G$. Let us show that the omission of the F here is justified. Suppose that E is an extension field of F. Put $S = E[X_{11}, X_{12}, \ldots, X_{nn}]$, regarded as an extension ring of the ring R above, and denote the annihilator ideal of G^0 in S by $\mathfrak{b}$. We have to show that the transcendence degree, $\text{tr} \cdot \deg_F R/\mathfrak{a}$, of $R/\mathfrak{a}$ over F is equal to that of $S/\mathfrak{b}$ over E.

Choose a basis $\{\alpha_i : i \in I\}$ of E over F. Then $S = \bigoplus_i \alpha_i R$, the direct sum being as F-space. It follows easily from $G \subseteq \text{GL}(n, F)$, cf. p. 73, that $\mathfrak{b} = \bigoplus_i \alpha_i \mathfrak{a}$. Let $x_1, x_2, \ldots, x_r$ be elements of R whose images in $S/\mathfrak{b}$ are algebraically dependent over E. Then there exists a non-zero polynomial f of $E[X_1, X_2, \ldots, X_r]$ such that $f(x_1, \ldots, x_r) \in \mathfrak{b}$. For some elements f_i of $F[X_1, X_2, \ldots, X_r]$ almost all, but not all, of which are zero, $f = \sum \alpha_i f_i$. The independence of the α_i implies that $f_i(x_1, \ldots, x_r) \in \mathfrak{a}$ for each i and since at least one f_i is non-zero it follows that the x_i are algebraically dependent over F. Therefore $\text{tr} \cdot \deg_F R/\mathfrak{a} \leq \text{tr} \cdot \deg_E S/\mathfrak{b}$.

Conversely, since $(R + \mathfrak{b})/\mathfrak{b}$ spans $S/\mathfrak{b}$ as E-space the ring

$$(R + \mathfrak{b})/\mathfrak{b} \cong_F R/\mathfrak{a}$$

contains a transcendence basis T of $S/\mathfrak{b}$ over E; and clearly T is also algebraically independent over F. Therefore $\text{tr} \cdot \deg_F R/\mathfrak{a} = \text{tr} \cdot \deg_E S/\mathfrak{b}$, and we have proved our contention, namely that $\dim G$ is invariant under ground-field extension.

14.6. Theorem. *Let G be a connected subgroup of $\text{GL}(n, F)$ and H a closed subgroup of G. Then $\dim H \leq \dim G$ with equality if and only if $H = G$.*

Proof. Let $\mathfrak{p}$ denote the annihilator of $G = G^0$ in $R = F[X_{11}, X_{12}, \ldots, X_{nn}]$ and $\mathfrak{q}$ the annihilator of H^0 in R. Then $\mathfrak{p}$ and $\mathfrak{q}$ are prime by 14.1 and 14.3 and trivially $\mathfrak{p} \subseteq \mathfrak{q}$. Hence by 14.5 we have $\dim H \leq \dim G$. Suppose that $\dim H = \dim G$. Then $\mathfrak{p} = \mathfrak{q}$ by 14.5 again, so $\mathscr{A}_F(H^0) = \mathscr{A}_F(G)$. Hence $\mathscr{A}_F(H) = \mathscr{A}_F(G)$ and so $G = G \cap \mathscr{A}_F(H) = H$ since H is closed in G. $\quad\square$

A linear group G satisfies the minimal condition on closed subgroups. The above theorem says in particular that G also satisfies the maximal condition on closed connected subgroups. It is an elementary exercise to show that $\dim \mathrm{GL}(n, F) = n^2$.

The concept of the dimension of a linear group gives us yet another way of setting up an induction argument for linear groups. If G is any linear group put

$$l(G) = (\dim G, (G : G^0)).$$

Order such pairs lexicographically, that is if G_1 is a second linear group write $l(G) \leq l(G_1)$ if either $\dim G < \dim G_1$ or, $\dim G = \dim G_1$ and $(G : G^0) \leq (G_1 : G_1^0)$. This is clearly a well-ordering so we may induct on $l(G)$. If H is a closed subgroup of G it is easily seen that $l(H) \leq l(G)$ with equality if and only if $H = G$. In [64] the author introduced the term "*level of G*" for $l(G)$ but it has not caught on.

One further result on dimension is perhaps worth mentioning, see [3a] p. 88 for a proof.

14.7. *If G is a closed subgroup of $\mathrm{GL}(n, F)$ and ϕ is a polynomial homomorphism of G then $\dim G = \dim (\ker \phi) + \dim (G \phi)$.*

Images of Algebraic Groups

Perhaps one of the more surprising properties of a polynomial homomorphism ϕ of an algebraic group G is that the image under ϕ of a closed subgroup of G is closed. The object of the present section is to indicate a proof of this result and to discuss some of its consequences and ramifications.

Let X be a topological space. A subspace U of X is *locally closed* in X if it is open in its closure in X. This is easily seen to be equivalent to saying that U is the intersection of an open set and a closed set (of X). A subspace of X is called *constructible* if it is the union of a finite number of locally closed subsets of X. (The collection of all the constructible subsets of X together with union and intersection is just the Boolean algebra generated by all the open and all the closed subsets of X.) It is simple to show that a constructible subset of a constructible subset of a space X is a constructible subset of X.

The following lemma is basic to this section; see [36e] p. 97, Corollary 2, for a proof.

14.8. Lemma (Chevalley). *Let U be the m-row vector space over F, W the n-row vector space over F and X constructible subset of U. If $\phi\colon X \to W$ is a map such that for some polynomials $p_1, p_2, \dots, p_n$ in m indeterminates we have*

$$x\,\phi = \bigl(p_1(x), \dots, p_n(x)\bigr)$$

for every x in X, then $X\,\phi$ is a constructible subset of W.

14.9. Lemma. *Let X be a topological space that satisfies the descending chain condition on closed sets. If Y is a constructible subset of X with closure $\overline{Y}$ in X, then Y contains a dense open subset of $\overline{Y}$.*

Proof. For any subset Z of X denote the closure of Z in X by $\overline{Z}$. It is a simple consequence of the descending chain condition on closed sets that Y is the union of a finite number of closed irreducible subsets of Y, say, $Y_1, Y_2, \dots, Y_r$. The Y_i are also constructible subsets of X and clearly $\overline{Y} = \bigcup \overline{Y_i}$. If for each i the set Y_i contains a dense open subset U_i of $\overline{Y_i}$ then $\bigcup U_i$ is a subset of Y that is dense and open in $\overline{Y}$. Hence we may assume that Y, and hence that $\overline{Y}$, is irreducible.

Now Y being constructible is the union of a finite number of locally closed subsets L_j of X. Clearly $\overline{Y} = \bigcup \overline{L_j}$ and yet $\overline{Y}$ is irreducible. Hence for some j we have that $\overline{L_j} = \overline{Y}$. By definition L_j is open in $\overline{L_j}$, so L_j is a subset of Y that is dense and open in $\overline{Y}$. ⬜

14.10. Lemma. *A constructible subgroup H of a CZ-group G is closed.*

Proof. By 14.9 there exists a subset U of H that is dense and open in the closure $\overline{H}$ of H in G. Now $\overline{H}$ is a subgroup of G by 5.9, so if $x \in \overline{H}$ then U and xU are both dense open subsets of $\overline{H}$. Thus $U \cap xU \neq \varnothing$ and consequently

$$x \in UU^{-1} \subseteq HH^{-1} = H.$$

Therefore $H = \overline{H}$. ⬜

14.11. Theorem. *Let H be a constructible subset of $\mathrm{GL}(n, F)$ and ϕ a polynomial map of H into $\mathrm{GL}(m, F)$. Then $H\,\phi$ is a constructible subset of $\mathrm{GL}(m, F)$ and if $H\,\phi$ is actually a subgroup of $\mathrm{GL}(m, F)$ then $H\,\phi$ is closed in $\mathrm{GL}(m, F)$.*

Proof. Denote by π_n the map of $\mathrm{GL}(n, F)$ into $\mathrm{SL}(n+1, F)$ given by

$$g\,\pi_n = \mathrm{diag}\bigl(g, (\det g)^{-1}\bigr), \qquad g \in \mathrm{GL}(n, F).$$

Then $(\pi_n|_H)^{-1}\,\phi\,\pi_m$ is a polynomial map of $H\,\pi_n$ into $\mathrm{SL}(m+1, F)$. Suppose for the moment that H is connected. Then there exist $(m+1)^2$ polynomials p_{ij} in $(n+1)^2$ indeterminates over F such that

$$h \phi \pi_m = (p_{ij}(h \pi_n)) \quad \text{for each } h \in H.$$

By 14.8 applied with $U = F_{n+1}$ and $X = H \pi_n$,

$$H \phi \pi_m = H \pi_n (\pi_n|_H)^{-1} \phi \pi_m$$

is a constructible subset of F_{m+1} and hence also of $\mathrm{SL}(m+1, F)$. Therefore $H \phi$ is a constructible subset of $\mathrm{GL}(m, F)$. In general the image of each connected component of H is a constructible subset of $\mathrm{GL}(m, F)$ and consequently $H \phi$ is also. Finally if H is a subgroup of $\mathrm{GL}(m, F)$ then $H \phi$ is closed in $\mathrm{GL}(m, F)$ by 14.10. $\quad\square$

14.12. Corollary. *If G is a closed subgroup of $\mathrm{GL}(n, F)$ and ϕ is a polynomial homomorphism of G into $\mathrm{GL}(m, F)$ then $G \phi$ is a closed subgroup of $\mathrm{GL}(m, F)$.*

Proof. $G \phi$ is a subgroup of $\mathrm{GL}(m, F)$ since ϕ is a homomorphism. Now apply 14.11 with $G = H$. $\quad\square$

14.13. Corollary. *If H and K are closed subgroups of $\mathrm{GL}(n, F)$ such that $HK = KH$, then HK is a closed subgroup of $\mathrm{GL}(n, F)$.*

Proof. HK is a subgroup of $\mathrm{GL}(n, F)$. Put

$$G = \{\mathrm{diag}(h, k): h \in H, k \in K\} \subseteq \mathrm{GL}(2n, F).$$

G is easily seen to be closed in $\mathrm{GL}(2n, F)$. Hence the corollary will follow from 14.11 if we can produce a polynomial map ϕ of G into $\mathrm{GL}(n, F)$ with $G \phi = HK$. Define ϕ by the n^2 polynomials p_{ij}, where for $i, j = 1, 2, \ldots, n$ we put

$$p_{ij} = \sum_{k=1}^{n} X_{ik} X_{n+k, n+j}. \quad\square$$

14.14. Lemma. *If $\{W_i: i \in I\}$ is a family of constructible irreducible subsets of $\mathrm{GL}(n, F)$ each containing 1, then $H = \mathscr{A}_F(\bigcup_i W_i)$ is connected and equal to some finite product*

$$W_{i_1}^{e_1} W_{i_2}^{e_2} \ldots W_{i_r}^{e_r},$$

where the e_i are ± 1.

Proof. Since inversion is a homeomorphism fixing 1 we may assume that for each i in I there exists a j in I with $W_i^{-1} = \{w^{-1}: w \in W_i\} = W_j$. If $\alpha = (i_1, i_2, \ldots, i_r)$ where each $i_k \in I$ write $W_\alpha = W_{i_1} W_{i_2} \ldots W_{i_r}$. Similarly for $\beta = (j_1, j_2, \ldots, j_s)$ where each $j_k \in I$, we put

$$W_{(\alpha, \beta)} = W_\alpha W_\beta = W_{i_1} \ldots W_{i_r} W_{j_1} \ldots W_{j_s}.$$

If X and Y are irreducible non-empty subsets of $GL(n, F)$ then XY is also irreducible: for let $x \in X$, $y \in Y$ and suppose that S and T are closed sets with $XY \subseteq S \cup T$. Assume that $x y \in S$. Then since $X y$ is irreducible we have that $X y \subseteq S$. But now $z Y$ is also irreducible for any z in X and $z Y \cap S \neq \emptyset$. Therefore $z Y \subseteq S$ for every $z \in X$, that is $XY \subseteq S$. It follows that XY is irreducible. A simple induction argument shows that each W_α is irreducible. Further W_α is the image under the obvious map of

$$D = \{\mathrm{diag}(w_1, w_2, \ldots, w_r):\ w_k \in W_{i_k}, k = 1, 2, \ldots, r\} \subseteq GL(r n, F)$$

and this map is clearly given by polynomials, cf. the previous proof. Further D is easily seen to be constructible, so W_α is also constructible by 14.11.

Choose α so that W_α has the maximum possible dimension. For any subset X of $GL(n, F)$ write $\overline{X}$ for its closure in $GL(n, F)$. Now $W_\beta W_\gamma = W_{(\beta, \gamma)}$ so just as in the proof of 5.9 we obtain $\overline{W_\beta} \, \overline{W_\gamma} \subseteq \overline{W_{(\beta, \gamma)}}$. Since $1 \in \bigcap W_i$ we have

$$\overline{W_\alpha} \cup \overline{W_\beta} \subseteq \overline{W_\alpha} \, \overline{W_\beta} \subseteq \overline{W_{(\alpha, \beta)}} = \overline{W_\alpha},$$

the final equality following from the maximal choice of α and the irreducibility of the $\overline{W_\gamma}$. In particular, $\overline{W_\alpha} \, \overline{W_\alpha} \subseteq \overline{W_\alpha}$ and $\overline{W_\alpha^{-1}} = \overline{W_\alpha}^{-1} \subseteq \overline{W_\alpha}$. Therefore $\overline{W_\alpha}$ is a subgroup. It now follows that $H = \overline{W_\alpha}$; in particular H is connected.

Finally W_α is constructible so by 14.9 there exists a subset U of W_α such that U is open and dense in H. Then for any x in H we have $U \cap x U \neq \emptyset$, so $x \in U U^{-1}$. Therefore

$$H = W_\alpha W_\alpha^{-1} = W_{i_1} \ldots W_{i_r} W_{i_r}^{-1} \ldots W_{i_1}^{-1}$$

where $\alpha = (i_1, i_2, \ldots, i_r)$. This completes the proof of the lemma. □

14.15. Corollary. *If G is an algebraic group then the subgroup of G generated by a set of closed connected subgroups of G is both closed and connected.*

This corollary is an immediate consequence of 14.3 and 14.14. □

14.16. Corollary. *Let G be a closed connected subgroup of $GL(n, F)$ and X any subset of $GL(n, F)$. Then*

$$[G, X] = \langle [g, x]:\ g \in G, x \in X \rangle$$

is a closed connected subgroup of $GL(n, F)$.

Proof. For each x in X the map ϕ_x of G into $GL(n, F)$ given by $g \phi_x = [g, x]$, where $g \in G$, is clearly a polynomial map. Hence by 14.3 and 14.11 the image of G under ϕ_x is irreducible and constructible; and trivially it contains 1. Hence

$$[G, X] = \langle G\phi_x : x \in X \rangle$$

is closed and connected by 14.14. □

14.17. Corollary. *Let G be a closed subgroup of $\mathrm{GL}(n, F)$ and H a closed subgroup of G. Then $[G, H]$ is a closed normal subgroup of G.*

Proof. That $[G, H]$ is a normal subgroup of G is elementary and has nothing to do with linear groups, cf. [50] 3.4.6. By 14.16 the groups $[G, H^0]$ and $[G^0, H]$ are both closed and connected in G, and therefore 14.15 implies that

$$L = \langle [G, H^0]^g, [G^0, H]^g : g \in G \rangle$$

is closed (and connected) in G. If we can show that $[G, H]/L$ is finite than it will follow that $[G, H]$ is closed in G. Thus the following lemma completes the proof of 14.17.

Lemma (R. Baer). *Let G be a group and H a subgroup of G such that both the indices $(G : C_G(H))$ and $(H : C_H(G))$ are finite. Then $[G, H]$ is finite.*

Proof. Note first that if ϕ is any homomorphism of G then $(G\phi : C_{G\phi}(H\phi))$ and $(H\phi : C_{H\phi}(G\phi))$ are also finite. Choose right transversals S of $C_G(H)$ to G and T of $C_H(G)$ to H. Then

$$X = \{[g, h] : g \in G, h \in H\} = \{[s, t] : s \in S, t \in T\}$$

is a finite generating set of $[G, H]$ normalized by $C_G(H)$. Hence $C_G(X) = C_G([G, H])$ has finite index in G. In particular the centre of $[G, H]$ has finite index in $[G, H]$ and consequently $[G, H]'$ is finite by Schur's Theorem ([24a] 8.1, [50] 15.1.13 or [27] 8.19). In view of our first remark we may factor out by $[G, H]'$, that is we may assume that $[G, H]$ is an abelian group.

For each h in H the map $x \mapsto [x, h]$ is a homomorphism of $[G, H]$ whose image lies in the finite set X. Clearly

$$[G, H, H] = \langle [G, H, t] : t \in T \rangle.$$

Therefore $[G, H, H]$ is a finite group. But $[G, H]$ is a finitely generated abelian group, so its periodic subgroup is finite. Hence $[G, H, H]$ generates a finite normal subgroup of G which by the first remark again we may assume is $\{1\}$.

We now have that H centralizes $[G, H]$ so for each g in G the map $h \mapsto [g, h]$ of H into X is a homomorphism of H. Thus $[G, h]$ is a finite group. But

$$[G, H] = \langle [G, t] : t \in T \rangle$$

and consequently $[G, H]$ is also finite. □

14.18. Corollary. *If G is an algebraic group then for each finite ordinal i, the i-th terms of the derived series and the lower central series of G are closed subgroups of G.*

Proof. Induct on *i* and apply 14.17.   ☐

14.19. Corollary. *If G is an algebraic group and H is a closed subgroup of G then the normal closure $\langle H^G \rangle$ of H in G is closed in G.*

Proof. $\langle H^G \rangle = H[G, H]$ and $[G, H]$ is a closed normal subgroup of *G* by 14.17. Therefore $\langle H^G \rangle$ is closed in *G* by 14.13.   ☐

Exercise 14.3. If *H* and *N* are closed subgroups of GL(*n, F*) such that *H* normalizes *N*, prove that $[N, H]$ is closed in *G*. (See [3a] p. 108 for a proof.)

Jordan Decomposition

We have already shown in 7.3 that if *G* is a closed subgroup of GL(n, *F*) then *G* is splittable, that is $g_u \in G$ and $g_d \in G$ for each *g* in *G* (recall we are assuming that *F* is algebraically closed). The other basic result on Jordan decomposition in algebraic groups is the following generalization of 6.5 and 6.6.

14.20. Theorem. *Let G be a closed subgroup of GL(n, F) and ϕ a polynomial homomorphism of G into GL(m, F). Then:*

   a) *If g is a unipotent element (resp. d-element) of G then $g\phi$ is a unipotent element (resp. d-element) of $G\phi$.*

   b) *If $g \in G$ then $g_u \phi = (g\phi)_u$ and $g_d \phi = (g\phi)_d$.*

   c) *If h is a unipotent element (resp. d-element) of $G\phi$ then there exists a unipotent element (resp. d-element) g of G with $g\phi = h$.*

Proof. a) Let $g \in G$ and put $h = g\phi$. There exist elements *x* of GL(*n, F*) and *y* of GL(*m, F*) such that g_d^x and h_d^y are diagonal. Also ϕ is determined by a finite set *S* of polynomials with coefficients in *F*. Let *R* be the subring of *F* generated by the coefficients of the elements of *S* and the entries in the matrices g, g_d, h_d, x, y and their inverses (so the entries of g_u, h, h_u and their inverses also lie in *R*). Then *R* is a finitely generated integral domain, so if $\{\mathfrak{m}_i : i \in I\}$ is the set of maximal ideals of *R*, then $\bigcap \mathfrak{m}_i = \{0\}$ by 4.1.

   For each $i \in I$ set $K_i = R/\mathfrak{m}_i$ and let π_i denote the natural projection of *R* onto K_i. Then π_i induces ring homomorphisms π_{in} of R_n onto $(K_i)_n$ and π_{im} of R_m onto $(K_i)_m$. If we apply π_i to the coefficients of each of the elements of *S*, then the resulting polynomials over K_i define a map ϕ_i of $\left(\mathrm{GL}(n, R) \cap G\right) \pi_{in}$ into $(K_i)_m$. It is simple to check that the following diagram commutes.

$$\begin{array}{ccc}
\left(\mathrm{GL}(n, R) \cap G\right) & \xrightarrow{\ \pi_{in}\ } & \left(\mathrm{GL}(n, R) \cap G\right) \pi_{in} \\
\Big\downarrow{\phi} & & \Big\downarrow{\phi_i} \\
R_m & \xrightarrow{\ \pi_{im}\ } & (K_i)_m
\end{array}$$

Since $1 \phi \pi_{im} = 1$, it follows easily that ϕ_i is a (group) homomorphism whose image lies in $\mathrm{GL}(m, K_i)$.

Suppose g is unipotent. Then $(g-1)^n = 0$; and π_{in} is a ring homomorphism, so $g \pi_{in}$ is also unipotent. But K_i has positive characteristic, p_i say; thus $g \pi_{in}$ has finite order a power of p_i. Therefore so does $g \pi_{in} \phi_i$ and consequently $g \pi_{in} \phi_i$ is unipotent. Using the above commutative diagram we obtain

$$(g \phi - 1)^m \pi_{im} = (g \pi_{in} \phi_i - 1)^m = 0$$

for each i, so

$$(g \phi - 1)^m \in \bigcap_i \ker \pi_{im} = \{0\}.$$

Therefore $g \phi$ is unipotent.

Now suppose that g is diagonalizable while $h = g \phi$ is not. Then h_u is non-trivial. There exists i therefore such that $h_u \pi_{im} \neq 1$. We have

$$h \pi_{im} = h_u \pi_{im} \cdot h_d \pi_{im} = h_d \pi_{im} \cdot h_u \pi_{im}. \qquad (*)$$

Now $h_u \pi_{im}$ is unipotent since π_{im} is a ring homomorphism. Also $y \in \mathrm{GL}(n, R)$ by the construction of R and clearly

$$(y \pi_{im})^{-1} (h_d \pi_{im}) (y \pi_{im}) = (y^{-1} h_d y) \pi_{im}$$

is diagonal. Hence $(*)$ exhibits the Jordan decomposition of $h \pi_{im}$. We have chosen i so that $h_u \pi_{im} \neq 1$, so $h \pi_{im}$ is not a d-element and so, being of finite order, it is not a p_i'-element. Again using the commutative diagram and the fact that ϕ_i is a homomorphism we find that $g \pi_{in}$ is not a p_i'-element. But $x \in \mathrm{GL}(n, R)$. Thus

$$(x \pi_{in})^{-1} (g \pi_{in}) (x \pi_{in}) = (x^{-1} y x) \pi_{in}$$

is diagonal, so $g \pi_{in}$ is a p_i'-element. This contradiction completes the proof of part a).

b) Now $g_u \in G$ and $g_d \in G$ by 7.3. By part a) above $g_u \phi$ is unipotent and $g_d \phi$ is a d-element. Also $g \phi = g_u \phi \cdot g_d \phi = g_d \phi \cdot g_u \phi$ so by the uniqueness (7.2) of the Jordan decomposition of $g \phi$ we obtain

$$(g \phi)_u = g_u \phi \quad \text{and} \quad (g \phi)_d = g_d \phi.$$

c) For any h in $G\phi$ there exists $g \in G$ with $g\phi = h$. By part b) above we have that $h_u = g_u \phi$ and $h_d = g_d \phi$, from which part c) follows. □

The Structure of the Algebraic Groups

In this section we give no proofs but merely try to summarize some of the more important results.

A *torus* of an algebraic (linear, as always) group G is a closed connected diagonalizable subgroup of G.

14.21. Theorem. *A torus of* $\mathrm{GL}(n, F)$ *of dimension r is isomorphic to* $D(r, F)$, *that is to the direct product of r copies of* F^*. (See [3a] p. 205.)

A closed diagonalizable subgroup D of $\mathrm{GL}(n, F)$ is simply the direct product of the torus D^0 and a finite subgroup ([3a] 8.7). Thus the isomorphism classes of diagonalizable closed subgroups of $\mathrm{GL}(n, F)$ can be described completely (in terms of F). Perhaps one further result on tori is worth mentioning: any subtorus of a torus is a direct factor (in a strong sense) of the torus, see [3a] p. 206.

It follows easily from 7.11 and 7.3 that any closed connected nilpotent subgroup of $\mathrm{GL}(n, F)$ is the direct product of a closed connected unipotent subgroup of $\mathrm{GL}(n, F)$ and a torus. We describe now the closed connected soluble subgroups of $\mathrm{GL}(n, F)$. We know already from the Lie-Kolchin Theorem (5.8) that such a group is triangularizable.

14.22. Theorem. *Let G be a closed connected soluble subgroup of* $\mathrm{GL}(n, F)$. *Then:*

a) *G_u the set of unipotent elements of G is a closed connected subgroup of G.*

b) *The maximal tori of G are all conjugate in G and complement G_u (that is $G = TG_u$ and $T \cap G_u = \{1\}$ for each maximal torus T of G).*

c) *Every d-subgroup of G is abelian and lies in a maximal torus of G. (Thus the maximal tori of G are exactly the maximal d-subgroups of G.)*

d) *If D is any d-subgroup of G then $C_G(D)$ is connected and equals $N_G(D)$.*

e) *G is nilpotent if and only if G_d is a subgroup of G, and in this situation G_d is the torus of G and $G = G_u \times G_d$.*

The situation of 14.22 is thus an almost classical "Schur-Zassenhaus type" situation. For a proof of 14.22 see [3a] 10.6.

A *Borel subgroup* of an algebraic group G is a maximal connected soluble subgroup of G. The Borel subgroups of G are necessarily closed in G by 5.9 and 5.11. We now state in full the conjugacy theorems already referred to in Chapter 12.

14.23. Theorem. *If G is a closed subgroup of* $GL(n, F)$, *then:*

a) *The Borel subgroups of G are all conjugate.*

b) *The maximal tori of G are all conjugate and are just the maximal tori of the Borel subgroups of G.*

c) *The maximal connected unipotent subgroups of G are all conjugate and are just the maximal unipotent subgroups of the Borel subgroups of G.*

The basic part here is a), see [3a] 11.1 for a proof. The other two parts follows from it and 14.22, see [3a] 11.3.

If char $F = 0$ then every unipotent subgroup of $GL(n, F)$ is connected (this is a very easy exercise, use 6.4 and 6.6). Thus by 14.23 for char $F = 0$ the maximal unipotent subgroups of a closed subgroup G of $GL(n, F)$ are all conjugate. This is also true if char $F > 0$ although here of course the unipotent subgroups of G need not be connected. In fact even more is true. If G is any closed subgroup of $GL(n, F)$, with no restriction on char F, and if p is any prime then the maximal p-subgroups of G are all conjugate in G, see 4.6 of [44], where other conjugacy theorems for algebraic groups are also given. Perhaps it is worth remarking that if G is also connected then the Borel subgroups of G are maximal soluble subgroups of G but not every maximal soluble subgroup of G need be a Borel subgroup, see [3a] p. 278.

Let G be a closed connected subgroup of G. The centralizer of a maximal torus of G is called a *Cartan subgroup* of G. It turns out that the Cartan subgroups of G are connected nilpotent and conjugate, the latter statement following at once from 14.23. A subgroup C of G is a Cartan subgroup of G if and only if it is a closed connected nilpotent subgroup of G such that $C = N_G(C)^0$. Another equivalent definition of Cartan subgroup is given by, C is a Cartan subgroup of G if and only if C is a maximal nilpotent subgroup of G such that every subgroup of finite index in C has finite index in its normalizer in G. For these results see [3a] 11.7, 11.12 and 12.6.

A number of sufficient conditions can be given for an algebraic group to be soluble. Again let G be a closed connected subgroup of $GL(n, F)$. If G contains a Borel subgroup B that is a d-group, then $G = B$ is a torus. If G has a unipotent Borel subgroup B then $G = B$ is unipotent. If $\dim G \leq 2$ then again G is soluble. If $\dim G = 1$ then G is isomorphic to either F^* or F^+, that is to either $GL(1, F)$ or $Tr_1(2, F)$. For these results see [3a] 10.9, 11.5 and 11.6.

By 3.8 any linear group G contains a unique maximal soluble normal subgroup R and clearly R^0 is the unique maximal connected soluble normal subgroup of G. Every soluble normal subgroup of G/R^0 is finite and thus if ϕ is any representation of G with kernel R^0 then $\{1\}$ is the maximal connected soluble normal subgroup of $G\phi$. A connected algebraic group is said to be *semisimple* if $\{1\}$ is its only connected soluble normal subgroup. In a semisimple algebraic group the Cartan subgroups coincide with the maximal tori, [3a] p. 316.

For each of the connected Dynkin diagrams (with weights) A_n, B_{n+1}, C_{n+2}, D_{n+3}, E_6, E_7, E_8, F_4 and G_2, $n \geq 1$, there exists an algebraic linear group over F that is simple modulo a finite central subgroup such that the following holds: if G is any semisimple algebraic group then there exists a group H with a finite central subgroup Z such that G is isomorphic to H/Z and H is a direct product of a finite number of the "nearly simple" groups associated with the Dynkin diagrams. In this association of Dynkin diagrams and nearly simple groups, A_n corresponds to $SL(n+1, F)$, C_n to $Sp(2n, F)$ and B_n and D_n essentially to the special orthogonal groups (essentially here meaning modulo finite central subgroups). The classification of the semisimple algebraic groups is much deeper than the other results quoted above and is due to Chevalley, see [8] or [48a].

Suggestions for Further Reading

We mention below a small selection of works from the bibliography. These items have been chosen solely on the grounds of how well they supplement this work. Either they contain many results that extend and develop ideas already in this text, or they contain techniques that we have not really touched upon and which have (or might have) a bearing on infinite group theory. All our suggestions are in English (or English translations are available).

Merzljakov's papers [36 b] and [36 d] extend the embedding theorems for soluble groups given in the first half of Chapter 2. Kegel has refined Mal'cev's local theorem (2.7) in the following way. If a group G has a local system of subgroups isomorphic to $C_n(K)$ for various finite fields K, where C_n is some fixed (but not arbitrary) Chevalley group, then for some field F, G is isomorphic to $C_n(F)$. The easiest place to study this work of Kegel's is, I would suggest, §4.B of [31 a]. (This section of [31 a] is independent of the preceding ones.) [69 f] discusses at some length the possibility of extending Nisnevič's Theorem (2.14) to generalized free products and contains many examples.

In Chapter 3 we gave no proof that required an investigation of the symplectic groups arising from 3.3. For examples of proofs that require such investigations and for improvements to the bounds obtained in 3.6 and 3.7 see Dixon [12], Newman [37 a] and Frick and Newman [14 b]. Suprunenko's book [54] on soluble and nilpotent linear groups, although now very much out of date being originally published in 1958, nevertheless contains very explicit discriptions of the maximal soluble subgroups of a general linear group and, for certain special fields, the conjugacy classes of such objects in the full general linear group. For soluble periodic linear groups, which have barely been touched upon in this book, see [65]. A. D. Gardiner, M. J. Tomkinson and particularly B. Hartley have shown that the results of [65] are just special cases of theorems concerning a certain abstract class of periodic locally soluble groups and have cleared up a number of points raised in [65].

Readers interested in devouring more on finitely generated linear groups really ought to digest Tits' paper [57 a] whose contents we summarized in Chapter 10. Dixon's little book [13 b] contains a wealth of material on finite linear groups, as does [10] of course. Perhaps someone should complement this present book by writing an up to date and comprehensive account of the theory of finite linear groups. (If someone has I apologise profusely for my ignorance.)

Serious students of infinite linear groups will have to pick up more that just a nodding acquaintance with the theory of algebraic groups at some time or another. There is no better place to do this that I know of, than Borel's book [3 a]. (Before studying [3 a] some preliminary reading on algebraic geometry, for example Chapter 1 of Mumford's book [36 e], might prove invaluable.) Those then catching the bug and wishing to delve into the classification theory of semisimple algebraic groups, but in English, there is Satake's book [48 a]. Otherwise there is always Chevalley's seminar [8] of course.

Platonov's paper [44] contains a wealth of diverse and interesting results, some of which have been given above, and also many examples. Much of the remainder concerns algebraic groups. Factorizations of linear groups have not been touched upon in this text. They are discussed by Kegel in [31]; his proofs depend heavily on the structure of algebraic groups.

Readers curious about "linear groups" over non-commutative rings might be interested to consult Zalesskii's paper [71 a].

Bibliography

Russian language entries are marked by an asterisk. For a fuller list of Russian papers prior to 1963 the reader is referred to the bibliography of [54].

1. Artin, E.: Geometric algebra. Interscience 1957.
2. Auslander, L.: On a problem of Philip Hall. Ann. Math. (2) **86**, 112–116 (1967).
2a. Baer, R.: Überauflösbare Gruppen. Abh. Math. Sem. Univ. Hamburg **23**, 11–28 (1957).
3. Borel, A.: Groupes linéares algébriques. Ann. Math. (2) **64**, 20–83 (1956).
3a. Borel, A.: Linear algebraic groups. W. A. Benjamin 1969.
4. Brauer, R., Feit, W.: An analogue of Jordan's Theorem in characteristic p. Ann. Math. (2) **84**, 119–131 (1966).
5. Brenner, J.L.: Quelques groupes libres de matrices. C.R. Acad. Sci. Paris **241**, 1689–1691 (1955).
6. Burnside, W.: The theory of groups of finite order, 2nd ed. C.U.P., 1911. Reprinted Dover Inc. 1955.
6a. Carter, R., Fischer, B., Hawkes, T.: Extreme classes of finite soluble groups. J. Algebra **9**, 285–313 (1968).
6b. Cassels, J.W.S., Fröhlich, A.: Algebraic number theory. Academic Press 1967.
7. Chevalley, C.: Théorie des Groupes de Lie, Actualités Sci. Indust. Paris; Tome II, (1951), no. 1152; Tome III (1955), no. 1226.
8. Chevalley, C.: Classification des Groupes de Lie Algébriques (2 vols). Seminaire Chevalley, École Norm. Sup. 1956–58.
9. Cohn, P.M.: Universal algebra. Harper and Row 1965.
10. Curtis, C.W., Reiner, I.: Representation theory of finite groups and associative algebras. Interscience 1962.
11. Dixon, J.D.: Complete reducibility in infinite groups. Canad. J. Math. **16**, 267–274 (1964).
12. Dixon, J.D.: The Fitting subgroup of a linear soluble group. J. Austral. Math. Soc. **7**, 417–424 (1967).
13. Dixon, J.D.: Normal p-subgroups of soluble linear groups. J. Austral. Math. Soc. **7**, 545–551 (1967).
13a. Dixon, J.D.: The soluble length of a soluble linear group. Math. Z. **107**, 151–158 (1968).
13b. Dixon, J.D.: The structure of linear groups. Van Nostrand Reinhold Co. 1971.
14. Dornhoff, L.: M-groups and 2-groups. Math. Z. **100**, 226–256 (1967).
14a. Dornhoff, L.: Jordan's Theorem for solvable groups. Proc. Amer. Math. Soc. **24**, 533–537 (1970).
14b. Frick, M., Newman, M.F.: Soluble linear groups. Bull. Austral. Math. Soc. **6**, 31–44 (1972).
15. Garăšcuk, M.S.: On the theory of generalized nilpotent linear groups*. Dokl. Akad. Nauk BSSR **4**, 276–277 (1960).

16. Garăščuk, M. S., Suprunenko, D. A.: Linear nilgroups*. Dokl. Akad. Nauk BSSR **4**, 407–408 (1960).

16a. Garăščuk, M. S., Suprunenko, D. A.: Linear groups with an Engel condition*. Dokl. Akad. Nauk BSSR **6**, 277–280 (1962).

16b. Gardiner, A. D., Hartley, B., Tomkinson, M. J.: Saturated formations and Sylow structure in locally finite groups. J. Algebra **17**, 177–211 (1971).

17. Gruenberg, K. W.: The Engel elements of a soluble group. Illinois J. Math. **3**, 151–168 (1959).

18. Gruenberg, K. W.: The upper central series in soluble groups. Illinois J. Math. **5**, 436–466 (1961).

19. Gruenberg, K. W.: The Engel structure of linear groups. J. Algebra **3**, 291–303 (1966).

20. Gruenberg, K. W.: The hypercentre of linear groups. J. Algebra **8**, 34–40 (1968).

20a. Gruenberg, K. W.: Cohomological topics in group theory. Springer Lecture Notes **143**, 1970.

20b. Gupta, C. K.: A faithful matrix representation for certain centre-by-metabelian groups. J. Austral. Math. Soc. **10**, 451–464 (1969).

20c. Gupta, C. K.: On free groups of the variety, $\mathfrak{AN}_2 \wedge \mathfrak{N}_2\mathfrak{A}$. Canad. Math. Bull. **13**, 443–446 (1970).

20d. Gupta, C. K., Gupta, N. D.: On the linearity of free nilpotent-by-abelian groups. J. Algebra, **24**, 293–302 (1973).

21. Hall, M.: The theory of groups. New York: Macmillan 1959.

22. Hall, P.: Theorems like Sylow's. Proc. London Math. Soc. (3) **6**, 286–304 (1956).

23. Hall, P.: The Frattini subgroups of finitely generated groups. Proc. London Math. Soc. (3) **11**, 327–352 (1961).

24. Hall, P.: A note on $\overline{SI}$-groups. J. London Math. Soc. **39**, 338–344 (1964).

24a. Hall, P.: Nilpotent groups. Lectures given to Canadian Mathematical Congress, 1957. Reissued as: The Edmonton Notes on Nilpotent Groups. Queen Mary College Mathematics Notes, 1969.

25. Huppert, B.: Monomiale Darstellung endlicher Gruppen. Nagoya Math. J. **6**, 93–94 (1953).

26. Huppert, B.: Lineare auflösbare Gruppen. Math. Z. **67**, 479–518 (1957).

26a. Huppert, B.: Endliche Gruppen I. Springer 1967.

26b. Jacobson, N.: Structure of rings. Amer. Math. Soc. Colloq. Pub. **37**, 1964.

27. Kaplansky, I.: An introduction to differential algebra. Paris: Herman 1958.

28. Kaplansky, I.: Notes on ring theory. Univ. Chicago 1965.

29. Kaplansky, I.: Commutative rings. Boston: Allyn and Bacon 1970.

30. Kargarpolov, M. I.: On periodic matrix groups*. Sibirsk. Mat. Ž. **3**, 834–838 (1962); erratum, Sibirsk. Mat. Ž. **4**, 198–199 (1963).

30a. Kargarpolov, M. I.: On finitely generated linear groups*. Algebra i Logika **6** (5), 17–20 (1967).

31. Kegel, O. H.: On the solvability of some factorized linear groups. Illinois J. Math. **9**, 535–547 (1965).

31a. Kegel, O. H., Wehrfritz, B. A. F.: Locally finite groups. North Holland Pub. Co. 1972.

31b. Kopytov, V. M.: Matrix groups*. Algebra i Logika **7** (3), 51–59 (1968). = Algebra and Logic **7**, 162–166 (1968).

31c. Kopytov, V. M.: Solubility of the embedding problem for finitely generated soluble matrix groups over number fields*. Algebra i Logika **10**, 169–182 (1971).

32. Kuroš, A. G.: The theory of groups (2 vols). Transl. of 2nd. Russian ed. by K. A. Hirsch, New York: Chelsea 1956. Transl. of 3rd. Russian ed. in preparation.

32a. Lennox, J. C., Roseblade, J. E.: Centrality in finitely generated soluble groups. J. Algebra **16**, 399–435 (1970).

32b. Levič, E. M.: Representation of soluble groups by matrices over a field of characteristic zero*. Dokl. Akad. Nauk SSSR **188**, 520–521 (1969). = Soviet Math. Dokl. **10**, 1146–

1148 (1969). See also Proc. Riga Seminar on Algebra, p. 74–97; Latv. Gos. Univ. Riga, 1969.

32c. Levič, E. M.: On the representation of 2-step soluble groups by matrices over fields of characteristic zero*. Mat. Sb. **81**, 352–357 (1970).

32d. Lyndon, R. C., Ullman, J. L.: Groups generated by two parabolic linear fractional transformations. Canad. J. Math. **21**, 1388–1403 (1969).

33. Magnus, W., Karrass, A., Solitar, D.: Combinatorial group theory. Interscience 1966.

34. Mal'cev, A. I.: On faithful representations of infinite groups of matrices*. Mat. Sb. **8**, 405–422 (1940). = Amer. Math. Soc. Transl. (2) **45**, 1–18 (1965).

35. Mal'cev, A. I.: On certain classes of infinite soluble groups*. Mat. Sb. **28**, 567–588 (1951). = Amer. Math. Soc. Transl. (2) **2**, 1–21 (1956).

36. Merzljakov, Ju. I.: Central series and commutator series in matrix groups*. Algebra i Logika **3** (4), 49–59 (1964).

36a. Merzljakov, Ju. I.: Verbal and marginal subgroups of linear groups*. Dokl. Akad. Nauk SSSR **177**, 1008–1011 (1967). = Soviet Math. Dokl. **8**, 1538–1541 (1967).

36b. Merzljakov, Ju. I.: On matrix representations of automorphisms, extensions and soluble groups*. Algebra i Logika **7** (3), 63–104 (1968). = Algebra and Logic **7**, 169–192 (1968).

36c. Merzljakov, Ju. I.: Matrix representations of groups of outer automorphism of Černikov groups*. Algebra i Logika **8**, 478–482 (1969).

36d. Merzljakov, Ju. I.: Integral representations of the holomorph of a polycyclic group*. Algebra i Logika **9**, 539–558 (1970).

36e. Mumford, D.: Introduction to algebraic geometry. Cambridge, Mass.: Harvard Univ. 1967.

37. Neumann, H.: Varieties of groups. Springer 1967.

37a. Newman, M. F.: The soluble length of soluble linear groups. Math. Z. **126**, 59–70 (1972).

38. Nisnevič, V. L. (Nisnewitsch, V. L.): Über Gruppen die durch Matrizen über einem kommutativen Feld isomorph darstellbar sind (Russian with German summary). Mat. Sb. **8**, 395–403 (1940).

39. Platonov, V. P.: Periodic subgroups of algebraic groups*. Dokl. Akad. Nauk SSSR **153**, 270–272 (1963). = Soviet Math. Dokl. **4**, 1653–1656 (1963).

40. Platonov, V. P.: Periodic and locally nilpotent subgroups of toplogical groups*. Dokl. Akad. Nauk SSSR **158**, 524–527 (1964). = Soviet Math. Dokl. **5**, 1261–1265 (1964).

41. Platonov, V. P.: The structure of periodic linear groups and algebraic groups*. Dokl. Akad. Nauk SSSR **160**, 541–544 (1965). = Soviet Math. Dokl. **6**, 144–148 (1965).

42. Platonov, V. P.: Automorphisms of algebraic groups*. Dokl. Akad. Nauk SSSR **168**, 1257–1260 (1966). = Soviet Math. Dokl. **7**, 825–829 (1966).

43. Platonov, V. P.: The Frattini subgroups of linear groups and finite approximability*. Dokl. Akad. Nauk SSSR **171**, 798–801 (1966). = Soviet Math. Dokl. **7**, 1557–1560 (1966).

44. Platonov, V. P.: The theory of algebraic linear groups and periodic groups*. Izv. Akad. Nauk SSSR, Ser. Mat. **30**, 573–620 (1966). = Amer. Math. Soc. Transl. (2) **69**, 61–110 (1968).

45. Platonov, V. P.: Linear groups with identical relations*. Dokl. Akad. Nauk BSSR **11**, 581–582 (1967).

45a. Platonov, V. P.: Several problems on linear groups*. Mat. Zametki **4**, 635–638 (1968).

45b. Platonov, V. P.: On a problem of Mal'cev*. Mat. Sb. **79**, 621–624 (1969).

45c. Platonov, V. P.: On one problem of finitely generated groups*. Dokl. Akad. Nauk BSSR **12**, 492–494 (1968).

45d. Platonov, V. P.: Adele groups and finitely approximable linear groups w.r.t. conjugacy*. Dokl. Akad. Nauk BSSR **14**, 777-779 (1970).

46. Platonov, V. P., Suprunenko, D. A.: On a theorem of Schur*. Dokl. Akad. Nauk BSSR **7**, 510-512 (1963).

46a. Platonov, V. P., Zaleskii, A. E.: On a problem of Auerbach*. Dokl. Akad. Nauk BSSR **10**, 5-6 (1966).

46b. Remeslennikov, V. N.: Representation of finitely generated metabelian groups by matrices*. Algebra i Logika **8**, 72-75 (1969).

47. Robinson, D.J.S.: Infinite soluble and nilpotent groups. Queen Mary College Mathematics Notes, 1968. For book based on these notes see: Finiteness conditions and generalized soluble groups (2 vols.). Springer 1972.

48. Samuel, P.: A propos du théorème des unités. Bull. Sci. Math. **90**, 89-96 (1966).

48a. Satake, I.: Classification theory of semi-simple algebraic groups. New York: Marcel Dekker, Inc. 1971.

49. Schur, I.: Über Gruppen periodischer Substitutionen. Sitzungsb. Preuss. Akad. Wiss. 1911, 619-627.

50. Scott, W. R.: Group theory. Prentice Hall 1964.

50a. Seitz, G. M.: *M*-groups and the supersolvable residual. Math. Z. **110**, 101-122 (1969).

50b. Seitz, G. M., Wright, C. R. B.: On finite groups whose Sylow subgroups are modular or quaternion-free. J. Algebra **13**, 374-381 (1969).

50c. Seitz, G. M., Schacher, M.: π-groups that are *M*-groups. Math. Z. **128**, 43-48 (1972).

51. Selberg, A.: On discontinuous groups in higher dimensional symmetric spaces, Contributions to Function Theory, Internat. Colloq. Function Theory. Tata Inst. of Fund. Research, Bombay 1960, 147-164.

51a. Serre, J.-P.: Cours d'Arithmétique. Paris: Presses Univ. de France 1970.

52. Speiser, A.: Die Theorie der Gruppen von endlicher Ordnung, 2nd ed., Berlin 1937. Reprinted, Dover Inc., 1945.

53. Suprunenko, D. A.: Soluble subgroups of the multiplicative group of a field*. Izv. Akad. Nauk SSSR, Ser. Mat. **26**, 631-638 (1962). = Amer. Math. Soc. Transl. (2) **46**, 153-161 (1965).

54. Suprunenko, D. A.: Soluble and nilpotent linear groups. Transl. of Math. Monographs **9**, Amer. Math. Soc. (1963).

55. Suprunenko, D. A.: Two theorems on matrix groups*. Dokl. Akad. Nauk BSSR **8**, 491-494 (1964).

55a. Suprunenko, D. A.: On the theory of soluble linear groups*. Dokl. Akad. Nauk SSSR **184**, 47-50 (1969). = Soviet Math. Dokl. **10**, 42-46 (1969). Proofs published under same title in Sibirsk. Math. Ž. **10**, 1161-1172 (1969). = Siberian Math. J. **10**, 859-867 (1969).

55b. Suprunenko, D. A., Matjuhin, V. I.: On soluble matrix groups over a Euclidean ring*. Izv. Akad. Nauk BSSR, Ser. Fiz.-Mat. Nauk 1965, No. 3, 5-9.

56. Suprunenko, D. A., Tyškevič, R. I.: Reducible locally nilpotent linear groups*. Izv. Akad. Nauk SSSR, Ser. Mat. **24**, 787-806 (1960).

56a. Suprunenko, D. A., Tyškevič, R. I.: Contractable subgroups*. Dokl. Akad. Nauk BSSR **12**, 861-862 (1968).

57. Swan, R. G.: Representations of polycyclic groups. Proc. Amer. Math. Soc. **18**, 573-574 (1967).

57a. Tits, J.: Free subgroups in linear groups. J. Algebra **20**, 250-270 (1972).

58. Tôgô Shigeaki: On splittable linear groups. Rend. Circ. Mat. Palermo(2) **8**, 49-76 (1959).

58a. Vapne, Ju. E.: On the representability as matrices, of the wreath product of groups*. Mat. Zametki **7**, 181-189 (1970).

58b. Vapne, Ju. E.: The criterion for the representability of wreath products of groups by matrices*. Dokl. Akad. Nauk SSSR **195**, 13-16 (1970). = Soviet Math. Dokl. **11**, 1396-1399 (1970).

59. Hsien-Chung Wang: Discrete subgroups of solvable Lie groups I. Ann. Math. (2) **64**, 1–19 (1956).

59a. Ward, H. N.: Automorphisms of quaternion-free 2-groups. Math. Z. **112**, 52–58 (1969).

60. Wehrfritz, B. A. F.: A note on d-groups. Canad. J. Math. **19**, 410–412 (1967).

61. Wehrfritz, B. A. F.: Conjugacy theorems in locally finite groups. J. London Math. Soc. **42**, 679–686 (1967).

62. Wehrfritz, B. A. F.: Conjugacy theorems in locally finite groups II. Arch. der Math. **18**, 470–473 (1967).

63. Wehrfritz, B. A. F.: A note on periodic locally soluble groups. Arch. Math. (Basel) **18**, 577–579 (1967).

64. Wehrfritz, B. A. F.: Sylow theorems for periodic linear groups. Proc. London Math. Soc. (3) **18**, 125–140 (1968).

65. Wehrfritz, B. A. F.: Soluble periodic linear groups. Proc. London Math. Soc. (3) **18**, 141–157 (1968).

66. Wehrfritz, B. A. F.: Locally nilpotent linear groups. J. London Math. Soc. **43**, 667–674 (1968).

67. Wehrfritz, B. A. F.: Frattini subgroups in finitely generated linear groups. J. London Math. Soc. **43**, 619–622 (1968).

68. Wehrfritz, B. A. F.: Transfer theorems for periodic linear groups. Proc. London Math. Soc. (3) **19**, 143–163 (1969).

69. Wehrfritz, B. A. F.: Groups of automorphisms of soluble groups. Proc. London Math. Soc. (3) **20**, 101–122 (1970).

69a. Wehrfritz, B. A. F.: Supersoluble and locally supersoluble linear groups. J. Algebra **17**, 41–58 (1971).

69b. Wehrfritz, B. A. F.: Remarks on centrality and cyclicity in linear groups. J. Algebra **18**, 229–236 (1971).

69c. Wehrfritz, B. A. F.: 2-generator conditions in linear groups. Arch. Math. (Basel) **22**, 237–240 (1971).

69d. Wehrfritz, B. A. F.: Wreath products and chief factors in linear groups. J. London Math. Soc. (2) **4**, 671–681 (1972).

69e. Wehrfritz, B. A. F.: Wreath products of linear groups – the characteristic p case. Bull. London Math. Soc. **3**, 331–332 (1971).

69f. Wehrfritz, B. A. F.: Generalized free products of linear groups. Proc. London Math. Soc., to appear.

69g. Wehrfritz, B. A. F.: Two examples of soluble groups that are not conjugacy separable. J. London Math. Soc., to appear.

69h. Wehrfritz, B. A. F.: Conjugacy separating representations of free groups. Proc. Amer. Math. Soc., to appear.

70. Weir, A. J.: The Sylow subgroups of the symmetric groups. Proc. Amer. Math. Soc. **6**, 531–541 (1955).

71. Winter, D. J.: Representations of locally finite groups. Bull. Amer. Math. Soc. **74**, 145–148 (1968).

71a. Zalesskii, A. E.: The structure of several classes of matrix groups over a division ring*. Sibirsk. Math. Ž. **8**, 1284–1298 (1967). = Siberian Math. J. **8**, 978–988 (1967).

72. Zariski, O., Samuel, P.: Commutative algebra (2 vols.). Van Nostrand 1958.

73. Zassenhaus, H.: Beweis eines Satzes über diskrete Gruppen. Abh. Math. Sem. Univ. Hamburg **12**, 289–312 (1938).

74. Zassenhaus, H.: On linear Noetherian groups. J. Number Theory **1**, 70–89 (1969).

75. Žavrid, G. P.: Maximal soluble subgroups of GL(4, P)*. Vesci Akad. Navuk BSSR, Ser. Fiz.-Mat. Navuk 1968, No. 3, 30–37.

Index

227

Jordan decomposition 91–93, 214–216
Jordan's Theorem 112

Level of a linear group 209
Lie-Kolchin Theorem 77
linear group 1
locally algebraic group 176, 179
— closed subspace 209
— completely reducible 93
— finite group, see periodic group
— nilpotent group 7, 80, 94, 96–108, 121, 125, 168, 182, 188, 191–194, 198
— π-separable group 119
— soluble group 46, 80, 123, 191, 192
— supersoluble group 7, 47, 48, 126, 162, 164–173, 200
— theorem (of Mal'cev) 27

Mal'cev's Theorem (on soluble groups) 44, 45
marginal subgroup 81, 149, 150
Maschke's Theorem 4
matrix representation 1
maximal closed subgroup 184
— condition on subgroups 21, 146, 196
— condition on abelian normal subgroups 107
— p-subgroup, see Sylow subgroup
— subgroup 70, 133, 166–168, 182–184, 196, 199
— π-subgroup 116; see also Hall subgroup
metabelian group 7, 26, 33, 56, 57, 62, 66, 127, 165
minimal condition (on subgroups) 21, 114, 123, 197
modular subgroup 94
monomial group 6–8, 155

Nilpotent group 23, 25, 33, 41, 47, 48, 60, 61, 78, 83, 94, 96–100, 103, 139, 140, 142, 216, 217
normal complement 180, 181, 184

P-adic linear group 108, 132, 162
paraheight 156–162, 169–173, 190, 200
parasoluble group 164, 166
periodic group 4, 15, 57, 83, 93, 112–134, 179–185, 196, 197
p-group 27, 83
p-normal group 181
polycyclic group 21, 26, 162, 196
poly $\pi\pi'$-group 123

polynomial homomorphism 203
— map 203–207
primitivity 4, 42–44, 48
Prüfer rank 25, 26, 141, 200

Quasi-Hall subgroup 119; see also Hall subgroup
quasi-linear group 186, 187

Rank 17, 26; see also Prüfer rank
rational linear group 108, 132, 162
rational mapping 84, 204–207
reducible 1
regular p-group 180–182
relatively free group 33, 34, 40
representation 1
residually a finite group 16, 51, 196
— a finite p-group 55

Schur's Lemma 3
semisimple algebraic group 218
π-separable group 119, 124
serial subgroup 128
series 128–130
simple group 52, 129–132, 218
soluble group 7, 16, 21–27, 33, 41–49, 65, 70, 77, 78, 79, 94, 123, 133, 140, 142, 145, 191, 192, 195, 199, 216
splittable linear group 93–100
subnormal subgroup 5, 130
super-residually 30
supersoluble group 7, 47, 48, 155–173
Sylow subgroup 114, 115, 133, 138, 177–181, 196, 197, 217

Thompson subgroup 177, 180
Tits' Theorem 145
torsion-free group 23, 25, 26, 56, 62, 68, 107, 108, 194
torsion group, see periodic group
torus 216
triangularizable 13, 16, 45, 47, 57, 77

Ultrafilter 28
ultraproduct 28
unipotent over commutative rings 187, 189–191, 197–199
— over fields 13, 14, 59, 90, 93, 158, 163, 174–176
unitriangularizable, see unipotent
upper central factors 109, 193
— central series 99, 101–111, 192–194

228

Ergebnisse der Mathematik und ihrer Grenzgebiete